AF828606

BEST OF
ARDUINO PROJECTS FOR HAM RADIO

Glen Popiel, KW5GP

ARRL The National Association for Amateur Radio®
225 Main Street, Newington, CT 06111-1400 USA
www.arrl.org

Printed in the USA

ISBN: 978-1-62595-147-2

First Edition
28 27 26 25 24   1 2 3 4 5

We strive to produce books without errors. Sometimes mistakes do occur, however. When we become aware of problems in our books (other than obvious typographical errors), we post corrections on the ARRL website. If you think you have found an error, please check **arrl.org/product-notes** for corrections. If you don't find a correction there, please let us know by sending email to **pubsfdbk@arrl.org**.

Sketches, libraries, diagrams, and other files for the projects in this book are available online from **kw5gp.com**.

Contents

Preface

I started out on my Arduino adventure 11 years ago, and four books later I'm still having fun. The sheer number of projects I could have included excited and amazed me, but many had to be left on the drawing board. There were 44 additional Arduino projects that didn't make the cut for this book. They weren't left out because they couldn't be made to work, it's actually quite the opposite. There simply wasn't enough room to fit them all in. Does that mean there will be another Arduino book? I certainly hope so because my own personal Arduino adventure is far from over. Rest assured that as soon as this book heads off to print, I'll start playing with a whole bunch of new Arduino project ideas, and who knows what could happen.

One of my favorite things to do with the Arduino is to research and create projects that highlight the devices and modules that are not well documented or find new uses for existing devices and modules. As part of this, I have also chosen some aspects of the Arduino that remain relatively unexplored territory, such as USB, ISM band communication, UART technology, lightning detection, and GPS technology. It is my goal to demystify and simplify the integration of these devices and technologies into your own Arduino projects.

Since my last Arduino book, there have been several new Arduino-compatible CPU boards, modules, and components. Part of my own personal Arduino adventure has been to take some of these newer parts and develop a project to demonstrate their usefulness in ham radio-related projects.

While the projects in this book were initially introduced in one of my previous books, each project has been redesigned, updated, and enhanced. There is also still plenty of room for you to add your own personal touches to enhance and customize your finished project beyond the scope of this book. What has also been part of the fun of this process is to see others pick up where I leave off and create some truly wonderful projects of their own. To me, this is what the open-source community and the Arduino are all about, sharing ideas and projects for others to expand upon.

In keeping with the open-source philosophy of sharing knowledge, in this book I have included projects designed to interface to real-world devices such as antenna rotators and transceivers. Don't be afraid to void those warranties and add an Arduino to your rotator controller or use an Arduino to talk to your rig's Computer Assisted Transceiver (CAT) interface. In many cases, you'd be surprised to find out just how easy it is to inexpensively add some very powerful and fun features to your existing equipment.

It has been a wonderful experience to see the art of homebrewing make a comeback in the ham radio world, with the Arduino being a large part of that resurgence. When I first started out homebrewing, way back in the Dark Ages before cell phones and the Internet, we didn't have things like the Arduino, and instead had to bury ourselves in the manufacturer's component databooks, then design and solder (or wirewrap) everything from scratch, with no guarantee of success. Now, with the Arduino, projects that would take weeks, or even months, can be built in a matter of days for a fraction of the cost. It's been this kind of thinking and homebrewing that has even led to groups of doctors and hams using the Arduino to power a COVID-19 patient ventilator that is currently under evaluation for federal approval and patient use.

So, who knows where your next Arduino project will take you? Let your imagination be your guide. My favorite words throughout this adventure have been, "Wouldn't it be cool if…?" That has led to the creation of many fun projects for me. Hopefully, it will be the same for you.

And whatever you do, please, share your ideas, projects, and knowledge with the rest of the open-source community. Without the philosophy of freely sharing knowledge and ideas, many things in the Arduino universe would not have been possible.

73,

Glen Popiel, KW5GP

Spring 2024

Acknowledgements

To my Dad, thanks for teaching me that by learning the fundamentals of how things work, you can do just about anything.

I would also like to thank you, my fellow Makers and Hams, for making my personal Arduino adventure so much fun. Without your support, this fourth ARRL Arduino book would never have happened. I would also like to thank Tim Billingsley (SK), KD5CKP, for introducing me to the Arduino, providing ideas and key insights during my Arduino learning process, and providing invaluable help, advice, and trying to keep me on track and focused on what I was actually trying to accomplish. I would also like to thank Craig Behrens (SK), NM4T, and the QRP Skunkwerks team, Rob Suggs, KB5EZ, and Joe Large, W6CQZ, who helped turn what started out as a "Wouldn't it be cool if…" moment into a reality. The JT65 project in my second book would never have happened without the efforts of Craig and the QRP SkunkWerks team.

There are many others who have helped along the way, and I apologize in advance to anyone I may have omitted. Thanks to the Olive Branch Mississippi and Helena Arkansas Amateur Radio Clubs, the Chickasaw Amateur Radio Association, along with the Orlando Hamcation and the Huntsville Hamfest for allowing me to present Arduino forums to help spread the word on the Arduino and how it can be applied to amateur radio.

I also need to thank retired ARRL Publications Manager, Steve Ford, WB8IMY, retired editor, Mark Wilson, K1RO, ARRL Publications Manager Becky Schoenfeld, W1BXY, and the staff at ARRL for the opportunity to work with them. Don't let anyone tell you otherwise, putting together a book like this is an arduous task, and they've helped to make the whole process enjoyable.

And a special thanks to the Open Source Community. Their sharing of knowledge and ideas helped make this book become a reality. It is through their efforts that the Arduino has become the popular microcontroller platform that it is today and we all owe them a huge debt of gratitude. Thank you.

About the Author

Glen Popiel is the author of ARRL's "Arduino for Ham Radio" series of books, along with ARRL's *High Speed Multimedia for Amateur Radio*. He is a retired network engineer and Technology Consultant for HTC Global Services. First published in *Kilobaud Microcomputing* in 1979 for circuits he designed for the RCA 1802 microprocessor, he continues to work with microcontrollers and their uses in amateur radio and has written numerous articles on computers and amateur radio.

Always taking things apart (and sometimes even getting them to work again afterwards), he discovered electronics in high school and has never looked back. As a teenager, he had one of the first true home computers, a Digital Equipment (DEC) PDP-8 minicomputer (complete with state-of-the-art Model 35 Teletype) in his bedroom that he and his friends salvaged from the scrap heap. Over his 40+ year career, he has worked for various aerospace, computer manufacturers, and technology companies, including radio and military turbojet research data acquisition and control systems.

Since discovering the Arduino over ten years ago, he has developed a passion for this powerful, inexpensive microcontroller and has given a number of seminars and hamfest forums on the subject of the Arduino and Open Source. He is a member of the Olive Branch Amateur Radio Club (OBARC), the Chickasaw Amateur Radio Association, QRP Amateur Radio Club International (QRP-ARCI), and the QRP Skunkwerks, a design team of fellow hams and Arduino enthusiasts who have succeeded in getting the JT65 digital mode working natively on a CW-only (so they thought) QRP Transceiver.

Glen is also a former cat show judge and has exhibited Maine Coon cats all over the country, with the highlight being a Best in Show at Madison Square Garden in 1989. He now lives in Southaven, MS, where he continues to create fun and exciting new Arduino projects for amateur radio with his two champion Maine Coon show cats and lab companions, Angel and Shadow.

About This Book

Welcome to the collection of what I feel are the best projects from my *Arduino for Ham Radio* series of books. I hope that you have as much fun building the projects in this book as I had creating them.

The projects in this book were chosen based on your feedback along with projects that are my personal favorites. Part of the project selection process was also based on uniqueness, simplicity, innovativeness of approach, and an all-around "cool" factor. Many of my Arduino projects start out from the simple thought of "Wouldn't it be cool if…?"

In this book I also focus on peeling away the layers of complexity on how to interface with the various rotator controllers, transceivers, communication protocols, and specialized Arduino components and functions, while at the same time trying to keep each project interesting, fun, easy to build, and as inexpensive as possible. Just to let you in on a little secret, I create and build these projects as much for myself and my ham shack as I do for you and the book. What I end up with is more of a chronicling of my own Arduino adventures in a format that allows others to join in on the fun as well as a bunch of fun projects to play with. And, for those of you that have been here before with my other books, isn't it fun to look back and see how much we've all painlessly learned in this process?

As with my other Arduino books, this book assumes that you have a basic working knowledge of electronics, components, and construction techniques. Don't be afraid. You don't have to be an expert, as most of these projects require only a handful of wires and components and are easy to build. However, you will need to know how to solder and feel comfortable building electronic projects. For those of you who are new to electronics, there are a number of good books and online tutorials to help you along the way, many of them available from the ARRL Store. Among the books that I recommend are *Electronics for Dummies* by Cathleen Shamieh; *Understanding Basic Electronics*, Second Edition, by Walter Banzhaf, WB1ANE; *Beginner's Guide to Reading Schematics* by Stan Gibilsco; and the *ARRL Handbook for Radio Communications*, published by the ARRL. Every ham should have a copy of the *ARRL Handbook* in their library. I find myself constantly referring to mine for research or when I want to learn some new aspect of electronics or ham radio. Many of these books and other useful resources are available from amateur radio dealers or at **arrl.org/shop**.

You will also need a working knowledge of the Arduino and the Arduino Integrated Development Environment (IDE). Entire books have been written about how to learn and use the Arduino and the IDE, and there is simply not enough space to cover those basics and still have room for all the fun projects. While the Arduino is easy to learn how to use and program, it's best that you take the time to become familiar with it. In particular, concentrate on the various methods of installing libraries. Library installation issues account for nearly 90% of all the Arduino questions I receive. Installing libraries is not really that difficult, but it is an often misunderstood and overthought process. The more you can work to understand Arduino libraries, the more it will help you further down the road.

There are several excellent introductory books and online tutorials to help you along the way. The ones I have found most useful include *Beginning Arduino* by Michael McRoberts and *Arduino Cookbook* by Michael Margolis. Another excellent book for learning the Arduino, especially the hardware interfacing aspect, is *Exploring Arduino: Tools and Techniques for Engineering Wizardry* by Jeremy Blum. There are also some outstanding Arduino tutorials available online at **arduino.cc**, **learn.adafruit.com**, and **learn.sparkfun.com**. Recently, I have discovered several other very good online courses and tutorials available for the Arduino. The Programming Learning Academy (**programmingelectronics.com**) offers a free 12-lesson Arduino Video Course and an inexpensive e-book PDF download of the *Arduino Course for Absolute Beginners* by Michael James. Both are excellent Arduino learning tools.

Since some of the projects in this book may require that you have a ham radio operator's license, now would be a good time get or upgrade your license if you haven't already. There is something for everyone in the ham radio community, and there is absolutely no reason for you to go it alone as you start out on your Arduino adventures. I recommend that you find a local ham radio club at **arrl.**

org/find-a-club and attend a few meetings. That's what got me started out on the Arduino, and you will not meet a friendlier, more helpful group of people anywhere. The odds are that, like me, you will meet other Arduino enthusiasts with whom you can collaborate on your own Arduino projects. I also urge you to attend some of the forums presented at many hamfests. Several of the best ideas for my Arduino projects have come from my Arduino forums and club presentations. Forums are also a great place to meet others who share similar interests and to share knowledge and ideas. For more information on how to become a radio amateur, check out the ARRL *Ham Radio License Manual*.

All the Arduino project sketches in this book were written and compiled on Windows 7, 8, and 10, using version 1.8.19 of the Arduino IDE. If you have any problems compiling the sketches, please try compiling using the same version of the IDE that I used and verify that your libraries are properly installed. While sometimes a bit on the vague side, the error messages you may receive when experiencing problems with your sketches can point you in the right direction as to where the problem is. When in doubt, do an internet search for the error message itself. Quite often, someone else has experienced the same issue and has posted an explanation or possible solution. All the sketch files, libraries, and any errata information for the book projects is available online at **kw5gp.com/BestOfArduino**. While every effort has been made to eliminate any errors in this book, they always find a way to creep in and cause all kinds of havoc, even in the most carefully prepared book. So please, check the errata first if you have any problems. And by all means, feel free to contact me if you need help, I'll be glad to help as much as I can.

How This Book Is Organized

This book contains a collection of the best projects that were presented in my previous books, *Arduino for Ham Radio, More Arduino Projects for Ham Radio*, and *More Arduino for Ham Radio*.

While I understand your desire to skip ahead and dive right into the projects, I urge you to take the time to read the chapters leading up to the projects. These preliminary chapters will provide vital information and insights into the various modules, components, and technologies used in the projects that follow. They will also provide information and guidance on creating a good, functional work area to build and troubleshoot your Arduino projects. The more prepared and organized you are in these areas, the better things will often turn out. This doesn't necessarily mean your work area will stay clean and uncluttered for very long, but it at least helps to start out that way. After 20 minutes on a project, my work area tends to degenerate into a complete and total mess, but at least I had good intentions.

Chapter 1, "The Arduino Family – an Update," provides an update to the world of the Arduino, showcasing and discussing the Arduino processor boards used for the projects in this book, as well as the latest members of the Arduino processor family.

Chapter 2, "Arduino Shields, Modules, and Devices - An Update," discusses the various boards and components used for projects in this book, as well as other new devices of interest for ham radio-related projects.

Chapter 3, "Arduino I/O Methods," discusses in detail the I/O capabilities of the Arduino, including some of the newer methods now available, which method is best for communicating with the various shields and components, and how to best implement each I/O method. This chapter also discusses the Bluetooth, USB, and Serial RS-232 communication methods with the Arduino, including the use of human interface devices (HID) such as keyboards, mice, etc. with the Arduino.

Chapter 4, "Creating Arduino Sketches," introduces the Arduino Integrated Development Environment (IDE), creating sketches (programs), outlining and flowcharting sketches, and organizing your projects using Sketchbook. It also discusses Arduino simulators and alternatives to the Arduino IDE, as well as how to install and manage your Arduino libraries.

Chapter 5, "Tools, Construction Techniques, and Troubleshooting," covers tools, test equipment, and methods on how to troubleshoot your sketches, library problems, hardware, and other issues you may encounter while building Arduino projects.

Chapter 6, "Random Morse Code Practice Generator," introduces a simple Morse code trainer project that you can use to learn Morse code.

Chapter 7, "Iambic CW Keyer," shows how to interface a CW keyer to the Arduino.

Chapter 8, "1 – 30 MHz DDS VFO," shows how to construct a high-precision VFO capable of driving a vacuum tube-based radio.

Chapter 9, "CW Decoder," Uses audio from a receiver to decode the incoming CW signals into text on a display.

Chapter 10, "Talking GPS-based UTC Clock and Grid Square Display," shows how to interface a GPS module to the Arduino, calculate and display the Maidenhead Grid Square, and output the results using a text-to-speech module.

Chapter 11, "Talking SWR Meter," demonstrates how to use sensors and analog-to-digital conversion methods to determine SWR of an antenna system and uses a text-to-speech module to convert the output into speech.

Chapter 12, "Antenna SWR Analyzer," allows you to scan, measure, and graph the SWR curve for an entire frequency range.

Chapter 13, "Yaesu G-450 and G-800SA Rotator Controller Rebuild," shows how to completely rebuild a broken Yaesu G-450 and G-800SA series rotator controller and interface it to rotator control software such as *Ham Radio Deluxe (HRD)*.

Chapter 14, "Yaesu G-450, G-800/1000SA Rotator Controller Modification," shows how to modify a Yaesu G-450 and G-800SA/G1000SA series rotator controller and interface it to rotator control software such as *HRD*.

Chapter 15, "Modified CDE/Hy-Gain Rotor Controller," describes how to modify a CDE/Hy-Gain rotator controller, sense the position of the antenna, and allow PC software such *as HRD* to automatically control your antennas.

Chapter 16, "Yaesu Azimuth/Elevation Rotor Controller," describes how to build an interface to control the Yaesu G5400/5500 satellite rotor controller, sense the position of the antenna, and allow PC software such *as HRD* to automatically rotate your antennas and track satellites.

Chapter 17, "CDE/Hy-Gain Keypad Rotator Controller," interfaces an inexpensive analog 4 × 4 keypad to your CDE/Hy-Gain rotator controller for direct bearing entry and rotation control without the need for a PC or holding down the control unit's front panel switches.

Chapter 18, "Yaesu CAT Display," demonstrates how to interface an Arduino and a color graphic TFT display to the Computer Assisted Transceiver (CAT) interface for many Yaesu transceivers such as the FT-450, FT-950, FT-1200, and others. This project could also easily be adapted to many other manufacturers and transceivers.

Chapter 19, "USB CW Keyboard," revisits the popular CW keyboard from my first book, *Arduino Projects for Ham Radio*, and updates it for use with a USB keyboard and color TFT display. This project also demonstrates how to interface USB human interface devices (HID) to the Arduino using a small, inexpensive USB host module.

Chapter 20, "Lightning Detector," demonstrates how to sense and display lightning distance and intensity at nearly 25 miles.

Chapter 21, "RTTY Reader," demonstrates how you can use an Arduino to decode and display a standard RTTY signal on a color graphic TFT display.

Chapter 22, "40-Meter QRP CW Transceiver," demonstrates how to build an inexpensive DDS-based 40-meter QRP CW transceiver, complete with dual range tuning, RIT, and a built-in iambic keyer.

Chapter 23, "In Conclusion," discusses processors, devices, projects, ideas, and concepts not included in this book. This chapter provides concepts and ideas for other projects to encourage going beyond the scope of this book.

Introduction

If someone had told me eleven years ago that I would have written five books about ham radio technology, I would have called them crazy. In high school, I would have been voted the kid least likely to have anything published. Now here we are, eleven years later and still going strong. Creating all these Arduino projects, and writing these books, has been an unending amount of fun. With my first two Arduino books no longer in print, and with some projects getting outdated by changes in the availability of devices, modules, and in the Arduino IDE itself, earlier this year, ARRL and I agreed that it was time to do a *Best of Arduino for Ham Radio* book, highlighting the best projects from all three Arduino books. Presented here are 17 of those projects. The projects have been redesigned, upgraded to newer components, and have added features that were not in the original projects. I enjoyed revisiting these older projects and applying what I have learned through the years to them. The first time around, these projects were fun to design, but as I worked with them again, I thought of so many new things to be added and improved. Every time I finished a project, I found myself amazed at how much more interesting the finished project turned out to be when compared to the original design.

Even since my most recent Arduino book, many new boards and devices have become available. These new boards and devices will form the foundation of where I go next with the Arduino and its ever-growing family. I'll even cover some of my future prospective projects at the end of this book so that you can get a head start yourself with the newer, more-powerful Arduino boards and devices.

Building, or "homebrewing" as hams prefer to call it, has always been a cornerstone of amateur radio. This has always been one of the aspects of amateur radio that I have enjoyed the most. In the early days, the only way to get on the air was to build it yourself or bribe a friend to help you. Companies such as the Heath Company flourished with kits, called Heathkits, of all descriptions. I remember eagerly awaiting every Heathkit catalog to see what new and exciting kit I could dream about building next. I built many of those kits over the years, in fact, my very first radio as a novice was the venerable Heathkit HW-16. I didn't build that one, but it had been modified by its builder. That custom-built kit radio inspired me to build those Heathkit radios for myself. Not being satisfied with what I had, modifying the kits I had just finished building was the next order of business. I'll never forget the day when I finished installing that receive preamplifier in my trusty Heathkit HW-101 so that I could hear the newest amateur radio satellite, Oscar 7, on 10-meters. It would not surprise me one bit if that old HW-101 is still out there somewhere, with its current owner wondering what the function of that little switch on the front panel was, and why it was put there.

Sadly, with the advent of the modern era, computers, and the complexity of today's typical ham shack, homebrewing has taken a back seat to flashy new rigs and accessories. Personal computers are now an integral part of many ham shacks. Why would anyone want to take the time to build something, or spend months writing a program, when a tried-and-true version is just a credit card and a click away? And so, the art of kit building and homebrewing has begun to fade into the background.

Enter the Arduino: a small, inexpensive, easy to program microcontroller. A whole new world is now opened to the homebrewer. Those once complex projects that would've taken too much time, money, and knowledge to complete are now just pennies and a few small steps away. As you will see in this book, complex projects can be built inexpensively and still result in a fully functional project. Now it seems that nearly every entry at the QRP-ARCI homebrew contest, held at their annual Four Days in May event in conjunction with Dayton Hamvention®, was Arduino-based, where in prior years there were only one or two Arduino entries. It's nice to see that homebrewing is becoming popular again, and that the Arduino is among the devices leading the way. I've had the opportunity to pick up an HW-16 and an HW-8 along the way, and I've had fun concocting all sorts of ways to bring those rigs into the 21st century with Arduino technology.

The Arduino is not without a following of its own. Its open-source model comes with a whole community of developers and builders known as "makers." Now you have a simple, easy-to-use, development platform and a whole world of developers sharing the fruits of their labor with everyone — for free. Partially due to this community of developers, the interest in homebrewing is back on the rise, and there are now all sorts of new and exciting projects for ham radio using the Arduino and other microcontrollers.

And please, don't forget the cardinal rule of open-source. When you do add your own personal touches, and enhance a project beyond what it is, give back to the open-source community. Who knows, there may be someone out there looking for exactly what you have created. It's that spirit of sharing knowledge and designs that make the open-source world what it is, a truly unique place to build and share wonderful new toys.

73,

Glen Popiel, KW5GP

Spring, 2024

The Projects – What's New

Nothing ever stays the same in the technology world, and this is true in the Arduino world as well. Some of the projects in my earlier Arduino books had become outdated, components discontinued or difficult to find, and/or the sketches and libraries updated to the point where the sketch itself would have to be recoded to support the newer IDE and library versions.

Because of this, every project included in this book has been updated to compile with the current version of the Arduino IDE (1.8.19 as of this writing). The displays have been upgraded from the older 16 × 2-LCD and Nokia-type monochrome displays to higher resolution color thin-film-transistor (TFT) LCD displays. Since many of the newer Arduino processors, modules, and devices require +3.3 V dc for power and signal interfacing, level shifters have been added to the projects. Since some components have been discontinued, or are no longer available, those projects have been updated to use currently available components. The Arduino UNO used in many projects has been replaced by an Arduino Nano to reduce the projects' size.

Additionally, new features have been added to many of the projects, often based on the enhancement ideas I had included with the original project. For example, the lightning detector now includes a relay output that can be used to disconnect your antennas when lightning is detected and reconnect them when the storm has passed.

In my opinion, these updates bring a whole new level of "cool" to the projects, and it was often difficult to decide where to stop with the upgrades and enhancements. But, as with all of my Arduino books, I have left plenty of room for you to add your own personal touches and functionality to these projects. I hope you enjoy this collection of new and improved Arduino for ham radio projects.

About ARRL

We're the American Radio Relay League, Inc. — better known as ARRL. We're the largest membership association for the amateur radio hobby and service in the US. For over 100 years, we have been the primary source of information about amateur radio, offering a variety of benefits and services to our members, as well as the larger amateur radio community. We publish books on amateur radio, as well as four magazines covering a variety of radio communication interests. In addition, we provide technical advice and assistance to amateur radio enthusiasts, support several education programs, and sponsor a variety of operating events.

One of the primary benefits we offer to the ham radio community is in representing the interests of amateur radio operators before federal regulatory bodies advocating for meaningful access to the radio spectrum. ARRL also serves as the international secretariat of the International Amateur Radio Union, which performs a similar role internationally, advocating for amateur radio interests before the International Telecommunication Union and the World Administrative Radio Conferences.

Today, we proudly serve nearly 150,000 members, both in the US and internationally, through our national headquarters and flagship amateur radio station, W1AW, in Newington, Connecticut. Every year we welcome thousands of new licensees to our membership, and we hope you will join us. Let us be a part of your amateur radio journey. Visit www.arrl.org/join for more information.

ARRL The National Association for Amateur Radio®

225 Main Street
Newington, CT 06111-1400 USA
Tel: 860-594-0200
FAX: 860-594-0259
Email: membership@arrl.org

www.arrl.org

1 The Arduino Family — An Update

The Arduino family has grown so much that it seems like every day new Arduino-compatible processor boards or entire families of Arduino-compatible processors are being released. This can make choosing an Arduino board for your project a daunting task. Generally, all these new boards are supported by the Arduino IDE. I usually try starting out with an Arduino Uno or Nano simply because of their versatility, low cost, and ease of use. If an Uno or Nano isn't right for the project, then I'll move up the chain, looking for the Arduino-variant that has the power and features the project will need. Amazingly, the Uno and Nano can handle about 90% of the projects I come up with, which just goes to show how versatile they are.

Since the Arduino was first introduced in 2005, dozens, if not hundreds, of new Arduino and Arduino-variant boards have been developed, each with its own set of features and enhancements. Some of the older boards have gone by the wayside, fallen out of use, or even been discontinued. But, due to the open-source nature of the Arduino, many compatible versions of these older boards can still be found from online suppliers such as eBay, **AliExpress.com**, and **Banggood.com**, in addition to the more mainstream Sparkfun, Adafruit, and similar Arduino suppliers. The following list of Arduino and Arduino-compatible processor boards is by no means complete. Many of the boards not listed here have already been extensively covered in other books, including my own *More Arduino Projects for Ham Radio*. Rather than repeat that information, I'll focus on the boards used in this book's projects, along with some of the newer Arduino processor boards.

Arduino Boards

When you talk about Arduino boards, the Arduino Uno (see **Figure 1.1**) is the first one that comes to mind. The Uno is considered the current mainstream Arduino board and is the board that most people start with. Most existing libraries and sketches are designed to work with the Uno, so you have a wide variety of code to work with as you start your Arduino adventures. I still prefer to use the Uno in many of my projects, primarily for its cost, simplicity, and ease of use.

The Arduino Uno

The Uno uses an 8-bit ATmega328 series microcontroller running at a clock speed of 16 MHz. It has 14 digital I/O pins and six 10-bit analog-to-digital pins that can be used for digital I/O if desired. Six of the digital I/O pins support *pulse width modulation* (*PWM*). This modulation allows you to control the pulse width of a square wave on the output pin to do things like dim an LED or play an audio tone. Two of the digital I/O pins can be configured to support external interrupts for hardware program con-

Figure 1.1 — Arduino UNO R3.

Figure 1.2 — Arduino Nano.

Figure 1.3 — FTDI.

trol, and all 14 I/O pins can be configured to provide a program interrupt when the I/O pin changes state. Interrupts can be pretty handy, and we'll talk more about them in the **USB CW Keyboard** project chapter. That project would never have worked properly without interrupts. Hopefully, seeing how it came together will help you understand how interrupts work and demonstrate how you can put them to interesting use in your own projects.

The Arduino Uno has three types of onboard memory: Flash, static random-access memory (SRAM), and electrically erasable programmable read-only memory (EEPROM). Flash memory is rewritable memory that is used primarily to store your Arduino programs, known as sketches. Flash memory is semipermanent and retains its contents even when the Arduino's power is turned off. You can rewrite the contents of flash memory approximately 100,000 times, meaning that you can use the same Arduino in dozens — even hundreds — of projects simply by loading a different sketch. Flash memory can also hold data that doesn't change, such as lookup tables, text, and other constants, to save valuable SRAM space using the PROGMEM keyword. SRAM holds variables and other dynamic data, and its contents are volatile, meaning they are lost when the Arduino is reset or powered off. The onboard EEPROM can be used to retain data such as calibration values and similar settings between resets or power cycles. As with flash memory, the Arduino EEPROM has a lifetime of approximately 100,000 write cycles.

The Arduino Uno can be powered through the USB port or by 7 – 20 V dc on either its onboard dc power jack or through the Vin pin. The ATmega-series microcontrollers support the industry-standard Serial Peripheral Interface (SPI) and Inter-Integrated (I^2C) bus communication protocols. The Uno, as with many other Arduino boards, has a standard 2.7 × 2.1-inch footprint. Female headers on the edges of the board allow for the stacking of add-on interface boards, also known as Shields, without additional wiring.

Good Things Come in Smaller Packages

Not all Arduinos come in the standard footprint that allows the use of shields. Some, such as the Arduino Nano (see **Figure 1.2**), are much smaller. These boards allow you to build more compact projects and still use the Uno's standard code base and libraries. These "baby" Arduinos — some not much larger than a postage stamp or two — are often based on the ATmega328 and have the same features and functionality as the Arduino Uno. While many have a smaller mini or micro-USB connector for programming, some may require a Serial-to-USB interface module (also known as an FTDI interface — see **Figure 1.3**) for programming and interfacing with the Arduino IDE.

Figure 1.4 — Sparkfun Microview - OLED.

To take the concept of tiny Arduino to the next level, SparkFun has combined the ATmega328 with a 64 × 48-pixel Organic LED (OLED) display. The SparkFun Microview – OLED Arduino Module (see **Figure 1.4**) is not much larger than a digital watch. Once it is programmed using the external USB programming adapter (see **Figure 1.5**), all the Microview needs is 3.3 – 16 V dc power and you're off and running.

Figure 1.5 — SparkFun Microview USB Adapter.

Wearable Arduinos

Wearable electronics are one of Arduino's newer applications. While you may not immediately think wearable electronics have a place in ham radio, these small boards are suitable for projects like electronic name badges and other interesting ham "wearables." Using conductive thread, these wearable Arduinos can even be sewn into clothing and powered by small rechargeable Lithium-polymer (LiPo) batteries. While some of the original "wearable" Arduino boards have been officially discontinued, these boards are still available from various online sources, including Sparkfun, Adafruit, and eBay, among others. Many of the boards in this category are based on the ATmega328 or ATmega32u4 processors, with the major differences between them being size, available I/O pins, and type of power or battery and programming connections. These boards are programmed using the FTDI interface adapter or via an onboard micro-USB connector. The LilyPad Arduino SimpleSnap (see **Figures 1.6** and **1.7**) is unique in that it has an onboard LiPo battery and charging circuit and is designed using conductive snap connectors, allowing you to remove the board from the project for washing or to use in other projects. The Adafruit FLORA also has matching Bluetooth and GPS modules available.

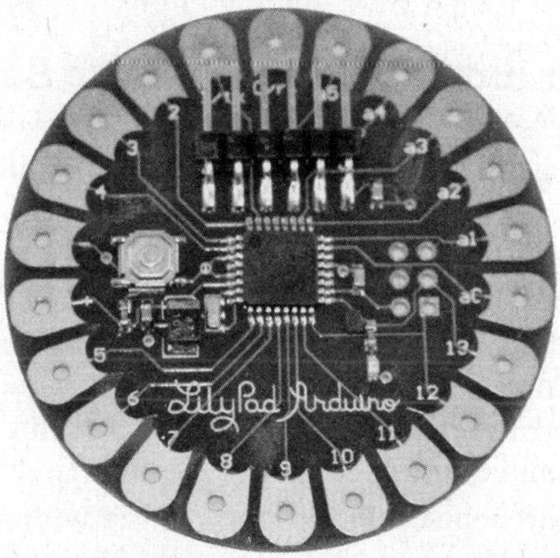

Figure 1.6 — LilyPad Arduino.

Figure 1.7 — LilyPad Arduino Simple Snap. (Photo courtesy of arduino.cc)

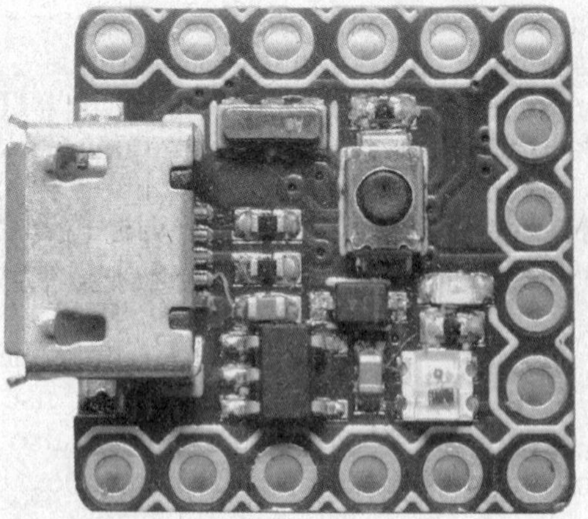

Figure 1.8 — Arduino Pico.

Other Arduino Uno/Nano Style Boards

Development on boards that use the ATMega328-style chips continues, including the Pico (see **Figure 1.8**), Mini Uno (see **Figure 1.9**), the Uno Mini (see **Figure 1.10**), and the Tiny 88 Nano (see **Figure 1.11**). It is important to note that the Tiny 88 Nano uses the Atmel ATtiny88 CPU, and many versions of these boards may not have the amount of Flash and memory found in the Uno and Nano. The T88 is also not pin-compatible with the Arduino Nano.

The Arduino Mega Series

With the advent of the larger touchscreen color graphic thin-film transistor (TFT) displays and the discontinuance of the Arduino Due, other Arduino processors with larger numbers of digital I/O pins, such as the Mega2560 (see **Figure 1.12**), are experiencing a resurgence. Still part of the mainstream Arduino line (and the first Arduino I purchased back in 2011), the Arduino Mega series offers substantially more memory, I/O, and features. Based on the 16 MHz ATmega2560 processor, the Mega2560 packs a punch with 54 digital I/O pins, 16 10-bit analog inputs, and four hardware serial ports on a larger 4 × 2.1-inch footprint. Six of the digital I/O pins can be configured for external hardware interrupts, and 15 of the digital I/O pins can provide PWM output. The I/O headers are compatible with most of the shields designed for the Uno and similar Arduinos, although you must match the desired shield to the Mega I/O pin layout. The Mega2560 ups the program's Flash memory to 256 KB, the SRAM to 8 KB, and the EEPROM to 4 KB. The Mega2560 ADK adds a USB Host port to enable communication with Android phones for use with Google's Android Accessory Development kit.

Recently, the Mega2560 has been downsized into an inexpensive, smaller, roughly ¾ Uno-sized footprint and designated the Mega2560 Mini (see **Figure 1.13**). The Mini has the same ATmega2560 processor and all the features in the full-size Mega2560. This board could easily become my "go to" Arduino board for when I need more memory or I/O than a standard Uno or Nano can provide.

A New Batch of Boards

Moving beyond the somewhat limited processing and memory capability of the ATmega328 and ATmega32u4, newer Arduino-compatible boards are using faster, more powerful processors with more memory and

Figure 1.9 — Mini Uno.

Figure 1.10 — Uno Mini.

are adding additional features. On these newer boards, the Arduino IDE can now install board-specific software plug-ins. This means that you can create your sketch for one type of processor, and with a simple change in your IDE settings, you can compile the same sketch for another processor type without having to modify your sketch other than to take advantage of features inherent in the new processor type. Care must be taken when working with these processors, as many of the new boards are powered by 3.3 V, and their I/O pins are generally not 5 V tolerant, meaning you can burn out your board if you apply 5 V on any of the I/O pins. Be sure to check the power and voltage specifications when using these boards in your projects.

The Arduino Every (see **Figure 1.14**) is a Nano-sized board with a beefier 20 MHz 8-bit ATMega4809 processor onboard. The Every is 5 V compatible and has 48 K of Flash memory, 6 K of SRAM, and 256 bytes of onboard EEPROM. The Every has 14 digital I/O pins (five can do PWM), eight analog pins, and can be used as a drop-in physical replacement for the Nano with slightly higher processor speed and significantly more memory.

The Arduino Nano 33 BLE boards (see **Figure 1.15**) are 3.3 V boards using a 32-bit Nordic Semiconductors nRF52840, containing an ARM Cortex-M4 processor running at 64 MHz. The Nano 33 BLE has 1 MB of Flash memory and 256 K of SRAM along with 14 digital I/O pins (all can be PWM) and eight analog pins, also making it an ideal replacement for the Nano, with significantly higher processor speed and more memory. In fact, the Nano 33 BLE is a physical drop-in replacement for the original Nano. The standard BLE comes with a nine-axis Inertial Measurement Unit (IMU) containing an accelerometer, gyroscope, and a magnetometer onboard, along with Bluetooth BLE that allows it to be both a BLE and Bluetooth client and host device.

The Arduino Nano 33 BLE Sense version has all the features of the standard Nano 33 BLE. It adds a series of embedded sensors, including humidity, temperature, barometric pressure, gesture, proximity, light color, and intensity sensors, along with a built-in microphone.

Figure 1.11 — Tiny88 Nano.

Figure 1.12 — Arduino Mega2560.

Figure 1.13 — Mega2560 Mini.

The PJRC Teensy 4.0 and 4.1

The Teensy 4.0 and 4.1 (see **Figure 1.16**) boards from PJRC are small footprint boards that use an iMXRT1062, containing a 32-bit ARM Cortex-M7 processor running at a whopping 600 MHz, with 2 MB of Flash (8 MB for the 4.1), and 1 MB of SRAM. These Teensy boards include two USB ports, three CAN bus ports, two I^2S digital audio ports, and one S/PDIF digital audio port. It also has SD card I/O capability, along with three SPI and three I^2C bus interfaces. The 4.0 supports seven serial interfaces, 40 digital I/O pins, and 14 analog

Figure 1.14 — Arduino Every.
(Photo courtesy of arduino.cc)

Figure 1.15 — Arduino Nano 33 BLE.
(Photo courtesy of arduino.cc)

Figure 1.16 — Teensy 4.1.

input pins. It also includes cryptographic acceleration, random number generation, and a real-time clock onboard. The Teensy 4.1 has all the features of the 4.0 but also increases the Flash memory to 8 MB. There are also locations on the 4.1 board to expand the Flash and SRAM memories with additional chips. The Teensy 4.1 increases the number of I/O pins to 55 and includes a 10/100 Ethernet interface onboard. One thing to be aware of with the Teensy boards is that not all of the pins are brought out to through-hole solder connections. Some of the signals are brought out to small solder pads on the board, which do not lend themselves very well to socket mounting.

The Espressif ESP32

The Espressif ESP32 (see **Figure 1.17**) family is an updated version of the earlier ESP8266. It offers Wi-Fi and Bluetooth functionality. The 32-bit ESP32 processor contains a dual-core Tensilica LX6 microcontroller running up to 240 MHz. In addition to the onboard 2.4 GHz IEEE 802.11 b/g/n Wi-Fi controller, the ESP32 also comes with integrated dual-mode Bluetooth, supporting both classic and BLE Bluetooth devices. The ESP32 has up to 16 MB of Flash and 520 KB of SRAM, along with 38–77 digital I/O pins, with 16 pins supporting PWM, and up to 18 analog input pins. The ESP32 also has an onboard Hall-effect sensor (for detecting magnetic fields and their intensity) and temperature sensor, with support for the SPI, I^2C, and automotive CAN buses. The ESP32 is a 3.3 V device, and the I/O pins are not 5 V tolerant.

The Adafruit Feathers

Adafruit has long been a leader in developing Arduino boards and modules. The Adafruit Feather series (see **Figure 1.18**) of Arduino-compatible boards supports a wide array of Wi-Fi, Bluetooth, and other methods of RF communications. The Adafruit Feathers are available with the Arduino 32u4 that is used in the Arduino Leonardo and other boards, as well as the ARM Cortex-M0, M0+, M3, and M4, the nRF52, the ESP-32, and the ESP8266, which are all on a smaller Arduino Nano-style footprint. Depending on the version, the Adafruit Feather supports Bluetooth LE, 2.4 GHz IEEE 802.11 b/g/n Wi-Fi, and more recently, the license-free 433 and 868/915 MHz industrial,

Figure 1.17 — Espressif ESP32.

Figure 1.18 — Adafruit Feather32u4 LoRa.

Figure 1.19 — The STM32.

medical, and scientific (ISM) bands, including 100 mW long range (LoRa), capable of communication over distances up to 20 kilometers. The memory and I/O specifications are different for each variant of the Feather; the best way to determine which Feather is right for you is to visit the Adafruit website at **adafruit.com**. The Feather also has a wide array of add-on boards available. These boards, known as "FeatherWings," are like the shields used on the larger Arduinos. All Adafruit Feathers are supported in the Arduino IDE simply by downloading the Adafruit plugins.

The STM32 Family

The STM32 (see **Figure 1.19**) is a family of 32-bit microcontrollers based on the ARM Cortex-M CPU. The STM32 "Blue Pill" has a 72 MHz ARM Cortex-M3 CPU with 64 K of Flash and 20 K of SRAM. It has 32 digital I/O pins with 12 pins capable of PWM and 14 analog input pins. It also has three UART ports, two SPI buses, and two I²C buses. Amazingly, the STM32 costs around $3.

The Raspberry Pi RP2040 Pico Family

The Raspberry Pi Pico (see **Figure 1.20**) is a MicroPython programming-language based microcontroller. Due to its ability to have the chip's "operating system" re-flashed with other programming languages, an operating system version that runs the Arduino's programming language and is programmed by using the Arduino IDE is now available. The Pi Pico has a dual-core 133 MHz ARM Cortex-M0+ CPU,

Figure 1.20 — The Raspberry Pi Pico.

Figure 1.21 — The RP2040 Development Board.

Figure 1.22 — The RP2040 Zero.

2 MB of Flash, 256 K of SRAM, 26 digital I/O pins (16 capable of PWM), and three 12-bit analog inputs. It also has two SPI buses, two I²C buses, USB Host capability, and eight programmable PIO "state machines," along with an accurate on-chip real-time clock and temperature sensors. The Pi Pico also includes a 2.4 GHz 802.11 b/g/n Wi-Fi chip.

An entire sub-family of processor boards (see **Figures 1.21** to **1.25**) has sprung up around the Raspberry Pi RP2040 Processor chip. Some of these variants even include on-board e-Ink or TFT displays.

Figure 1.23 — RP2040 with 0.96-inch OLED display.

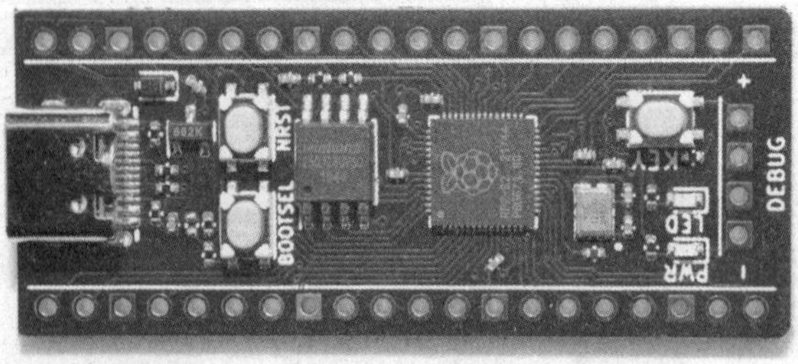

Figure 1.24 — The WeAct RP2040 Pico with 16 MB Flash.

Figure 1.25 — The Lilygo T-Display RP2040.

Summary

There is a wide array of Arduino boards to choose from for your Arduino projects, and we have only covered a small portion of them. Newer, more powerful Arduino boards are being developed all the time, and with the smaller footprint of these boards, creating smaller, portable Arduino projects has been made much easier. In this chapter, you have seen most of the different Arduino processor boards currently available that you may want to use in your own projects. In the next chapter, we will discuss the add-on boards, modules, and components you are most likely to use in your Arduino projects.

2 Arduino Shields, Modules, and Devices — An Update

Like Arduino-compatible processor boards and processor families, it seems like every day new shields, modules, and components for the Arduino come out. While researching the ideas for some of the projects in this book, I discovered even more new devices just begging to be used, and I've got a bunch of additional projects in mind that incorporate these new components. As with the Arduino processor boards, rather than discuss every possible shield, module, and component available, I'll focus on those that would commonly be used in ham radio-related Arduino projects. **Table 2.1** lists some additional Arduino parts sellers. I go to eBay, **AliExpress.com**, and **Banggood.com** in addition to Sparkfun and Adafruit for the majority of my Arduino parts. Parts on eBay are usually cheaper than anywhere else, and I have never had any problems with bad parts or quality. When it comes to the Arduino, at least at the hobby-project level, more often than not, "parts is parts," no matter where you choose to buy them.

Table 2.1

Arduino Parts Suppliers

Adafruit	**adafruit.com**
AliExpress	**aliexpress.com**
Arduino	**arduino.cc**
Argent Data Systems	**argentdata.com**
Banggood	**banggood.com**
DFRobot	**dfrobot.com**
Dr.Duino	**drduino.com**
EBay	**ebay.com**
HobbyPCB	**hobbypcb.com**
Modern Device	**moderndevice.com**
Nextion	**nextion.tech**
Solarbotics	**solarbotics.com**
SparkFun	**sparkfun.com**

The Arduino was designed to interface with all manner of shields, modules, and devices. The Uno and Mega's standard layout of I/O header pins allows for a wide array of shields to plug in to the top of the Arduino, and without any wiring at all, you can add display, networking, and all manner of additional features to your Arduino project.

With its built-in support for the industry standard SPI, I^{2}C, and One-Wire bus protocols, you can add any number of modules and components, including displays, temperature sensors, GPS modules, lightning detection, Ethernet modules, and all kinds of interesting sensors and control capabilities to the Arduino. The list of things you can add to your Arduino is nearly endless and growing daily. While it is not feasible to discuss every available Arduino-compatible shield, module, or device, assembled here is a list and discussion of the various Arduino add-ons that I have found most useful for ham radio-related projects.

Shields

As you start out creating projects with the Arduino, you may find it easier to use preassembled shields rather than figuring out how to work with and wire up individual modules and devices. Shields are boards that are designed to plug into the headers on top of the Arduino processor board, allowing instant access to the features on the shield without having to do any additional wiring. You can even stack shields one on top of the other, adding additional functionality with each shield; however, the shield stacking method can get rather cumbersome in a hurry, and you always run the risk of I/O pin usage conflicts between the various shields. As a general rule, you'll want to stack only one or two shields on top of the Arduino before decid-

Figure 2.1 — HobbyPCB RS-UV3 shield.

ing that it may be best to use a protoshield (prototyping shield) or some other method to move your project onto something more functional and aesthetically pleasing. Shields are usually supported by program libraries and example code, which allow you to quickly and easily develop working sketches. A thorough list of available Arduino shields can be found at **shieldlist.org**.

Figure 2.2 — Argent data radio shield. (Photo courtesy argentdata.com)

Figure 2.3 — Adafruit LCD shield. (Photo courtesy Adafruit.com)

The HobbyPCB RS-UV3 Radio Shield

The HobbyPCB RS-UV3 (see **Figure 2.1**) is a complete 144/220/450 MHz FM transceiver shield that allows you to create 1,200 and 9,600 baud packet radio, repeaters, Echolink stations, and other VHF/UHF radio projects. The RS-UV3 can also be used as a stand-alone radio, without the need for an Arduino. While it is essentially a stand-alone board, the RS-UV3 can be used as an Arduino shield, and it has pin headers to match the Arduino Uno and Mega board shield footprint. The RS-UV3 has built-in level shifters allowing the use of either 3.3 or 5 V Arduino or Raspberry Pi boards, with an on-board 7.2 V lithium ion (Li-ion) battery charger and a power output of 200 mW. The RS-UV3 also includes software to allow you to control the RS-UV3 from a PC, with a virtual front panel application to give access to all the functions of the RS-UV3. The RS-UV3 is controlled via the Arduino Serial port, and because it was designed as a stand-alone transceiver, there are no specific Arduino libraries needed to use the RS-UV3. The RS-UV3 has an optional aluminum case available. However, if you plan to use the RS-UV3 with this case, there is not enough room to stack it on top of an Arduino and fit everything inside the case. HobbyPCB also has an optional 5 W VHF/UHF amplifier that works with both the RS-UV3 and the HamShield.
A complete review of the HobbyPCB can be found in the April 2016 issue of *QST*.

Argent Data Radio Shield

The Argent Data Radio shield (see **Figure 2.2**) can be used to provide AX.25 packet radio send and receive capability to the Arduino. Packets are sent and received in AX.25 UI frames at 1,200 baud, allowing operation on the VHF APRS network. There are a number of projects that use this board in ARRL's *Ham Radio for Arduino and PICAXE* by Leigh Klotz, WA5ZNU. The Argent Data Radio shield also includes a prototyping area and an HD44780-compatible LCD interface.

Liquid Crystal Display (LCD) Shield

While it has largely been replaced by the newer display devices, one of the first displays that just about everyone uses when starting out with the Arduino is the 16-character by 2-line (16 × 2) LCD display (see **Figure 2.3**). Also available as a stand-alone module, the 16 × 2 LCD shield is a quick and easy way to add a display to your Arduino project. Many LCD shields also include several push-button switches, allowing you to add control functions to your Arduino sketches.

Color TFT Display Shield

To give your Arduino project a really nice look, you can use one of the many color graphic thin-film-transistor (TFT) displays, such as the one shown in **Figure 2.4**. These high-resolution color graphic displays come in various sizes, from 1.8 inches (128 × 160 pixels), all the way up to a monster 7-inch (and beyond) 800 × 480-pixel display. The Adafruit Color TFT shield features a 1.8-inch TFT display along with an onboard microSD card slot you can use to store images for display. The 1.8-inch version of this shield also includes a five-way joystick, while the 2.8-inch version incorporates a resistive touch screen. Using the powerful Adafruit GFX library, these displays allow you to rotate the entire screen via software, draw pixels, lines, rectangles, circles, triangles, and other shapes, along with the ability to display normal text.

Recently, some of the larger and inexpensive color TFT displays have become available from websites such as eBay. These displays

Figure 2.4 — Adafruit color TFT shield. (Photo courtesy Adafruit.com)

Figure 2.5 — 3-inch TFT and shield.

Figure 2.6 — 7-inch TFT and shield.

Figure 2.7 — DFRobot 4 channel relay shield. (Photo courtesy DFRobot.com)

Figure 2.8 — DFRobot rotor driver shield. (Photo courtesy DFRobot.com)

range in size from 3 – 7 inches (and even larger). Due to the number of I/O pins needed to interface with these displays, you will need a display shield similar to the ones in **Figures 2.5** and **2.6**. This shield is available for both the Arduino Uno and Arduino Mega footprint boards and will allow you to interface with many of the larger TFT displays now available. While these larger TFT displays can be used with the Arduino Uno, due to the number of I/O pins these larger displays require, you can't use the touch screen features of the display with the Uno-type boards. For that, you would need a board with more available I/O pins, such as the ATmega2560 and the matching LCD shield.

Relay and Motor Shields

Because controlling stuff is one of the things that the Arduino does best, a relay shield such as the one shown in **Figure 2.7**, or a motor shield (see **Figure 2.8**), is often the quickest, easiest, and safest way to interface your Arduino to high current and high voltage devices such as motor and power control systems. The DFRobot relay shield, for example, allows you to control four onboard relays, and it includes test buttons and indicator LEDs.

The motor driver shield allows you to control motors directly from your Arduino. The DFRobot motor driver shield can drive two 1-amp, bidirectional, dc motors.

Figure 2.9 — Ethernet shield.

Figure 2.10 — SparkFun USB host shield.

Ethernet Shield

The Ethernet shield (see **Figure 2.9**) gives your Arduino the ability to communicate over your home network and the internet. With its powerful libraries and example sketches, you can turn your Arduino into a web-enabled simple chat server, web server, FTP server, or telnet client. It can be used to allow your Arduino to make http requests from websites and much more. The Ethernet shield also has an onboard microSD card slot that can be used to store files and other data. Some Ethernet shields also include an option to install a Power-over-Ethernet (PoE) module, allowing your Arduino to be powered via the Ethernet cable using a PoE switch or power injector.

USB Host Shield

The USB Host Shield (see **Figure 2.10**) adds USB host capability to your Arduino project, allowing you to communicate with USB devices such as keyboards, mice, joysticks, game controllers, and many others.

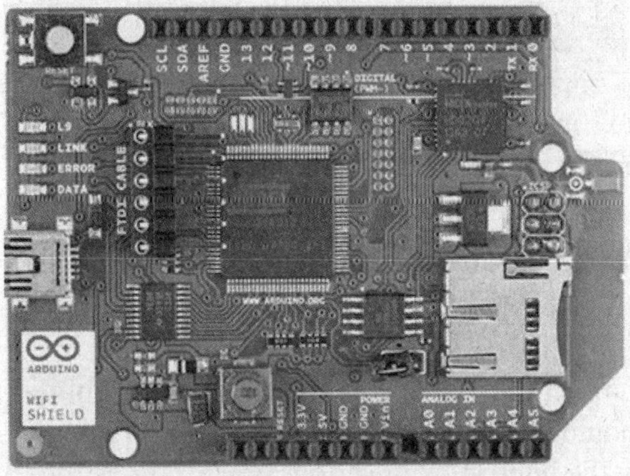

Figure 2.11 — Wi-Fi shield. (Photo courtesy Arduino.cc)

Wi-Fi Shield

While many of the newer Arduino processor boards have Wi-Fi built in, you may want to experiment with Wi-Fi by using something like an Uno and a Wi-Fi shield. The Arduino Wi-Fi shield (see **Figure 2.11**) allows the Arduino to connect to IEEE 802.11 b/g networks and supports the WEP and WPA encryption protocols. With the proliferation of Arduino boards and variants that include Wi-Fi support onboard, the

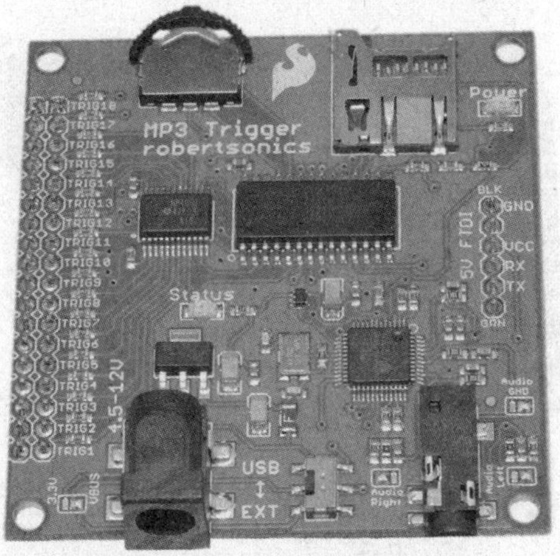

Figure 2.12 — SparkFun MP3 trigger.

Figure 2.13 — Adafruit Power boost shield.

Wi-Fi shield is not as popular as it once was, but it is still a quick and easy way to add Wi-Fi capability to your Uno-based projects. Arduino.cc and other suppliers no longer offer some of the original Wi-Fi shields, but they are still available from suppliers such as eBay.

Audio Shields

Some of your Arduino projects may involve making sounds or playing audio, as in the case of a repeater identifier or contest voice keyer. Using one of the various audio shields, you can quickly and easily add high-quality sound and even music to your Arduino projects. The SparkFun MP3 Trigger (see **Figure 2.12**) has 18 external trigger pins that allow you to directly play a selected MP3 file from the onboard microSD memory card. The MP3 Trigger shield has the capability for remote triggering of up to 256 separate MP3 files with up to a 192 kHz data rate.

Figure 2.14 — Uno I/O sensor shield.

Power Boost Shield

With its low power consumption, the Arduino is ideal for portable, battery-powered projects. While 9 V batteries or AA battery packs do work — and I use them in a number of my projects — sometimes a self-contained rechargeable power source would do the job a whole lot better and a lot more efficiently. For that, there is the Adafruit PowerBoost shield shown in **Figure 2.13**. This shield allows you to attach a high-capacity rechargeable lithium polymer (LiPo) battery directly to the Arduino. The shield contains a built-in battery charger as well as a dc-to-dc converter to provide the 5 V needed by the Arduino Uno-style boards. Capable of providing at least 500 mA of output current, with a peak output up to 1 A, the shield can use any 3.7/4.7 V Li-ion or LiPo battery, with the recommended battery being a 1,200 or 2,000 mAH capacity LiPo battery.

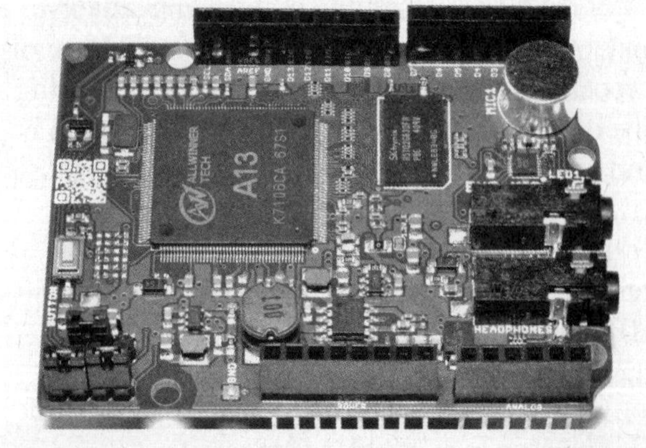

Figure 2.15 — The Audeme MOVI™ Arduino shield.

I/O Sensor Shields

Using an I/O sensor shield (see **Figure 2.14**) is a great way to begin creating your own Arduino projects. This shield brings all the Arduino pins out to groups of 3-pin headers, along with power and ground on each header group. This particular shield also has header sockets that can be used with either the Uno or Nano board footprint. This shield allows you to easily connect the I/O pins to your breadboard and prototyping boards using jumper wires while developing your project. Once development is complete, all you do is remove the connecting wires, and your Arduino is ready for the next project.

Figure 2.16 — The Dr.Duino Explorer shield.

Voice Recognition and Speech Shield

I have had some wonderful conversations with people who have contacted me about some of the projects in my books. I probably should have expected it, but I was surprised by the interest from various organizations for the vision-impaired, asking for more voice-related Arduino projects. The Audeme MOVI™ Arduino shield (see **Figure 2.15**) has some very interesting potential for use in ham radio projects. Adding voice recognition to Arduino projects would take them to the next level and allow you to create some truly practical and fun projects. The MOVI™ performs both speech synthesis and speech recognition. Adding voice recognition is definitely at the top of the list for any new projects.

The Dr.Duino Shields

The Dr.Duino Shields provide a great way to work with the Arduino. Several of the projects in this book started out as proof of concept and prototypes using the Dr.Duino Explorer shield (see **Figure 2.16**). The Dr.Duino shields have several of the most commonly used components when learning and working with the Arduino. For example, the Dr.Duino Pioneer (see **Figure 2.17**), has three LED outputs, three potentiometers for analog inputs, four push-button switches, a piezo buzzer, and an eight-LED addressable RGB LED strip, along with connections for other sensors and devices such as a servo, an ultrasonic sensor, a light-dependent resistor, and a passive IR sensor, all on a shield that fits the Uno-type boards. The Pioneer is an excellent starter shield for learning how to use the Arduino. For more information, please see the Dr.Duino Pioneer product review in the November 2020 issue of *QST*.

The Dr.Duino Explorer shield is targeted to be a mini-development platform. The Explorer has two breadboard areas: the standard temporary wire-jumper type, and a second area with solder pads for a more permanent use. The Explorer features four LED outputs, four push-button inputs, three potentiometers for analog input, a light-dependent resistor, an HC-05 Bluetooth module, a 128 × 32 Organic LED display, a 5 V 3 A dc-to-dc converter, and an eight RGB-addressable LED strip, along with additional headers for external connections for an atmospheric sensor, an ultrasonic proximity sensor, a servo, and other external device connectors. The Explorer also has all the Arduino I/O pins brought out to jumpers on the board,

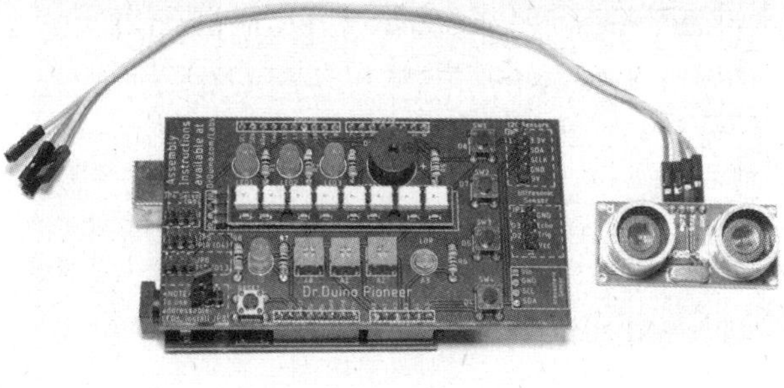

Figure 2.17 — The Dr.Duino Pioneer shield.

Figure 2.18 — The prototyping shield.

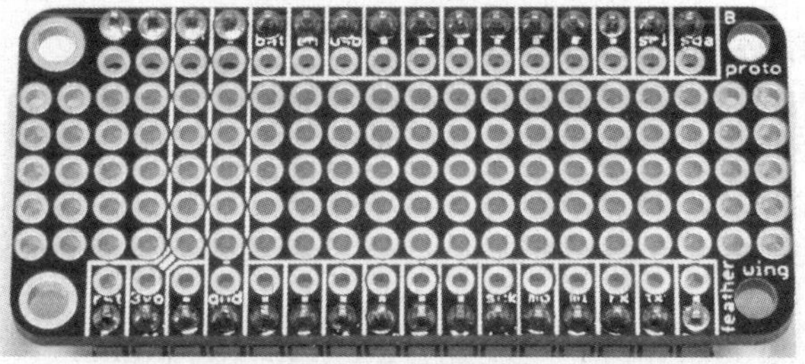

Figure 2.19 — Adafruit FeatherWing Proto.

allowing you to reroute one or more I/O pins to the Explorer onboard components or pass these signals on to an Uno-type shield attached to the Explorer board. The Explorer has header sockets for either an Uno or Nano processor. The Explorer is an ideal tool to build and test your project prototypes. You can read a full product review on the Dr.Duino Explorer in the January 2021 issue of *QST*.

Prototyping Shield

Once you have finished breadboarding and testing your project, you may want to create a more permanent version that involves wiring and soldering components. I have found the easiest method to interface my projects to the Arduino is with a prototyping shield (see **Figure 2.18**). For their Feather line, Adafruit has the FeatherWing Proto (see **Figure 2.19**) that can be stacked on top of a Feather board like a standard Arduino shield. The protoshield brings all the Arduino pins and voltages to solder pads, with the center of the prototyping shield laid out like a standard prototyping board. I like to solder DuPont-style

2.54-millimeter header pins in the prototyping area so that I can wire the Arduino pins to the header and then use a connector to interface to my offboard modules and components. This also allows for easy troubleshooting, reconfiguration, and replacement of the external parts as needed. An added benefit to using prototyping shields is that you can quickly swap shields and sketches as needed and use the same Uno-style board for many different projects.

Arduino Modules

While shields provide a quick and easy way to interface with the Arduino, the real fun begins when you interface your Arduino with external modules and components. With just a handful of modules and parts, you can wire up some pretty amazing projects. Most of the modules for the Arduino include software libraries, which greatly simplify working with these modules, and you can create fully functional and powerful Arduino projects in just a few short hours.

Using individual modules and components also helps lessen the pain if and (in my case) when you do horrible things to your project and challenge Ohm's Law (hint: Ohm's law usually wins and signifies its victory by sending out smoke signals). Shields are usually more expensive than individual modules and components. With many shields, when something goes horribly wrong, the entire shield no longer works as intended and you have to buy a new one. With the lower cost of individual modules and components, you shed fewer tears, and if you're like me, you probably have more than one willing victim in your now rapidly growing supply of parts. Many of the various modules and components can be mounted to your project using header sockets, allowing for quick and easy replacement if bad things should happen.

In reality, I have found that the Arduino, shields, and modules are very forgiving when you miswire them. As long as you pay attention to the 3.3 and 5 V differences between boards and components, you have to really work to damage them. I'm not saying it can't be done, and it's amazing just how much smoke a tiny surface-mount component has inside of it, but you really have to do something wrong to get things to that point.

Another big advantage of using modules instead of shields is that projects using shields tend to be bulky and not very aesthetically pleasing. Modules give you control over the size and shape of your project, allowing you to determine what the finished project will look like. To this end, there are a number of enclosures designed for the Arduino and Arduino projects that can really spice up the look of your completed project. Also, you are not limited to enclosures designed for the Arduino. I've found a number of suitable enclosures at craft stores that are ideal for Arduino projects. In this next section, we will introduce and briefly discuss the various modules and components that I have found to be most useful for creating Arduino-powered ham radio projects. Again, this is by no means an exhaustive list, as new modules and components for the Arduino are being released every day. The modules and devices shown here are the ones most suited for Arduino ham radio projects.

Direct Digital Frequency Synthesis (DDS) Module

Prior to the development of the direct digital frequency synthesis (DDS) module, the only way to generate a radio frequency (RF) signal was to use a crystal or an unstable (relatively speaking) resistive-capacitive (RC) oscillator. Using a DDS module such as the AD9850 DDS module shown in **Figure 2.20**, you can programmatically generate highly stable wave-

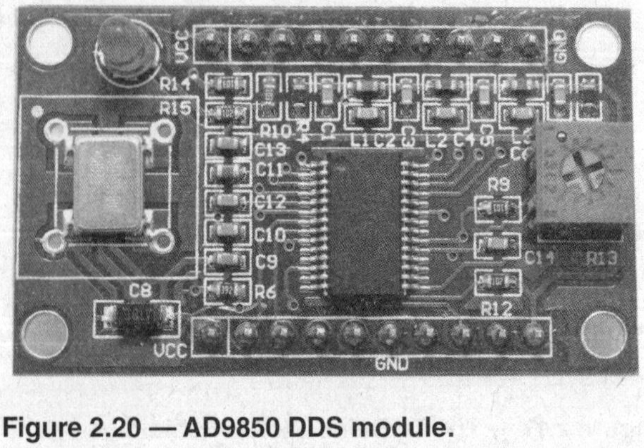

Figure 2.20 — AD9850 DDS module.

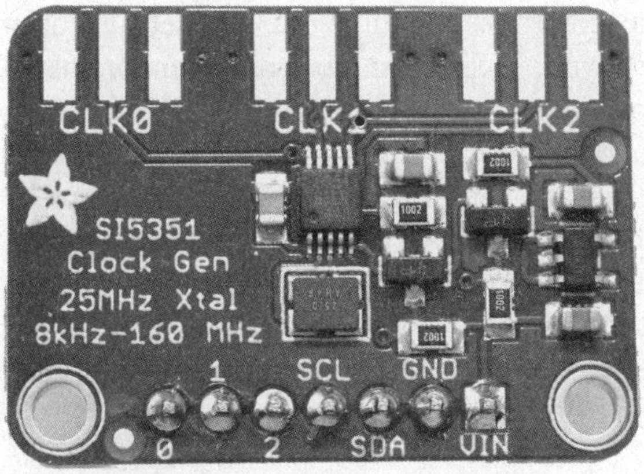

Figure 2.21 — Si5351 programmable clock generator module.

forms up to 62.5 MHz, with an accuracy of 0.0291 Hz. Other DDS modules support higher frequencies than the AD9850, but I have found the inexpensive AD9850 to be more than adequate for most Arduino ham radio projects. A DDS generates a smooth sine or square wave, while some DDS modules, such as the AD9833, also allow you to generate triangle waves. One of the major advantages of using a DDS module over processor-generated waveforms is that the DDS chip offloads all the frequency and waveform generation from the processor and, in the case of the AD9850, uses a 10-bit digital-to-analog converter on the chip to produce clean, stable waveforms across the entire operating range. The AD9850 DDS also has the capability of shifting the phase of the output waveform from 0 – 2π (0 – 720°), and interfaces to the Arduino using the SPI bus. There are a number of libraries supporting the DDS modules, including a very simple and easy-to-use library for the AD9850 written by Paul Darlington, MØXPD.

Figure 2.22 — Mini USB host module.

Si5351 High-Frequency Programmable Clock Generator Module

The Si5351 (see **Figure 2.21**) High-Frequency Programmable Clock Generator Module is similar to a DDS in that both generate highly precise frequencies. In the case of the Si5351, it can simultaneously generate up to eight different frequencies from 8 KHz to 160 MHz, although many of the inexpensive Si5351 modules currently available use a three-output version of the chip. The Adafruit Si5351A module communicates with the Arduino using the I^2C bus and has three independent outputs, each capable of providing 3 V peak-to-peak square waves. The Si5351 is a 3.3 V device, so you will need level shifters on the Si5351 I/O pins if you are using a 5 V Arduino. One major difference between a typical DDS and the Si5351 is that the Si5351 only has a square wave output, as compared to the sine and square wave output of the AD9850 DDS. This means that an unfiltered signal generated by the Si5351 will also generate unwanted harmonics at multiples of the original frequency. For this reason, if you are going to use the Si5351 as part of a transmitter, for example, you will want to add a filtering section to smooth the square wave into a sine wave to reduce harmonic output. Adafruit and others have excellent libraries that support the Si5351 and allow you access to all the features of the Si5351 chip.

USB Host Module

While technically a mini shield designed for use with the Arduino Pro Mini, the Mini USB Host shield/module shown in **Figure 2.22** is ideal for adding USB host capability to your Arduino project. The Arduino USB Host library designed for the full-sized USB Host shield works with this module as well, allowing you to easily incorporate USB devices, including USB Human Interface Devices (HID) to your Arduino projects. The USB Host module also has eight general-purpose input and eight general-purpose output pins that are accessible through the USB host library.

Figure 2.23 — Adafruit Ferroelectric random access memory (FRAM) module.

Ferroelectric Random Access Memory (FRAM) Module

Back when computers were just getting started and took up whole rooms (or racks as they got smaller), they predominantly used ferrous core memory, which under magnification looks like a whole bunch of miniature toroids all woven together. With the advent of dynamic and static RAM chips, the good old core memory pretty much disappeared from use. Now it appears there is a new version of Ferroelectric Random Access Memory (FRAM) (see **Figure 2.23**) that is non-volatile, meaning it retains the stored information even after power is turned off, similar to the Flash memory used by the Arduino to store your sketches. Unlike Flash and EEPROM, however, FRAM is much faster to write to, has a much longer lifespan than Flash memory, and can be randomly accessed. Flash and EEPROM memory can typically be rewritten only about 100,000 times, whereas FRAM can be rewritten 10,000,000,000,000 (ten trillion) times and can retain data for up to 95 years at room temperature. FRAM comes in multiple sizes, typically from 64 kilobit (8 kilobyte) to 256 kilobit (32 kilobyte).

Relay Modules

There are many choices in relay modules for the Arduino. SainSmart has several interesting options for adding relay control to your project. Their iMatic series of relay control boards offers wired-Ethernet or wireless Wi-Fi-based relay control for up to 16 relays. They also offer a Network Web Server controller version that allows you to remotely operate the relays through an onboard web server. SparkFun also offers a high-power 40 A, 28 – 380 V ac 4-channel solid-state relay board with a preprogrammed Atmel ATTiny84 microcontroller onboard, and it allows you to daisy-chain more than 100 of these boards together for some serious ac power control configurations.

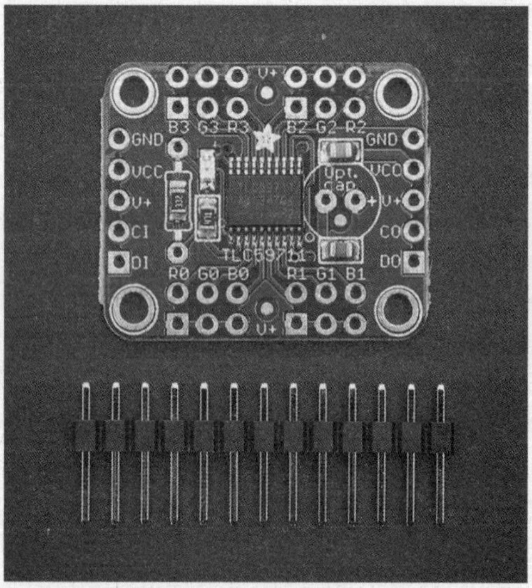

Figure 2.24 — Adafruit 12 channel module. (Photo courtesy Adafruit.com)

Figure 2.25 — Adafruit 24 channel module. (Photo courtesy Adafruit.com)

Stepper Motor Driver module

There are several stepper motor drivers available for the Arduino. One of these is the SparkFun ProDriver, based on the TC78H670FTG bipolar stepper motor driver chip. The ProDriver can be controlled either by clock-input stepping or serial commands, and has high resolution stepping control down to 1/128 of a step.

RobotDyn nRF24L01 2.4 GHz ISM Band Transceiver Module

The RobotDyn 2.4 GHz ISM band transceiver module is an ultra-low power 2.4 GHz transceiver that operates in the industrial, scientific, and medical (ISM) band and is capable of 2 Mbps data rates.

LED Driver Modules

With the advent of the new addressable RGB LEDs, you may find that you don't have enough Pulse-Width Modulation (PWM) pins on your Arduino processor board to build the project you have in mind. To solve that problem, rather than using a different Arduino board with more PWM pins, you can use a PWM LED driver board. Adafruit has a 12- and 24-PWM channel LED driver board (see **Figures 2.24** and **2.25**), capable of 16- or 12-bit control of the PWM duty cycle, respectively. These boards require just two or three SPI pins to operate, based on the board you select. You can even drive LEDs in series using these boards. SparkFun also has a 16-channel LED Driver module with a 12-bit PWM duty cycle control, a 6-bit current limit control, and can be daisy-chained with other boards.

Keypad Modules

In addition to the thin membrane keypads that have been associated with the Arduino from the early days, several new options for keypads are now available. Adafruit has several choices, including the original thin membrane keypad. The latest addition to their keypad lineup is the NeoTrellis 4 × 4 key matrix board with a silicone elastomer keypad cover and enclosure, as shown in **Figure 2.26**. This board is available in several styles; one version uses monochrome 3-millimeter LEDs to backlight each key, while the other uses addressable RGB LEDs. The NeoTrellis boards can also be connected to use up to 32 boards (512 keys) in a single configuration.

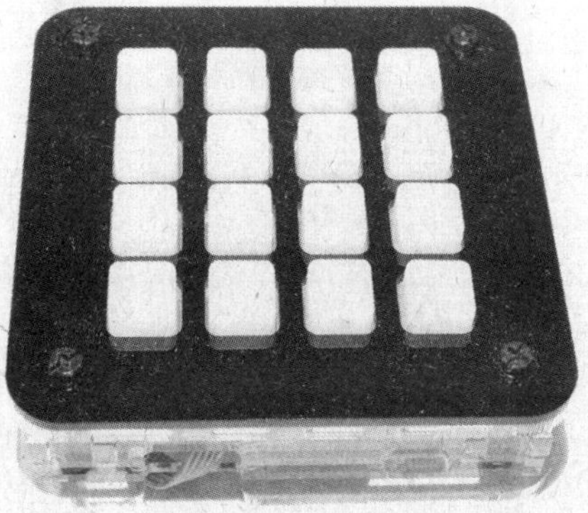

Figure 2.26 — Adafruit NeoTrellis 4 × 4 keypad and enclosure.

Figure 2.27 — 4 × 4 Keypad and Enclosure.

Banggood.com and other suppliers also offer a 4 × 4 keypad (see **Figure 2.27**) with removable keys and key covers that allow you to use your own key labels. It comes with blank keys, so you will need to create your own labels. I used laser-printed decal paper to create mine and transferred them to the keycap just like you would apply a decal to a model car or airplane. Unlike other keypads, this keypad outputs an analog voltage unique for each keypress, and then can be read using an analog input pin on the Arduino. It needs only power, ground, and an analog input pin to operate. It's available from many of your usual Arduino parts suppliers, including **Banggood.com** and eBay.

Figure 2.28 — The INA169 current sensor module.

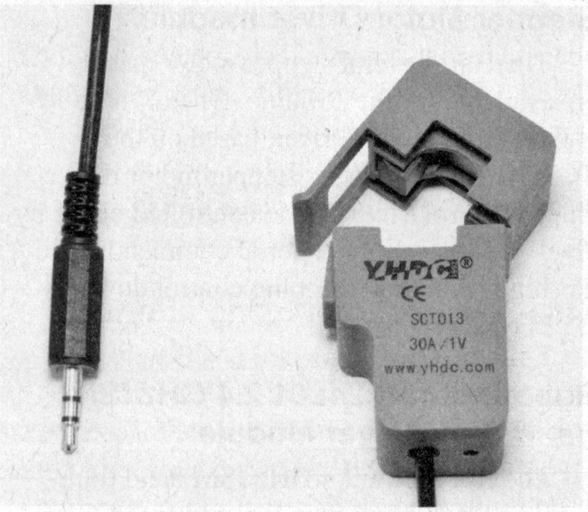

Figure 2.30 — Non-Invasive ac current sensor.

Current Sensors

There are now a number of current and voltage sensing modules, both invasive and non-invasive, for the Arduino. Starting out small, there is the INA169 (see **Figure 2.28**) that can measure the dc current in a circuit. A "high-side" current sensor, the Texas Instruments INA169 chip converts the dc current flow across a shunt resistor into an output voltage that corresponds to 1 V per amp, which can then be read using an analog input pin on the Arduino. Capable of monitoring up to 5 A at a maximum of 60 V continuously, the shunt resistor can be changed to increase or decrease the sensitivity of the sensor.

Another current sensor you can use is the Hall-Effect current sensor (see **Figure 2.29**). Available in multiple ranges from ± 5 A to ± 75 A, a Hall-Effect current sensor uses the electromagnetic field flowing through a conductor to measure current, thereby isolating the current source from the sensing electronics itself, providing a measure of protection for your measurement device.

Next, there is the non-invasive ac current sensor shown in **Figure 2.30**. The non-invasive current sensor clamps around one leg of an ac circuit, and through the use of a precision transformer, converts the current flowing in the

Figure 2.29 — 30 A Hall-Effect current sensor.

Figure 2.31 — CMJU-116 9-Axis IMU module.

ac circuit into a low voltage that can be rectified and read using an analog input pin on the Arduino. One of the benefits of a non-invasive current sensor is that you do not need to insert the sensor into the circuit path, providing a measure of isolation and protection for your measuring device.

Nine-Axis Motion/Position Sensor

The latest in motion and position sensing modules is the nine-axis, or 9 Degree of Freedom (9-DoF) sensor. Based on the InvenSense MPU-9250, the CMJU-116 module (see **Figure 2.31**) incorporates a three-axis gyroscope, a three-axis accelerometer, and a three-axis magnetometer in a single small module, allowing you to create projects that can include a built-in compass, as well as azimuth and elevation sensing, and even motion sensing. This would be an ideal module for a handheld satellite tracking antenna or high-altitude balloon telemetry projects, among others.

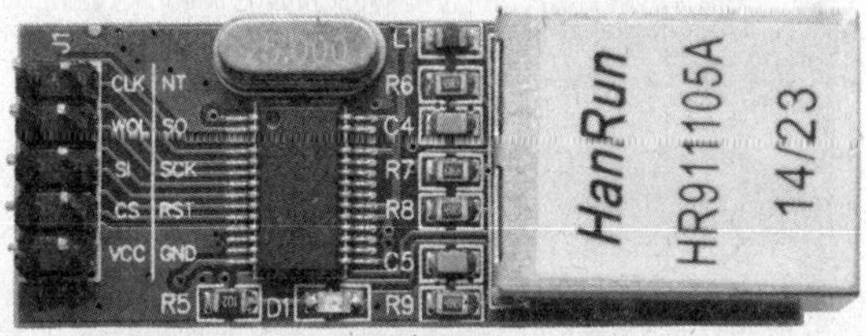

Figure 2.32 — ENC28J60 Ethernet module.

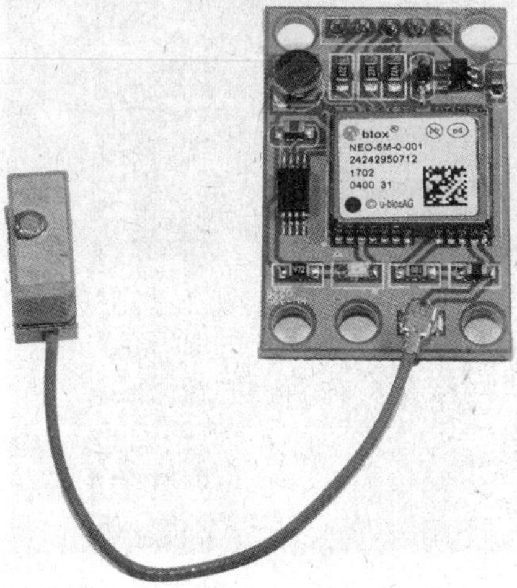

Figure 2.33 — ublox NEO-6M GPS module.

Figure 2.34 — ISD1820 voice recorder and playback module.

ENC28J60 Ethernet Module

The ENC28J60 Ethernet Module (see **Figure 2.32**) allows you to quickly add wired 10/100 Mb/s Ethernet networking capability to your Arduino projects. Easily interfaced to the Arduino via the SPI bus, the Ethernet module is supported by a number of Arduino libraries and allows you to turn your Arduino into a web server, FTP file server, and even a Network Time Protocol (NTP) server.

GPS Module

With a GPS module such as the one shown in **Figure 2.33**, you can add all the power of a standard GPS to your Arduino, providing such information as time accurate as 1/100 of a second, as well as latitude, longitude, altitude, speed, course, and other features. These GPS modules communicate with the Arduino via a standard TTL serial I/O port, typically with a default speed of 9,600 baud, and they output standard National Marine Electronics Association (NMEA) NMEA-0183 messages. The TinyGPS and TinyGPS+++ Arduino libraries by Mikal Hart at **arduiniana.org** parse the NMEA messages sent by the GPS into data that can be accessed using simple function calls from within your Arduino sketch.

Figure 2.35 — DFRobot speech module.

Figure 2.36 — HC-05 Bluetooth module.

ISD1820 Voice Recorder and Playback Module

The ISD1820 (see **Figure 2.34**) is capable of recording and playing back an 8- to 20-second message. The record duration and sampling frequency is selected by an external resistor. The module even includes an onboard microphone and LM386-driven speaker output, along with push buttons to activate the recording and playback functions.

DFRobot Gravity Text-to-Speech Module

The DFRobot Gravity Text-to-Speech Module (see **Figure 2.35**) allows you to add natural sounding speech to your Arduino projects. Capable of speaking in English and Chinese, the Gravity communicates with the Arduino via I^2C or by using a standard TTL serial I/O port, and includes an onboard audio amplifier and speaker.

Bluetooth Module

There are two basic modes for the standard Arduino Bluetooth modules: subnode and main. The HC-05-type Bluetooth module shown in **Figure 2.36** supports both modes. In subnode mode, you can pair your Arduino to a device such as a workstation or cell phone and have the Arduino act as a mouse, keyboard, or other Human Interface Devices (HID). In main mode, you can pair and connect multiple Bluetooth subnode modules to your Arduino. Be careful if you want to connect a Bluetooth keyboard or mouse to your Arduino, as it will need to be an HID-compliant unit, and the vast majority of the standard Arduino Bluetooth modules, such as the HC-05, do not support HID devices. You might consider using a USB host shield and Bluetooth dongle, or selecting an Arduino-variant processor board with onboard Bluetooth support for your Arduino Bluetooth HID projects.

Some of the currently available Bluetooth modules can be switched between main and subnode mode from within your Arduino sketch.

Figure 2.37 — Level Converter module.

The Arduino communicates with the Bluetooth module using standard TTL serial communications, usually at a speed of 9,600 baud.

Be careful and verify the specifications for your particular Bluetooth module, as some accept power in the range of 3.6 – 6 V, but allow only a maximum of 3.3 V on the data pins. You can use a simple resistor divider network, or a level converter module to protect the data pins from overvoltage.

Level Converter Module

With the increasing number of 3.3 V Arduino boards and modules, you may find yourself needing to convert signal levels from 3.3 V to 5 V and vice versa. Connecting 3.3 V pins to a 5 V source can often cause damage to the 3.3 V device. Fortunately, you can use a level converter module to convert between the two signal voltage levels. The Logic Level Converter shown in **Figure 2.37** can bi-directionally convert between 3.3 V and 5 V signal levels on up to eight pins.

Real-Time Clock Module

The DS3231 shown in **Figure 2.38** is an inexpensive but extremely accurate real-time clock (RTC) module for the Arduino. A vast improvement on earlier real-time clock modules, the DS3231 chip has an internal temperature-compensated crystal oscillator (TCXO). Since the Arduino does not have a real-time clock onboard, the DS3231 is ideal for those projects where you need to keep track of date and time. The DS3231 module communicates with the Arduino via the I²C bus and has an onboard coin-cell battery to maintain the clock when the power is off. As with earlier RTC modules, the DS3231 can keep track of time, day, month, and year, with built-in leap-year compensation up to the year 2100. The DS3231 also has a built-in temperature sensor accurate to ± 3° C. As with most Arduino modules, the DS3231 is supported by excellent libraries, including my personal favorite, the DS3231 library from Henning Karlsen at Rinky-Dink Electronics (formerly **henningkarlsen.com**).

Figure 2.38 — DS3231 real-time clock module.

Figure 2.39 — Lightning sensor module.

Lightning Sensor Module

The Sparkfun Lightning Sensor module shown in **Figure 2.39** is based on the Austriamicrosystems (now ScioSense) Franklin AS3935 Lightning Sensor chip and can detect lightning at a distance of up to 40 kilometers. The AS3935 is capable of detecting both cloud-to-ground and cloud-to-cloud lightning. The AS3935 includes a proprietary embedded algorithm to reject man-made electrical noise software-selectable threshold settings. The AS3935 will also statistically calculate the distance to the leading edge of the thunderstorm, along with the estimated strength of the lightning strike.

Figure 2.40 — ADS1115 16-bit A/D module.

Figure 2.41 — MCP4725 D/A module.

Analog-to-Digital Converter Modules

The analog-to-digital (A/D) converter used for the analog input pins on the Arduino is a single-ended 10-bit A/D, providing a count of 0 – 1,023 over a 5 V range with respect to ground. For many of your Arduino projects, this resolution is more than adequate; however, there will be occasions, as in some of the antenna rotator control projects, where you will want higher resolution in your analog voltage measurements than the A/D onboard the Arduino allows. In the case of an antenna rotator position sensor, you will want to read a position in degrees from 0° – 360° (or 450° in the case of some rotators). Using a 10-bit A/D, this leaves you with a resolution of about 1°. A 12-bit A/D would provide a resolution of about 1/10°, and a 16-bit A/D would provide you with a resolution of about 1/100°. As you can see, the more bits your A/D has, the higher the resolution.

The Texas Instruments ADS1015 12-bit A/D module can provide a count of 0 – 4,095 on four single-ended or two differential inputs, with a program-selectable sampling speed of up to 3,300 samples per second. The TI ADS1115 16-bit A/D module (see **Figure 2.40**) can provide a count range of 0 – 65,535 on four single-ended or two differential inputs, with a program-selectable sampling speed up to 860 samples per second. Both communicate with the Arduino using the I^2C bus and feature six program-selectable gain settings, from a full-scale reading of 6.144 V (do not ever exceed VDD + 0.3 V on any input pin; you will never be able to read more than the supply voltage at the 2/3 gain setting) all the way down to a full-scale reading of 0.256 V at the 16 x gain setting. The Arduino libraries for these modules allow access and control of all the internal parameters and make using these modules in your Arduino projects a breeze.

Digital-to-Analog Converters

Performing the reverse function of an A/D module, a digital-to-analog module (D/A) converts digital data into an analog voltage representation of the value of the digital data. The one thing the Arduino is missing is an onboard D/A converter, although some of the newer variants have this feature built-in. In trying to add this functionality to my Arduino projects, I have built and used several types of D/A converters, from a simple 8-bit resistive ladder all the way up to an I^2C D/A module. The results of these

Figure 2.42 — MCP23008 8-pin I/O expander.

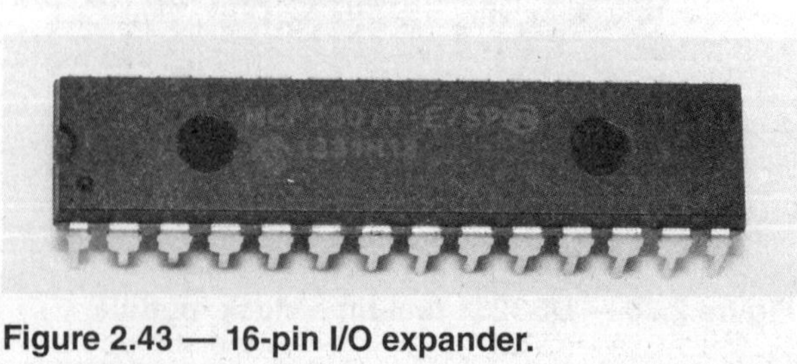

Figure 2.43 — 16-pin I/O expander.

tests showed that the Microchip MCP4725 12-bit I^{2}C D/A converter shown in **Figure 2.41** to be my D/A of choice, both for its simplicity and ease of use. Using the MCP4725 Arduino library and sample sketches, you can quickly have your Arduino outputting analog voltages, sine waves, and even triangle waves, among other waveforms.

I^{2}C Digital I/O Expanders

Sometimes, the 14 pins of digital I/O available on the standard Arduino aren't enough, and you don't want to splurge on a Mega or a variant with more I/O pins. Using an 8- or 16-pin I^{2}C I/O expander chip can get the job done and cost you a whole lot less. The Microchip MCP23008 I^{2}C serial I/O expander chip (see **Figure 2.42**) will add eight pins, and the Microchip MCP23017 (see **Figure 2.43**) will add 16 pins of digital I/O to your Arduino. Fully supported by Arduino program libraries, the serial I/O expander chips communicate with the Arduino via the I^{2}C bus, and they have programmable address settings, allowing for up to eight expander chips on a single I^{2}C bus. This means you can add up to 64 or 128 pins of digital I/O to your Arduino project to handle those bigger projects. The MCP23S08 and MCP23S17 provide similar capability but communicate with the Arduino using the SPI bus instead of the I^{2}C bus.

Figure 2.44 — HT9200B DTMF tone generator.

Figure 2.45 — MT8870D DTMF receiver chip.

Figure 2.46 — Close-up showing the WS2812B chip embedded in the LED itself.

Figure 2.47 — Addressable RGB LED stick.

I^{2}C Bus Extender

While the average usable length for an I^{2}C Bus connection is typically 9 – 12 feet, the Texas Instruments P82B715 I^{2}C Bus Extender chip can be used to buffer and extend the bus to a length of 50 meters (164 feet).

DTMF Tone Generator

Dual-tone multi frequency (DTMF) is a mainstay in amateur radio. Just about every modern mobile radio has a DTMF tone keypad, and many repeaters utilize DTMF for control and other functions. DTMF is a simple way to control many of your Arduino projects remotely, without a lot of complex circuitry. The Holtek HT9200 DTMF Generator shown in **Figure 2.44** allows your Arduino to generate the standard 16 DTMF tones, as well as 8 single tones. The HT9200B allows for both a serial and parallel mode interface and easily interfaces with the Arduino without the need for a separate library.

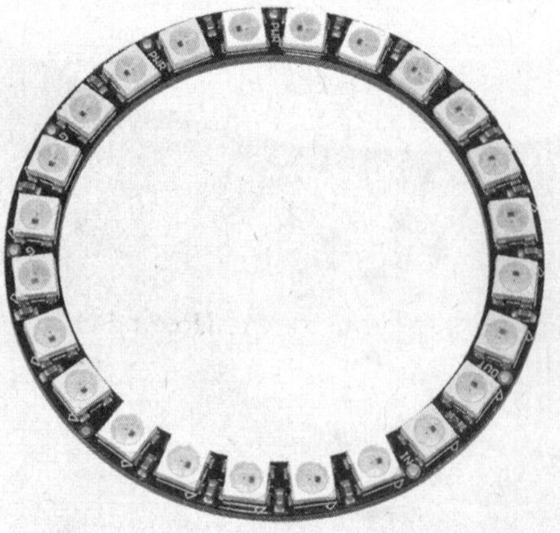

Figure 2.48 — Addressable RGB LED ring.

DTMF Receiver

Of course, if you're going to use a DTMF Generator, you most likely will also need a DTMF receiver. The Zarlink MT8870D shown in **Figure 2.45** is a DTMF receiver chip that converts the 16 DTMF tones into a 4-bit digital output. The only major external component required is a 3.79545 MHz crystal for the oscillator circuit portion of the chip. This is a very inexpensive crystal that was originally designed for use in extracting the colorburst portion of a television signal, and it is readily available online from places such as eBay and most electronics parts suppliers.

Addressable LEDs

Individually addressable LEDs are becoming very popular in Arduino projects. Adafruit refers to their version of these LEDs as NeoPixel LEDs. These LEDs can be wired together as individual LEDs or purchased in sticks or strips that can be up to 10 meters long, and they can be daisy-chained together to create even longer strips. The WS2812B Intelligent controller chip embedded inside each 5050-type surface-mount RGB LED often powers these LEDs, as you can see in **Figure 2.46**. While they do come with plain white LEDs, the RGB color versions are far more popular. These LEDs are available in sticks (see **Figure 2.47**), strips, rings (see **Figure 2.48**), and even quarter-circle arcs for the larger LED rings. The individual LEDs are usually pulse-width modulated at 400 Hz or higher, which keeps the power utilization much lower than you would expect. For example, a

Figure 2.49 — 16 × 2 LCD display.

10-meter (32.8 feet) flexible strip generally has 300 individual LEDs. The super-bright 5050-type LEDs used in most strips can draw up to 50 mA per LED. The math says if you have all 300 LEDs lit at full brightness, the current draw would be 15 A. In reality, you rarely operate at full brightness, and in actual operation, you'll see that the current consumption will be substantially lower. Still, you need to be aware of the power requirements for larger numbers of LEDs in your projects and should consider adding an external fused power source to power your LED strips.

Arduino Displays

Many different types of displays can be used with the Arduino. Some, such as the basic 16-character by 2-line LCD display, can be interfaced simply by connecting to the Arduino's digital I/O pins, while others can be interfaced using standard TTL serial communications, or the serial peripheral interface (SPI) and I^2C buses.

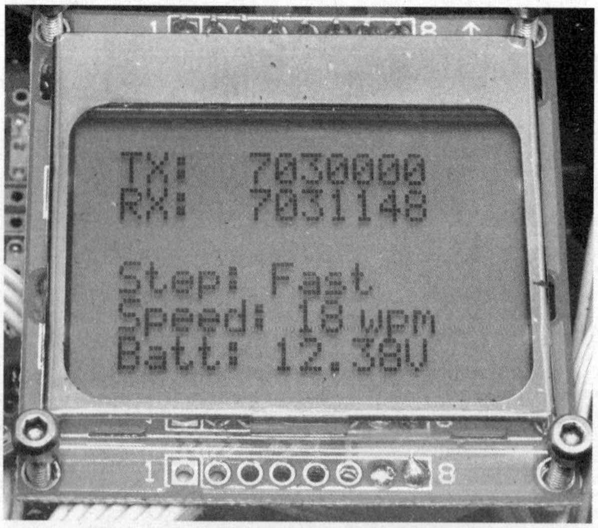

Figure 2.50 — The Nokia 5110 LCD display.

Figure 2.51 — Three types of the Nokia 5110 (front view).

Figure 2.52 — Three types of the Nokia 5110 (rear view).

The most commonly used display (although becoming a bit outdated) for the Arduino is the 16-character by 2-line (16 × 2) LCD shown in **Figure 2.49**. A larger 16 × 4 version of this display is also available. Both interface with the Arduino using six digital I/O pins, two for control and four pins for data. Libraries and example sketches for these displays are built into the Arduino IDE. Newer versions of these displays are also available that support I²C and/or TTL serial communication interfaces. I generally prefer the I²C version of these displays, as they only require a simple 4-wire connection of +5 V, ground, and the Arduino SDA and SCL I/O lines to function.

The Nokia 5110 Graphic LCD Display

Originally used in the older Nokia cell phones, the Nokia 5110 84 × 48-pixel graphic LCD shown in **Figure 2.50** is another one of the popular original Arduino displays that is becoming outdated. Easy to interface to the Arduino using five digital I/O pins, the Nokia 5110 is small, backlit, easily readable, and capable of displaying both graphics and up to six lines of 14-character-per-line text. While the Nokia 5110 is technically an SPI-bus capable display, more often than not it is connected to the Arduino directly to digital I/O pins and the SPI bus is not used. The Nokia 5110 is based on the Phillips PCD8544 LCD controller, and while specified for 3.3 V, I have used the Nokia 5110 with supply voltages from 2.7 – 5 V, without 3.3 – 5 V level shifters, and with no damage to the display. Contrast settings can vary widely from display to display, but most libraries allow you to adjust the contrast level via software. The Nokia 5110 also has a four LED backlight that can be controlled through the use of an additional digital I/O pin or a resistor from the backlight pin to ground. More recently, multiple variants of the Nokia 5110 have become available, with some minor electrical differences between them. **Figures 2.51** and **2.52** show three different versions of the Nokia 5110 display currently available. Note that the power pin is

Figure 2.53 — Adafruit 128 × 64 Organic LED Display. (Photo courtesy of Adafruit.com)

Figure 2.54 — Sparkfun Microview OLED Display.

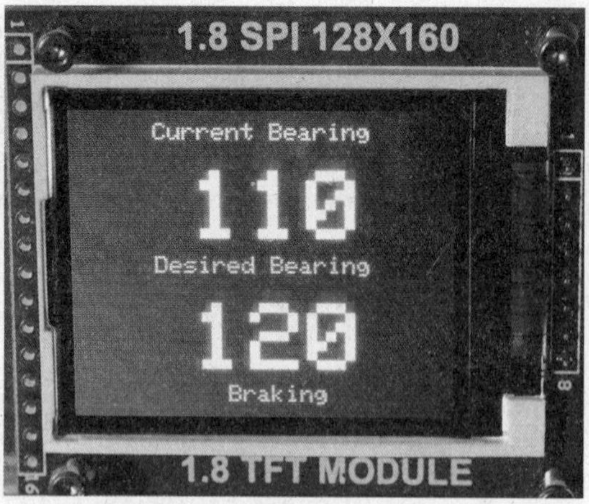

Figure 2.55 — 1.8-inch TFT Display.

different on the one in the center, and that the pin designations may not exactly match up with the ones in the project schematics. The LCD on the far left is the one that I use most often for projects that call for using the Nokia display. While most of the Nokia 5110 displays use a dropping resistor for the backlight LEDs, the Adafruit version uses a transistor and requires a logic-high level to enable the backlight. When wiring up the Nokia 5110 module, always use the pin labels and not the pin numbers, as these designations differ between the various Nokia 5110 versions. When in doubt, build a simple test project on your breadboard to determine the correct wiring for the version of the Nokia display that you have on hand.

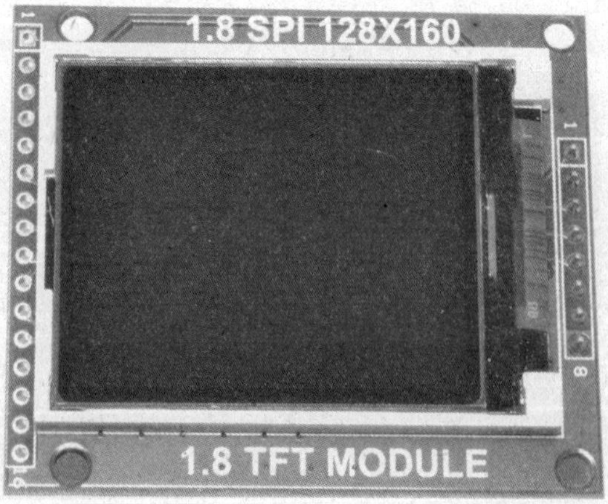

Figure 2.56 — A generic 1.8-inch 128 × 160-pixel Color TFT display module.

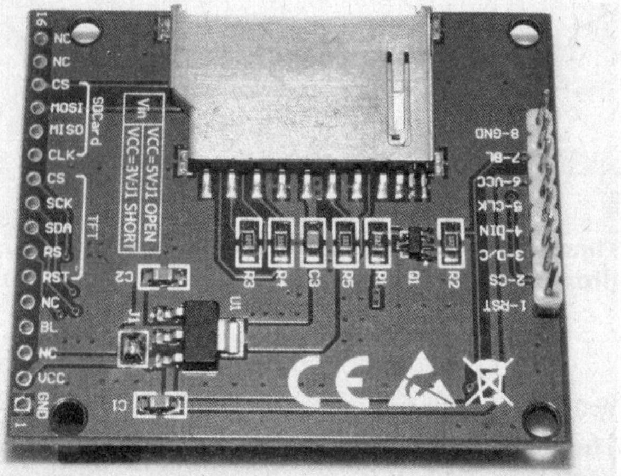

Figure 2.57 — Reverse side of a generic 1.8-inch TFT display showing the 3.3/5 V solder jumper.

OLED Displays

The organic LED (OLED) displays are small, graphic LED displays that can communicate with the Arduino using either the SPI or I²C bus. There are several versions of OLED displays available, primarily the 128 × 32 pixel and the 128 × 64 pixel (see **Figure 2.53**) versions. These are small displays, with the 128 × 32 module being about an inch wide by 3/8-inch high, and the 128 × 64 module is about an inch wide by 5/8-inch high. A newer OLED, the 64 × 48 pixel SparkFun Micro OLED (see **Figure 2.54**), is a mere 0.66 inch across and about 0.5 inch high. An OLED display is essentially comprised of tiny individual LEDs, and because it uses LED technology, it does not need a backlight. OLED displays are very bright and can be used in sunlit applications. The libraries supporting the OLED displays allow you to display both text and graphics. The OLED display is an ideal choice when a small, bright, clearly readable display is desired.

TFT Color Displays

If you really want to spice up your Arduino display, a colored thin film resistor (TFT) graphic display, as shown in **Figure 2.55**, is the way to go. TFT displays for the Arduino come in a variety of sizes, from 1.8 inches (128 × 160 pixels) all the way up to a massive 9-inch (800 × 480) pixel display, and they offer up to 262,144 shades of color. The smaller 1.8-, 2.2-,

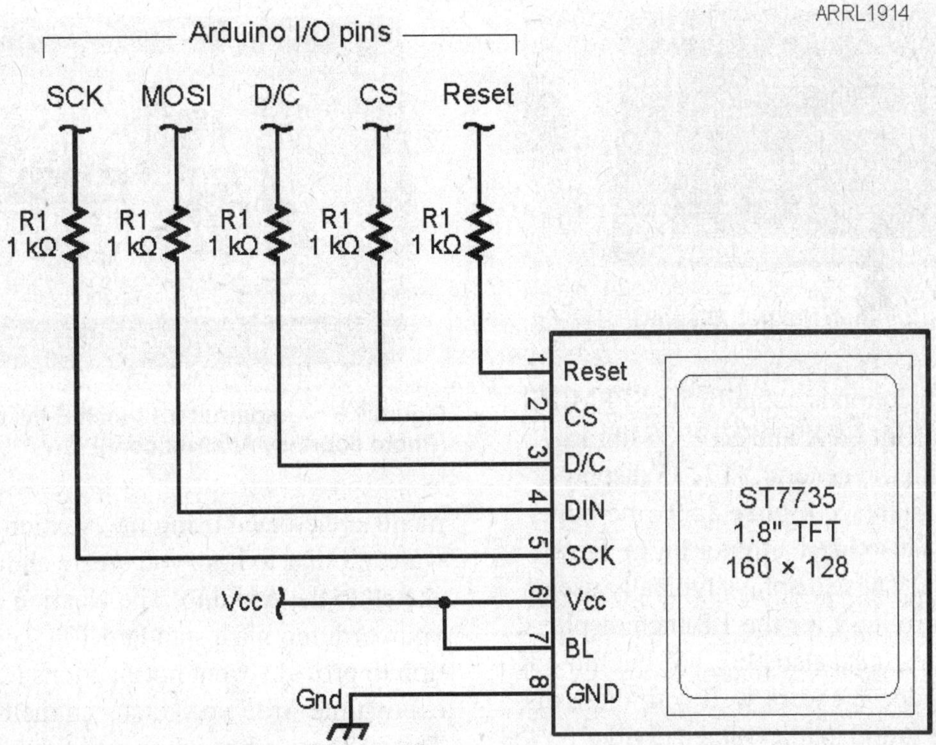

Figure 2.58 — Schematic for adding resistors to the signal lines for the 1.8-inch ST7735-type TFT display.

and 2.8-inch versions of these displays communicate with the Arduino using either the SPI or I²C bus. The 3.2-inch and larger displays require an adapter shield due to the larger number of I/O pins to support the touchscreen on these display sizes. Many of these display modules also include an onboard microSD card slot, allowing you to save bitmap images onboard. Many of these displays also include a touch screen, however, you will more than likely need a board such as the Arduino Mega to support both the display and touch screen simultaneously.

When purchasing the generic TFT displays from places such as eBay, there are a few things you need to be aware of, particularly the minor differences between the generic displays and the ones you get from places such as Adafruit and Sparkfun. Don't be afraid to use the generic TFT displays. I use them all the time in my projects without any problems. The Adafruit TFT libraries work just fine with the generic TFT displays.

The 1.8-inch generic displays (see **Figure 2.56**) are usually based on the ST7735 controller but are designed to operate from 3.3 or 5 V, depending on the presence of a solder jumper on the back of the module (see **Figure 2.57**). When using 5 V, the data lines still must be fed with 3.3 V. To achieve this, you can either use a voltage divider, level shifter module, or you can add a 1 K resistor in series with each signal line as shown in **Figure 2.58**. Also, while the generic board may be designated as an I²C and SPI device, it actually does not support I²C and is strictly an SPI device. When wiring the display, the SCL label is actually the SPI SCK pin, and

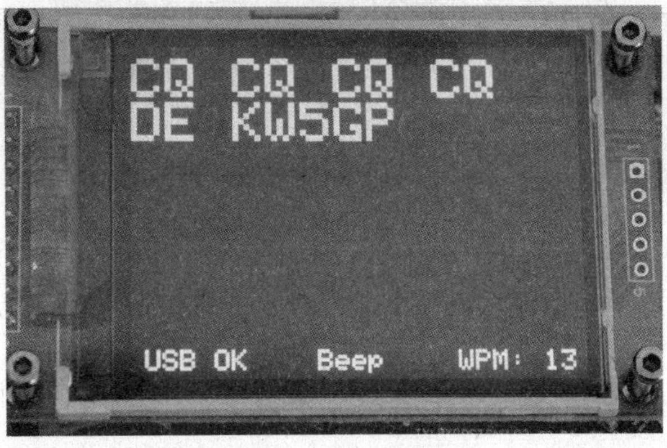

Figure 2.59 — A 2.2-inch ILI9341 generic TFT display.

Figure 2.60 — 2.8-inch Nextion Display.

the SDA connection is actually the SPI MOSI pin. The Adafruit GFX and ST7735 libraries work fine with the generic ST7735 displays.

The 2.2-inch (see **Figure 2.59**) and larger TFT displays also need a minor bit of extra consideration. These displays typically use an ILI9341 controller. Like the 1.8-inch displays, the signal lines must also be 3.3 V, but the 1 K resistor trick doesn't work well with these displays. I have found that a voltage divider on the signal lines works just fine. As with the 1.8-inch displays, the Adafruit GFX and ILI9341 libraries work with these generic TFT units.

Nextion TFT Touchscreen

The Nextion TFT touch screen display (see **Figure 2.60**) is a member of the next generation of TFT displays. The Nextion displays combine an onboard processor and touch screen in what they refer to as a Human Machine Interface (HMI). Unlike the standard TFT displays, the Nextion's screen display and touch screen elements are created using the Nextion Editor software, similar to how you create and upload your sketch to the Arduino. The Nextion connects to your Arduino via a standard TTL serial connection to provide event notifications (screen presses) that the Arduino sketch can then process. This offloads a huge amount of the display and touch screen processing load from the Arduino, making your sketches much easier to create and manage. The Nextion display is available in screen sizes from 2.4 inches all the way up to a huge 10.1-inch display. While creating displays is a bit more involved with the Nextion as compared to a standard TFT display, the results are well worth the effort and sketches perform graphic display operations much faster than its standard TFT counterpart. While a bit on the pricey side, especially compared to the standard TFT displays, the Nextion display is rapidly gaining popularity among developers.

E-Ink Displays

If you've ever used a Kindle or Nook E-reader, you've used an E-Ink display (see **Figure 2.61**). E-Ink displays are unique in that the

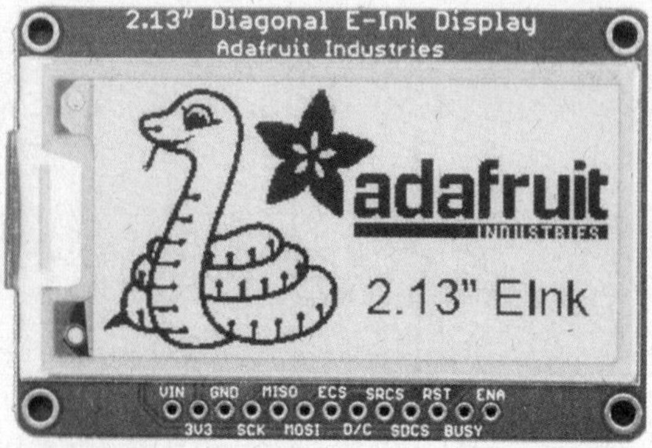

Figure 2.61 — Adafruit 2.13-inch E-Ink display. (Photo courtesy Adafruit.com)

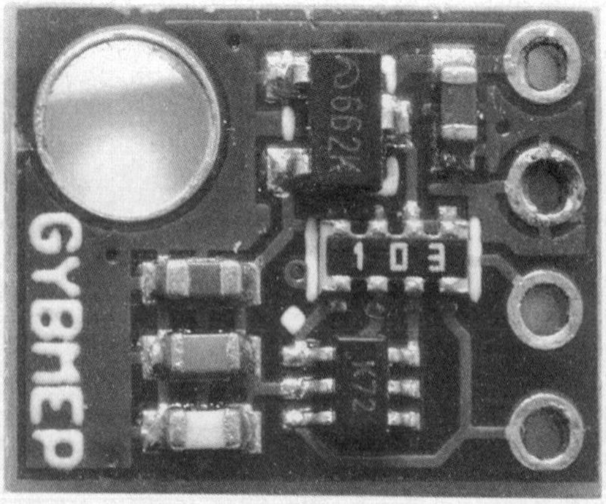

Figure 2.62 — GYBMEP temperature, humidity, and barometric pressure sensor.

Figure 2.63 — Wind speed sensor.

Figure 2.64 — SC16IS750 UART module.

Figure 2.65 — Arduino Uno in an Altoids mint tin.

image stays on the display even after the power is completely disconnected. The image is high-contrast and daylight-readable, and it has been said to look just like printed paper, hence the term E-Ink. E-Ink displays consume power only when they are active and are excellent for projects requiring low power consumption. The E-Ink displays are available in a wide variety of sizes. Some Arduino-compatible processor boards such as the STM32 and Raspberry Pi RP2040 variants available from suppliers such as **AliExpress.com** include an E-Ink display on the board.

Environmental Sensors

The BME280 Temperature, Humidity, and Barometric Pressure Sensor (see **Figure 2.62**) is a precision Bosch sensor that not only performs routine environmental measurements, but can also be used for precise altitude and elevation measurements.

The Modern Device Wind Speed Sensor (see **Figure 2.63**) doesn't require any moving parts to accurately measure wind speed. Instead, it uses a technology known as a hot-wire anemometer. This technology uses an onboard heater and thermistor combination to measure the power needed to maintain a constant temperature. The sensor comes in two versions, with the "C" model measuring wind speeds from 0 – 60 mph, and the "P" model capable of accurately reading wind speeds up to 150 mph.

Serial Communications

With some Arduino projects, it seems like you just don't have enough serial ports, or the software serial port just won't do the trick. Ordinarily, this would mean moving up to a more powerful and expensive Arduino processor board. Fortunately, there is the SC16IS750 UART module (see **Figure 2.64**). The term UART hearkens back to the early days of computer based serial communications and is short for "universal

Figure 2.66 — Solarbotics MEGA S.A.F.E.

asynchronous receiver-transmitter." Simply put, a UART allows you to inexpensively add a hardware serial communications port to your Arduino. Using a UART module, your Arduino can now support a wider range of serial communication protocols than you can with the Arduino's standard serial port. For example, it would be rather cumbersome to set up your Arduino to transmit and receive standard RTTY (radio teletype). You would have to do it using a software serial port and a library that supports the 5-bit 45.45 baud Baudot standard ham RTTY protocol. Instead, you can just add this UART module and configure it directly to support the bit length and speed required. You can then do the BAUDOT-to-ASCII conversion in a sketch and build an Arduino-powered teletype with only a few parts and not a whole lot of coding.

Figure 2.67 — Baseball display case.

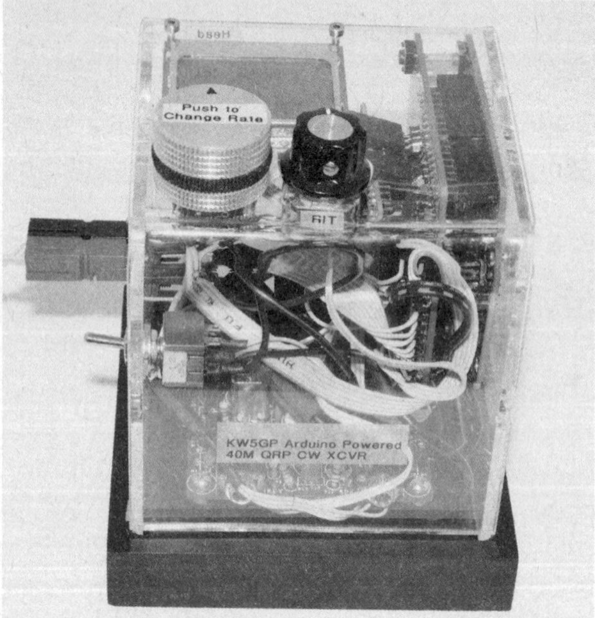

Figure 2.68 — Softball display case.

Enclosures

Once you've built your project, you have a number of enclosure options available. To add to the fun, I've even found some rather unusual places you can get some nice enclosures that work very well for Arduino-based projects.

For those tiny projects that can fit in your pocket, there's the good old Altoids mint tin (see **Figure 2.65**). The larger Altoids mint tin is a near-perfect fit for an Arduino Uno, and several 3D designs are available to print a plastic insert to insulate the Arduino from the metal case.

The major Arduino suppliers are always good places to find enclosures. Both SparkFun and Adafruit have a variety of Arduino enclosures. Another good source for Arduino enclosures is eBay. My personal favorite is the Solarbotics S.A.F.E and Mega S.A.F.E enclosures (see **Figure 2.66**). I particularly like the Mega S.A.F.E enclosure for use with an Arduino Uno. There's enough room for the Uno, a prototyping shield stacked on top, and leftover room for a 9 V battery with plenty of places to mount a display, switches, LEDs, rotary encoders, etc.

More recently, I have found that craft stores have clear acrylic golf ball, tennis, baseball, softball, and model car display cases (see **Figures 2.67** and **2.68**) that are easily adapted into nice-looking Arduino enclosures. I have started using these in a number of my Arduino projects and have been very pleased with the results. They're easy to work with and your finished Arduino project looks similar to one of those "ship in a bottle" models.

3 Arduino I/O Methods

The Arduino communicates in a variety of ways with attached modules, devices, and the outside world. These are known as input/output (I/O) methods. It is important to understand how each of these methods can be used in your Arduino projects. This will help you with adding your own features and enhancements.

The main purpose of a microcontroller such as the Arduino is to interface directly with sensors and devices that allow the Arduino to detect and control things like in the antenna rotator simulator in **Figure 3.1**. The primary methods of I/O on the Arduino are digital I/O, digital I/O with Pulse Width Modulation (PWM), analog input, and TTL/USB serial communication. The Arduino also supports several bus-type protocols, including the Serial Peripheral Interface (SPI), the Inter-Integrated Circuit (I^2C), and 1-Wire buses. In addition to these methods, many of the newer Arduino variants support Bluetooth and Wi-Fi.

Through the use of a shield or external module, your Arduino can also interface and communicate with USB devices. The Arduino also supports hardware and software interrupts, which allow program execution flow to be changed (interrupted) by an external event.

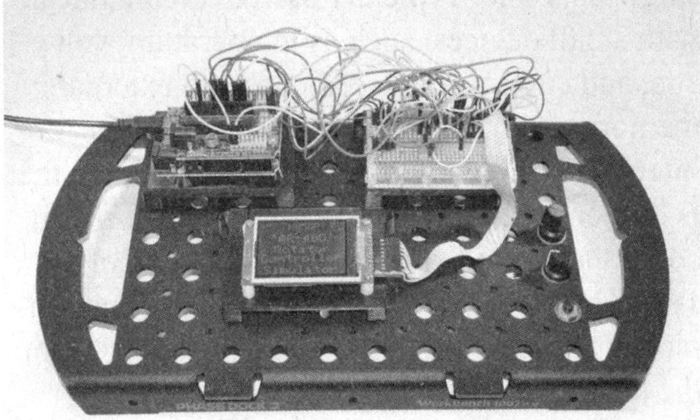

Figure 3.1 — A Hy-Gain AR-40 rotator simulator.

Digital I/O

The simplest, and most common form of I/O on the Arduino is performed using the digital I/O pins. These pins can be used to turn on LEDs, relays, and other external devices, as well as to sense the position of a switch, a pushbutton, or another form of on/off (digital) input. Through the use of several Arduino libraries, you can also use digital I/O pins to allow TTL serial, SPI, or I^2C communications using pins other than those normally designated for these communication protocols. The Arduino digital I/O pins also have an internal pull-up resistor (normally disabled) that can be enabled via software, saving you from having to add an external pull-up resistor, which is useful, for example, when sensing the position of a switch. Some Arduino variants also have internal pull-up resistors. When designing your project, be sure to verify that the Arduino variant you are using supports this feature. On the Arduino Uno, Mega 2560, and Leonardo, the value of the internal pull-up resistor is from 20 kΩ to 50 kΩ. The Arduino digital I/O pins are defined as being tri-state, meaning that the pin can either be at a logic-level high or low, or be in a high-impedance (disabled) state, which effectively removes the pin from the circuit.

Digital I/O with Pulse Width Modulation

Six of the digital I/O pins (pins 3, 5, 6, 9, 10, and 11) on the Arduino Uno and 14 digital I/O pins on the Mega 2560 can be configured to output data using PWM, which allows you to control the duty cycle of a square wave output on the digital I/O pin using an 8-bit value from 0 – 255. Using PWM, you can dim an LED, output an audio tone, control the speed of a dc

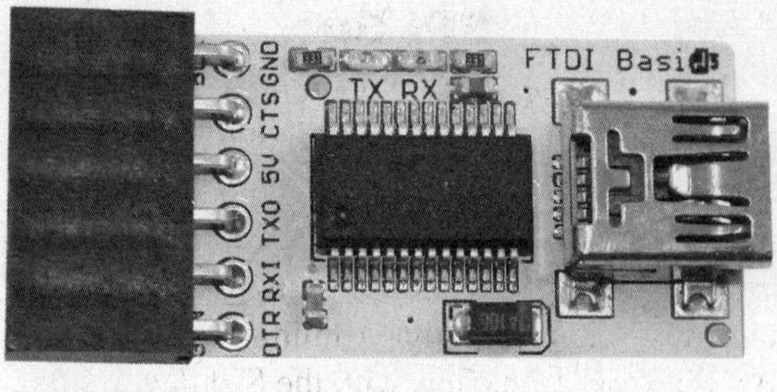

Figure 3.2 — An FTDI module.

motor, modulate a signal for an infrared LED remote, and perform other tasks where you need to modify the duty cycle of the output waveform. The number of PWM digital I/O pins varies between Arduino variants, so be sure to check the digital I/O pin specifications if you plan to use an Arduino variant in your project.

Analog Input and Output

The Arduino Uno has six analog input pins. Interestingly, even though it uses the same Atmel ATmega328 processor as the Arduino Nano, the Nano has eight analog input pins. These pins convert an analog input of 0 – 5 V (3.3 V on the 3.3 V Arduinos) into a 10-bit digital value from 0 – 1,023. This range can be modified either by using an external reference voltage or by selecting the Uno's internal reference of 1.1 V. The Arduino Mega 2560 has 12 analog input pins and also supports an internal reference voltage of 2.56 V. The analog input pins can also be used as digital I/O pins, and like the regular digital I/O pins, also have an internal pull-up resistor (normally disabled) that can be enabled via software. The value of the internal pull-up resistor on the analog pins is the same as for the standard digital I/O pins.

While the standard Arduinos, such as the Uno, Nano, and Mega 2560, do not have I/O pins to convert from a digital representation to an analog output voltage, some of the more recent Arduino boards and variants do have digital-to-analog (D/A) pins. Using the D/A pins, you can output an analog voltage, sine waves, and other waveforms directly from the Arduino itself, rather than having to use an external D/A module.

Serial I/O

The Arduino integrated development environment (IDE) on your workstation communicates with the Arduino via the hardware serial port. Most Arduinos communicate with the IDE via the USB connector. However, some Arduinos, such as the Solarbotics Ardweeny, require a USB-to-TTL serial converter, also known as an FTDI module (see **Figure 3.2**), to access the serial port on the Arduino. The USB port normally shares the I/O pins with the Arduino's hardware serial port. On the Arduino Uno and similar Arduinos, digital I/O pins 0 and 1 are used for serial communication. When designing your Arduino project, care must be taken if you use the dedicated hardware serial I/O pins in your project, as they can interfere with your ability to communicate and upload sketches to the Arduino.

The Arduino IDE can be used to upload sketches and has a Serial Monitor that allows you to view the serial port data, as well as send serial data to and from the Arduino. If your project requires a separate serial I/O port, you can use the SoftwareSerial library, which allows you to assign any pair of digital I/O pins as a software-driven serial port. Additionally, some Arduinos and variants, such as the Mega 2560 and others, have additional hardware serial ports on the board.

1-Wire Bus

The 1-Wire bus interface was designed by Dallas Semiconductor Corp. (now part of Maxim Integrated and Analog Devices) to provide low-speed data signaling and power over a single data line. Typically used to communicate with small devices, such as temperature, voltage, and current sensors, along with external memory and other devices, the 1-Wire bus interface is a bus-type interface architecture that is implemented using a single wire (two if you count the ground wire). With a data rate of up to 16.3 Kbps, the 1-Wire bus can reliably communicate with devices over 100 meters away from the host.

Each 1-Wire device has its own unique 64-bit serial number embedded in the device, allowing many 1-Wire devices to be attached to the same digital I/O pin, without the need for any additional configuration or wiring. Using an advanced algorithm, the host, also known as the bus master, can quickly identify all of the devices attached to the 1-Wire bus, and determine their type based on the device type information that is embedded in the lower 8 bits of each device's serial number. This algorithm can scan the 1-Wire bus and identify up to 75 sensors per second.

A unique feature of the 1-Wire interface is that many 1-Wire devices can be powered entirely from a single digital I/O pin used as the data line for the 1-Wire bus. Using this method, known as parasitic power, the device does not need a separate power supply, and instead, a small capacitor (typically 800 pF) is integrated inside the device to store enough of a charge to operate the device's interface to the bus.

MaxDetect also makes a series of relative humidity and temperature sensors that use a proprietary 1-Wire interface that is different than the 1-Wire bus described above. The MaxDetect 1-Wire interface, however, is not a bus architecture, and the MaxDetect devices do not have embedded addresses, meaning that you can only have one MaxDetect 1-Wire device attached to the digital I/O pin at a time. The MaxDetect 1-Wire interface does not support parasitic power mode, and must be supplied power on a pin separate from the device's data pin. The MaxDetect 1-Wire interface is not compatible with the Dallas Semiconductor Corp./Maxim Integrated 1-Wire bus, so care must be taken when mixing these two types of 1-Wire devices in your projects.

Both types of 1-Wire devices and interfaces are supported by Arduino sketch libraries and example sketches, making it easy to interface and use 1-Wire devices in your Arduino projects.

SPI Bus

The SPI bus protocol was developed by Motorola to be a high-speed, full-duplex communication, bus-type protocol between one main and multiple subnode devices. SPIs, along with I^2Cs, are essentially the workhorse buses that you will use most often with your Arduino projects. The Arduino communicates with the SPI devices on a single shared bus using four signal lines. These lines are designated as clock (SCLK, SCK, or CLK), subnode or chip select (SS or CS), main-out subnode in (MOSI), and main-in subnode out (MISO or DI). Each device attached to the SPI bus requires a separate subnode select line. Think of the subnode select line as a device select line, which tells the selected device that the information on the SPI bus is addressed to it.

SPI is a loosely defined standard and can be implemented in slightly differing ways between device manufacturers. Because SPI is considered to be a synchronous communications protocol, data is transferred using the clock line, with no formally defined upper limit on speed. Some SPI implementations can run at more than 100 Mbps. SPI has four modes (Modes 0, 1, 2, and 3) that define the clock edge on which the MOSI line clocks the data out, the SPI master device samples the MISO line, and the clock signal polarity. Fortunately, the Arduino sketch libraries for SPI and the various SPI devices handle the proper signaling required to communicate with the various SPI devices you may attach to your Arduino project.

On the Arduino Uno, the SCLK, MISO, and MOSI pins are permanently defined at pins 13, 12, and 11, respectively. Digital I/O pin 10 is often used at the subnode select pin for the first SPI device, however, any available digital I/O pin may be used as a subnode select pin. On the Arduino Mega 2560, the SCLK, MOSI, and MISO pins are defined as digital I/O pins 52, 51, and 50, respectively, with the subnode select pin for the first SPI device typically assigned to digital I/O pin 53. On some Arduinos, such as the Uno R3, Nano, and Mega 2560, among others, the SPI signals are also brought out to the six-pin In-Circuit Serial Programming (ICSP) header.

Because each SPI device attached to your Arduino requires a digital I/O pin assigned to the Subnode Select of each SPI device, your projects are limited to the number of digital I/O pins available. Some Arduinos and variants have multiple SPI buses on the board.

There are also several Arduino sketch libraries available that allow you to software-define the SPI bus interface to use regular digital I/O pins and communicate with the SPI devices via these pins by using software-based timing of the SPI signals (also known as bit banging). With this method, the digital I/O pins are turned on and off manually by the library to simulate the hardware timing needed to communicate with the SPI device. While this bit-banging method does work, it requires all of the bit timing and clocking to be performed in software, which is much less efficient than the hardware-based SPI method. It is usually better to design your projects to use hardware SPI, and only use software-based SPI communications when absolutely necessary.

As compared to the I²C bus, SPI is generally faster, and assuming you have the digital I/O pins available in your project, is the preferred method for interfacing most modules and devices that support both SPI and I²C communications.

I²C Bus

The I²C bus was developed by Philips (now NXP Semiconductors) for attaching low-speed peripherals to a host device. On the Arduino, the I²C bus is also known as the Two-Wire Interface (TWI) bus. I²C is a serial bidirectional 8-bit communications protocol used by many manufacturers who develop peripherals and devices for embedded systems and microcontrollers such as the Arduino.

The I²C bus requires only two communication lines: serial data (SDA) and serial clock (SCL). On the Arduino Uno and Nano, these are defined as analog pins A4 and A5, respectively. On the Arduino Mega 2560, SDA and SCL are assigned to pins 20 and 21, respectively. Some Arduino variants have a second I²C interface, with the pins designated as SDA1 and SCL1.

The I²C standard defines the speed of the I²C bus as 100 Kbps (Standard Mode), 10 Kbps (Slow Mode), 400 Kbps (Fast Mode), 1 Mbps (Fast Mode Plus), and 3.4 Mbps (High Speed Mode). The Arduino I²C bus defaults to a bus speed of 100 Kbps, but the bus speed can be changed by modifying the internal Arduino Two-Wire Bit Rate Register (TWBR), or by modifying the TWI speed definition in the Arduino Wire library. Unless your project requires it, and your devices can support it, it is best not to modify the Wire library, as it may cause issues when you compile other projects that use the same Wire library.

I²C devices have a unique 7- or 10-bit address, with some devices capable of having their I²C address reassigned using jumpers, switches, or another hardware method, allowing you to have multiple devices of the same type co-existing on the I²C bus. On the Arduino, I²C addresses 0 – 7 and 120 – 127 are reserved, leaving 112 7-bit addresses available for devices. Every I²C device connects to the bus using open-drain (which is the same as open-collector, except with MOSFET devices), requiring the use of pull-up resistors on the SDA and SCL bus lines. Typically, the value of the pull-up resistors is 4.7 kΩ, so the Arduino's digital I/O pin's internal pull-up resistor will not suffice as a viable I²C bus pull-up resistor value.

As with software SPI, there are several Arduino sketch libraries available that allow you to use regular digital I/O pins for the I²C bus and communicate with I²C devices on those pins using software-based timing of the I²C signals (bit-banging, as mentioned previously). With this method, the digital I/O pins are turned on and off manually by the library to simulate the hardware timing needed to communicate with the device. While this method does work, it requires that all the bit timing and clocking be done in software, which is much less efficient than the hardware-based I²C method. It is far better to design your projects using hardware I²C, and use software I²C communication only when absolutely necessary. Rather than using software to create a second I²C bus if needed, choose an Arduino variant that has multiple I²C bus interfaces available on the board.

Interrupts

While not often seen as being an Arduino I/O method, interrupts allow certain conditions and events to modify the way your Arduino sketches execute. The Arduino has two types of interrupts — hardware and timer. Hardware interrupts are triggered by an external event, such as a change in the logic level of a digital input pin. An interrupt will pause (interrupt) the current program execution and immediately execute a user-defined function, known as an interrupt handler or Interrupt Service Routine (ISR). When the ISR function is complete, program execution resumes the program normally at the command that was last executed prior to the interrupt. An interrupt can happen at any time and allow your Arduino sketch to immediately respond to external events without having to constantly check (or poll) to see if the desired condition exists. Without using interrupts, the USB CW Keyboard project in this book would not have been feasible.

There are four types of interrupt conditions that can be defined: Rising, Falling, Change, and Low. These interrupt conditions refer to the state of the digital I/O pin used to generate the interrupt. The Rising condition will generate an interrupt when the I/O pin goes from a logic-level low state to a high state; falling will generate an interrupt when the I/O pin goes from high to low. The Change condition will generate an interrupt when the I/O pin changes from either low to high or high to low. The Low condition will generate an interrupt when the I/O pin is low. Some Arduinos and Arduino variants also have an additional interrupt condition, High, which generates an interrupt when the pin is high.

The Arduino Uno has two interrupts assigned to digital I/O pins 2 and 3. The Arduino Mega 2560 has six interrupts, on pins 2 and 3, and 18 – 21. Some Arduinos and Arduino variants allow you to configure interrupts on all available I/O pins. The Arduino Uno can also handle Change interrupts on all I/O pins, but unlike hardware interrupts on the defined interrupt pins, the ISR function must decode the interrupt and determine which I/O pin generated the interrupt.

The Arduino Uno has three internal timers, defined as `Timer0`, `Timer1`, and `Timer2`, which can be used to generate `Timer` interrupts. `Timer0` is an 8-bit timer used by the Arduino for internal timing functions such as `delay()` and `millis()`. Because it can affect these functions, modifying the `Timer0` settings is not recommended. `Timer1` is a 16-bit timer often used by some Arduino sketch libraries such as the Servo library, and `Timer2` is an 8-bit timer used by the Arduino `tone()` function. As long as you are aware of any potential interaction with these functions, you can modify the settings on `Timer1` and `Timer2` for use in your sketches. The Mega 2560 has three additional 16-bit timers — `Timer3`, `Timer4`, and `Timer5` — which are not used by any Arduino internal functions. Other Arduino and variant boards may also have additional timers available.

You can configure the timers to generate a software interrupt on overflow, or when the timer count reaches a desired value. The timers are based on the Arduino CPU clock rate (i.e., 16 MHz on the Arduino Uno). You can use the Timer Counter/Control Register (TCCR) for each timer to control the timer clock setting. By modifying the Clock Select bits, you can control how fast the timer increments the counter. On the Arduino Uno, the available settings are `Clk/1` (clock speed), `Clk/8`, `Clk/64`, `Clk/256`, and `Clk/1024`. At the maximum setting of `Clk/1024`, you can have 16-bit `Timer1` generate a software interrupt approximately every 4,194 seconds. Using the Clear Timer on Compare Match (CTC) setting, you can adjust the time to generate an interrupt when the timer reaches a preset value. Using 15624 as the preset value will cause the timer to generate one interrupt per second. Using a 1-second interrupt in this manner, you can add precision timing to your sketches without having to manually keep track of time using the `millis()` function or other manual methods.

Implementing interrupts does add a level of complexity to your Arduino sketches, and you must remind yourself that an interrupt can occur at any time during program execution. You will have to remember and plan for this as you write and troubleshoot your sketch. You can enable and disable the interrupts as needed from within your sketches to allow for uninterrupted executions of critical or time-sensitive points in your sketch where you don't want the sketch execution to be interrupted.

Used properly, interrupts can be a powerful tool in developing your Arduino projects, and once you get comfortable using interrupts, you will find that they can greatly simplify and enhance your project development, because you no longer need to have your sketches running in timing loops waiting for an event to occur. Instead, you can have your sketch off doing other things and only respond when the actual event occurs.

Bluetooth Communication

Bluetooth in the Arduino context is often misunderstood and frustrating. Deciphering how Bluetooth technology can be integrated into your Arduino projects could easily take an entire book. I had hoped to include some Bluetooth projects in my books, but there was no proper way to come up with a viable project without some major additional research and time investment on my part, which unfortunately never seemed to be available. Rest assured, Bluetooth technology is at the top of the list for Arduino projects in the future.

When you think of Bluetooth, you typically envision the devices you use in conjunction with smartphones, such as wireless headsets and microphones, keyboards, mice, etc. These types of Bluetooth devices are often referred to as Human Interface Devices (HID).

The vast majority of Arduino projects involving Bluetooth communication use a non-HID version of Bluetooth, known as a Serial Port Profile (SPP). Some of the more recent Bluetooth modules can emulate the role of an HID device when communicating with a PC or smartphone, but you cannot connect a Bluetooth keyboard to your Arduino using one of these Bluetooth modules. So, when you think of Bluetooth and the Arduino, think along the lines of a wireless serial data connection from one Bluetooth module to another, or a Bluetooth module to a PC or smartphone. If you want to use a Bluetooth keyboard with an Arduino, more than likely the easiest way would be to use a USB Host Shield/module with a standard Bluetooth dongle and connect the Bluetooth keyboard that way.

To complicate matters even further, there are multiple versions of Bluetooth technology: Bluetooth and Bluetooth Low Energy (BLE), also known as Bluetooth 4.0. Both operate in the 2.4 GHz Industrial, Scientific, and Medical (ISM) band, but that's pretty much where the similarities end.

The older Bluetooth is often referred to as Bluetooth Classic, and incorporates the Bluetooth 1.x, 2.x, and 3.x standards. Often, you will hear the term EDR (Enhanced Data Rate) associated with this version. EDR refers to enhancements within this version of the Bluetooth protocol, allowing faster data rates (up to 3 Mbps). At 1 Mbps, Bluetooth BLE has a slower data throughput than Bluetooth Classic but operates at a much lower power level. Unfortunately, the two technologies don't often play well together. While a Bluetooth BLE smartphone can link and communicate with a Bluetooth Classic device, the reverse is not true. For simplicity's sake, when working with Bluetooth devices, always use the same version on both ends of the connection to avoid any incompatibility issues.

When using the Bluetooth SPP, one end of the connection is designated the main and the other end is designated the subnode. The HC-05 Bluetooth module (see **Figure 3.3**) is used the most often with the Arduino. While the firmware varies widely between the HC series of modules, typically the HC-05 can be configured for either the main or subnode role. The companion HC-06 module can only operate in the subnode role. The HC-05 does not support the HID Bluetooth profile and therefore can only be used for serial data links. And the fun's not over yet, there is a lot of variance in the firmware for the various HC-05 modules used with the Arduino, so you really need to play with these modules to get a

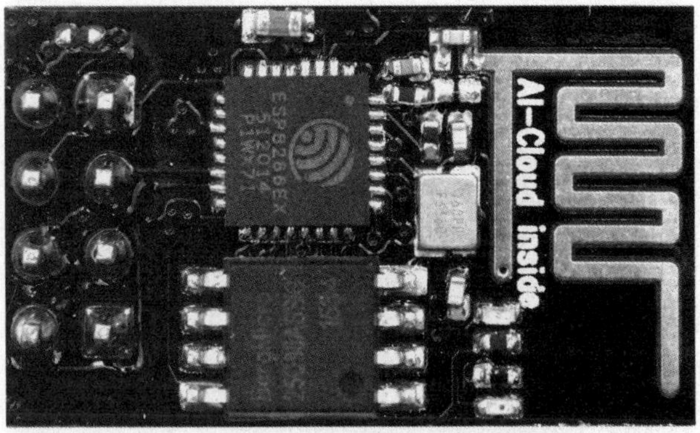

Figure 3.3 — The HC-05 Bluetooth module.

feel for how to integrate them into your projects.

Many of the HC-05-type modules can be powered by 5 V, but the signal lines operate at 3.3 V. To prevent damage to the module, you will need to use a voltage divider or level converter on the signal connections to the module.

The HC-05 has two operating modes: Data and AT Command modes. The AT Command mode is reminiscent of the old dial-up modem AT command prefixes. The command mode is used to perform any configuration required to communicate with the subnode unit. The HC-05 powers up into the default Data mode, and there are differing methods of entering the AT Command mode based on which version of firmware you are running. The most recent and most common version of the HC-05 board has a small pushbutton switch that will put the HC-05 into AT Command mode if the switch is pressed before the unit is powered up, and during operation. The LED on the HC-05 board should blink slowly on and off about once every two seconds to indicate that it is in AT Command mode. When in Data mode, the LED blinks faster, about 5 times per second, while waiting for a connection, and then gives a quick double flash about every two seconds when it is connected to another device.

There are other Bluetooth modules becoming popular for use with the Arduino, including the HC-10, HC-11, and RM-42. Again, it would take a whole book to just cover all the various Arduino Bluetooth modules and their individual intricacies. But once you get the hang of working with the HC-05, working with the others should be just as easy, if not easier.

USB Communications

In many ways, USB technology is like an iceberg — on the surface, it appears to be just a straightforward high-speed communication method, but it's that hidden 90% that will absolutely ruin your day, especially when it comes to the Arduino. There are two basic components of a USB connection, the USB Host and the device. This distinction is an important piece in understanding USB communication as it relates to the Arduino. In order to be programmed using the Arduino IDE, just about every Arduino communicates with your PC via the Arduino's onboard USB port (a handful of Arduino-variants use only Bluetooth or Wi-Fi), or through an FTDI module to convert between the Arduino's serial port to the PC's USB port. In these types of connections, the PC is the USB Host, and the Arduino is the device. Most USB devices are HIDs, such as keyboards, mice, external hard drives, etc. Here's where the bottom half of the iceberg comes into play.

The reason for this issue is the nature of the USB protocol. USB stands for Universal Serial Bus, meaning that it is designed to be the standard for interfacing all manner of devices to the host device. As such, a large part of determining (also known as enumerating) what device has been connected, as well as installing and/or configuring the device drivers needed to communicate with the device, is necessary for the attached device to operate properly. Because of the huge variety of USB devices, you can see that setting up to communicate with a USB device is not really all that simple, but this all happens behind the scenes on your PC or smartphone. That can take a lot of processing overhead, horsepower, and memory to handle everything that is needed to communicate with a USB HID.

Some Arduinos and Arduino variants can emulate a USB HID to a host, such as a PC or smartphone. But, if you want to connect a USB keyboard, mouse, or other HID device to your Arduino project, the Arduino must have USB Host functionality. There are only a few Arduino boards, such as the Arduino Due and Mega ADK, both of which are officially discontinued

Figure 3.4 — The SparkFun USB Host Shield.

Figure 3.5 — The mini USB Host module.

but are still available through suppliers such as eBay, that can function as a USB Host without the use of a module or shield. Some of the Teensy Arduino boards from **pjrc.com** can also function as a USB Host.

A far simpler and less expensive method to add USB Host capability to your Arduino project is to use a USB Host Shield (see **Figure 3.4**) or module. Both are supported by the USBHost library and function identically. In actuality, the mini USB Host module (see **Figure 3.5**) is a shield designed for the Arduino Pro Mini, but it works equally as well as a module with other Arduino boards, such as the Uno or Nano, as you'll see later in this book in the **USB CW Keyboard** chapter.

Amazingly, through the use of the USBHost library, you can connect all manner of USB devices easily. While I have not personally tried it yet, it should be feasible to also add Bluetooth HID capability through the use of a Bluetooth USB dongle plugged into the Arduino's USB Host Shield/module port. This is definitely at the top of the list of things to test, as it opens up a whole new world of simple and inexpensive wireless Arduino projects.

4 Creating Arduino Sketches

In this chapter, we will discuss how to plan and design the program, or sketch, as it's called in the Arduino world, for your Arduino project. We will also cover the basics of installing Arduino Libraries and using the Arduino Integrated Development Environment (IDE) (see **Figure 4.1**) to create your sketches.

Creating sketches is the heart of every Arduino project. Using the Arduino IDE, you can create, edit, upload, and test your Arduino sketches. But before you even start to write the sketch for your Arduino project, it helps to plan exactly what you want your sketch to do, what I/O pins and methods it will use, and what shields or modules you'll be using. I refer to this form of program planning and design as the "divide-and-conquer" method. Each major part of the sketch is broken into smaller chunks of code that can be coded and debugged, one piece at a time. The block diagram and flowcharts help you stay on track with what you're trying to accomplish with your Arduino project and sketch.

Generally, I like to start with a block diagram of the project hardware. A block diagram similar to the one in **Figure 4.2** is a very high-level overview of your project, giving you an idea of what shields, modules, and components you will need to assemble your project. A block diagram will also help identify what I/O pins and methods to use in your project. You can also use the block diagram to help determine which Arduino board or variant may be best suited for your project. This is also a good time to think about what type of enclosure you want to use.

After I create the block diagram, I draw out a rough schematic similar to the one in **Figure 4.3**. This rough schematic is usually a hand-drawn version that I use to construct the project, us-

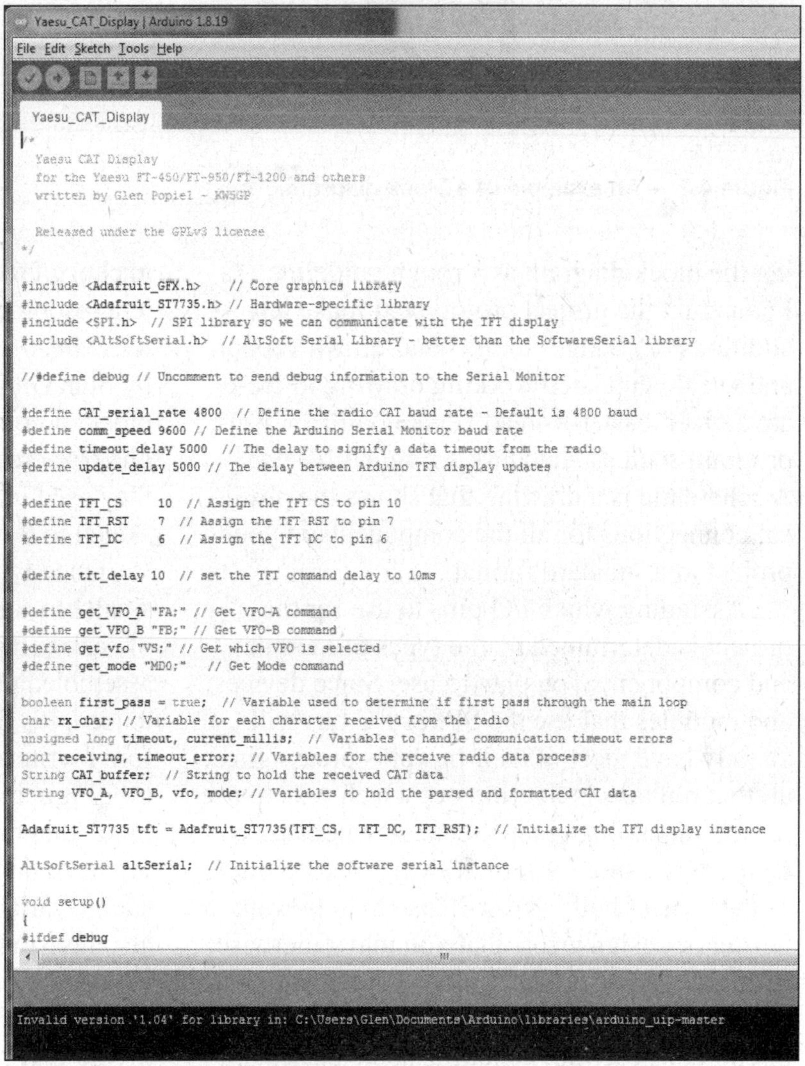

Figure 4.1 — The Arduino IDE.

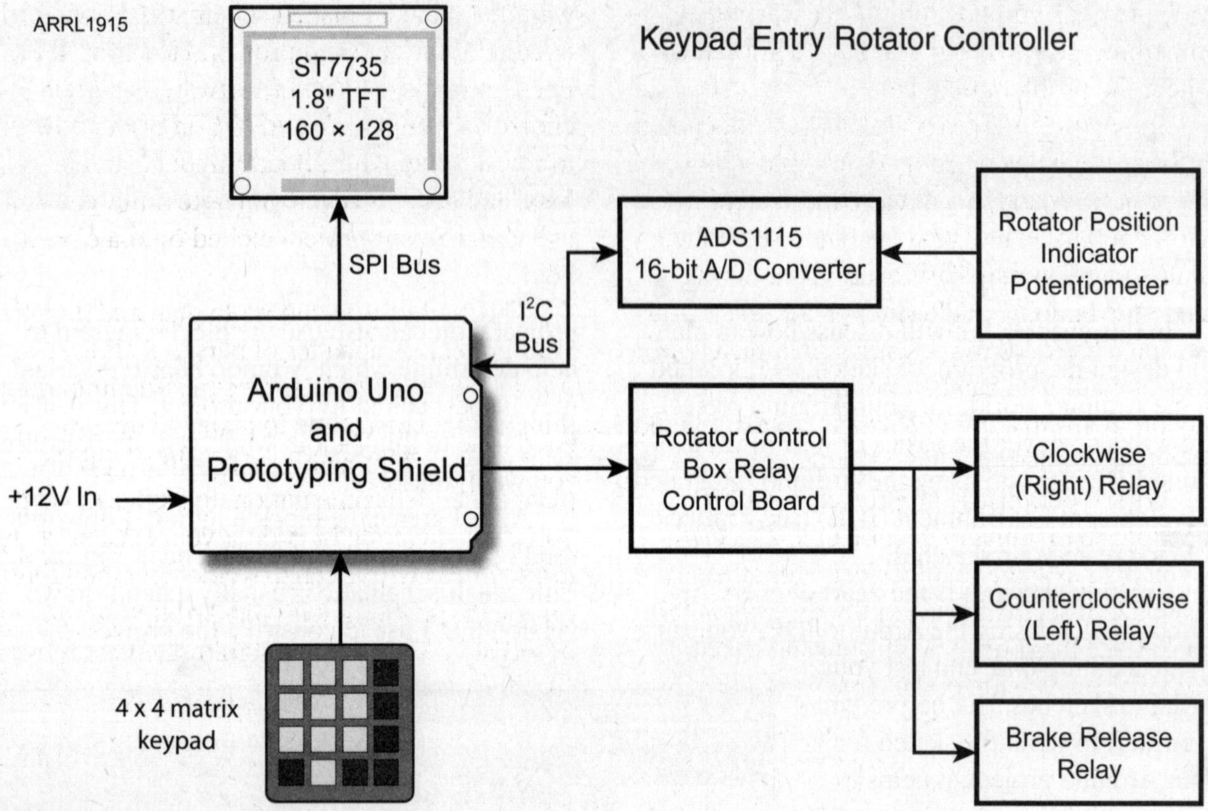

Figure 4.2 — An example of a block diagram.

ing the block diagram as a rough guideline. As I construct the project prototype, I make any additions or changes to the hand-drawn version and use this updated working drawing to create a nicer, easier-to-read version with a CAD program such as Autodesk's *Eagle* or *KiCad*. A schematic is a drawing that shows the physical connections for all the components in your project in a standard format.

Assigning which I/O pins to use for the project is determined by the types of modules and components you plan to use. Since devices and modules that use the SPI or I²C bus will already have most of their bus pins preassigned, all that remains is figuring out which pins to use for the remaining components. A schematic can be used as a step-by-step drawing to keep track of the project build and will also help prevent wiring or design errors that you may otherwise overlook. If you are unfamiliar with how to read a schematic, the ARRL website has a section on circuit construction, including some basics on how to read circuit diagrams (schematics) at **arrl.org/circuit-construction**. Another excellent resource for learning how to read schematics is *Beginner's Guide to Reading Schematics* by Stan Gibilisco. Finally, if I plan to use a block of code or library for a device that someone else created, I'll often use the same pins they used in their project so that I can use their sketch to test that piece of the project.

Once I finish building the project, I'll use small blocks of code to test the various functions of each component on the board. After I have an assembled project that I am sure will work as intended, it's time to start on the sketch itself. Before I write any code, I'll first create a flowchart (see **Figure 4.4**). Flowcharts are essentially a block diagram for your program. They help you organize how to write the sketch and what constants, variables and types you may need. They also help determine what libraries you need to include in your sketch. Flowcharting breaks down your sketch into smaller pieces using the divide-and-conquer method. This method allows you to write your sketch in blocks, testing each block as you go along. That way, when something goes wrong, you have a good idea of where the problem may be.

Taking the time to create this basic documentation really helps when it comes time to build the project prototype and write the sketch. It allows you to review your design to see if there is anything you may have overlooked during the initial building and testing phase. Since you did some basic testing when you completed the project build, you can be reasonably certain that any problems that crop up at this point are in the sketch itself and not with the hardware.

When you finish your project, it's a good idea to take the time to document it, especially if you plan to share your project with others. As you will see with each of the projects in this book, once the project is finished and tested, I used the free version of Autodesk's *Eagle* software to create a finished schematic drawing to document the finished project. However, the open-source, free *KiCad* software is a good choice as well. *Eagle* and *KiCad* both allow you to create schematics, board layouts, and parts lists, and they can even generate a file you can use to have your project etched onto a circuit board.

The method you choose to plan and organize your projects is a matter of personal choice, so use whatever works best for you. The important thing is to take the time to plan and document your work as thoroughly as possible. A little planning and documentation up front can keep you from soldering or coding yourself into a corner and getting frustrated. Remember, building things and working with the Arduino is supposed to be fun. Block diagrams, schematics, and flowcharts can help keep it that way.

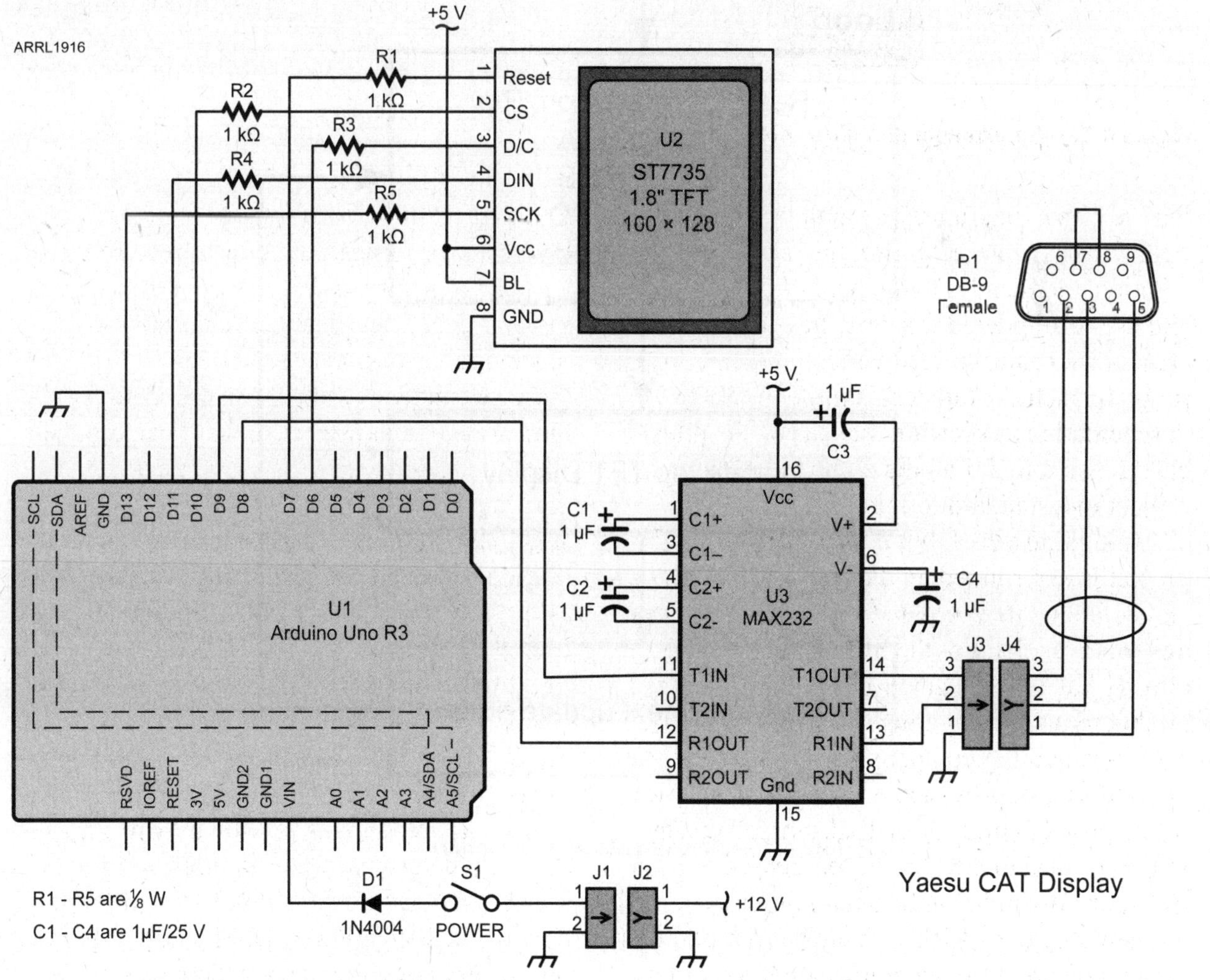

Figure 4.3 — An example of a schematic.

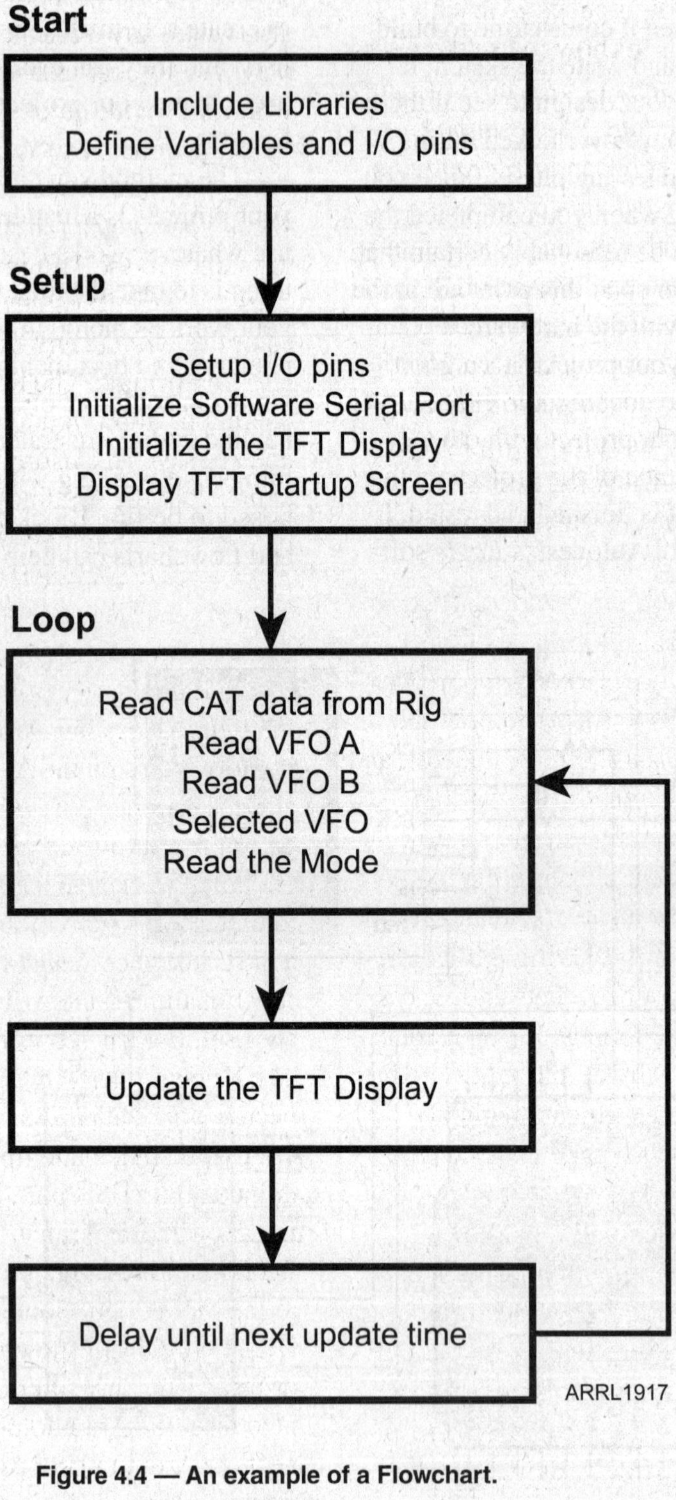

Figure 4.4 — An example of a Flowchart.

The Arduino IDE

Arduino sketches are created using the Arduino Integrated Development Environment (IDE). This book is not intended to teach you how to write programs, or how to use the Arduino IDE, as there are already a number of books available for this, including *The Arduino Cookbook* by Michael Margolis, and *Programming Arduino – Getting Started with Sketches* by Simon Monk, that can devote far more pages to teaching you how to use the Arduino IDE than is possible here. There are also some excellent online tutorials for installing and learning how to use the Arduino IDE at **docs.arduino.cc/software/ide-v1/tutorials/arduino-ide-v1-basics**, **learn.sparkfun.com/tutorials**, **learn.adafruit.com**, and other websites. In this book, we will focus on a few IDE basics and the things you will need to know to use the Arduino IDE with the projects in this book, such as installing Arduino libraries and how to add support for Arduino-variant boards to the Arduino IDE.

The Arduino IDE is used to create, edit, and upload your Arduino sketches. The Arduino IDE is very versatile and can support a range of different hardware and processor types. Using the IDE's text editor, you can create an Arduino sketch. To help you, the IDE will highlight keywords and match up your brackets and braces to help you identify the loop and if-then code blocks in your sketch. Under the **TOOLS** menu, the **AUTO-INDENT** feature will automatically indent your sketch text to make your sketch more readable. Preferences are set under the **FILES>PREFERENCES** menu option, allowing you to customize some important aspects of the IDE, including where the IDE will save your Arduino sketches. The location where the IDE saves your Arduino files is also known as the 'sketchbook,' and is used by the IDE to create a menu listing your available Arduino sketches. Please become familiar with the **PREFERENCES** menu, as you will be using this option to add support for additional Arduino boards and variants if they are not already built into the IDE.

The Arduino IDE is used for more than simply creating sketches. The IDE has a message area that is used to provide feedback while saving, compiling, and uploading sketches, it is also where any error messages or warnings display. The toolbar icons near the top of the IDE window provide quick access to the most used IDE functions: **VERIFY**, **UPLOAD**, **NEW**, **OPEN**, and **SAVE**. The **VERIFY** operation will compile and check your sketch for errors but does not upload the sketch to the Arduino. The **UPLOAD** operation does the same as **VERIFY**, except when the sketch compiles without any errors, it will then upload it to the Arduino. The **NEW**, **OPEN**, and **SAVE** operations allow you to create a new sketch and open or save an existing sketch. It is important to note that the Arduino IDE performs an auto-reset on the Arduino board as part of the upload process. Some Arduino boards do not support the auto-reset function. On these boards, you will have to manually press the Arduino's reset button immediately prior to selecting **UPLOAD** in the IDE. Also, some Arduino boards, particularly those based on the ATmega32U4, will actually lose connectivity with the workstation as part of the auto-reset process, which may cause the board to change the USB port it's connected to on the workstation. This is primarily due to a USB interface design that integrates the USB functionality on the Arduino (or variant) processor itself, which allows the Arduino to also act as a Human Interface Device (HID) to a connected workstation. As part of the reset process, your workstation attempts to identify the Arduino and its USB port, but may fail to do so in the allowed time frame. You may have to go into the IDE and reset the serial port selection back to the proper port. Sometimes, the Arduino's USB port doesn't show up at all after an upload or reset. The only solution I have found for this is to reset the Arduino or disconnect and reconnect the Arduino USB port until the IDE finally figures things out. This can get annoying sometimes, and is the primary reason why I prefer the standard Arduino Uno or Nano for many of my projects.

Interestingly, the Arduino sketch editor allows you to create your sketch using multiple tabs within the editor. In theory, this would let you break a large sketch into smaller, more readable sections, and allows you to incorporate other files usable by the Arduino IDE, such as .h, .cpp, and .c files, and the files used within a library. Using multiple tabs to create your Arduino sketch can be confusing since the IDE will compile your sketch in alphabetical order, starting with the main sketch and adding, or concatenating, the tabbed files so they effectively appear to be at the bottom of the main sketch. This can cause issues with any `#define` statements and variables defined in the tabbed files but used in the main sketch, among other things. Some developers have used this as a quick way to add a library to their Arduino sketch. While this may work, it is not the ideal method. Creating a sketch using multiple tabbed files is often more trouble than it's worth, and it's best to use a single tab for your Arduino sketch.

The **TOOLS** menu is where you select the type of Arduino board you are working with, along with the USB communications port the board is attached to. If you do not select the correct board and USB port, your sketch may verify and compile, but the upload process will fail. You can also add support for third-party hardware, such as Arduino-variant boards, by placing their board definitions, core libraries, bootloaders, and programming definitions into a sub-folder in the hardware folder of your Arduino sketchbook. You can create the hardware folder if it does not exist, and the IDE will incorporate the new board into the IDE the next time the IDE is started. Fortunately, with the newer versions of the IDE, you can also add support for additional boards by adding a third-party URL for the new boards in the IDE preferences. These will automatically be added to the IDE board manager, where you can select and download everything you need to support a new Arduino board or variant.

Also, on the **TOOLS** menu, you will find the **AUTO FORMAT** option. This will reformat the text in your sketch so that the curly braces all line up and the text is properly indented, making your sketches easier to read, which is a great help when you get around to troubleshooting your sketch. You can also save your Arduino sketch in compressed .zip format using the **ARCHIVE SKETCH** option. This will create a compressed .zip file of your Arduino sketch and save it in the sketchbook folder.

One of the most used options on the IDE **TOOLS** menu is the **SERIAL MONITOR**. The **SERIAL MONITOR** will display the serial data sent and received from the Arduino's USB/serial port. Typically, you will use the **SERIAL MONITOR** to show basic program output and any debugging information you have included in your sketch. You can also use the **SERIAL MONITOR** to send characters and commands to the Arduino via the USB/serial port. The Arduino's USB/serial port can also allow your Arduino to communicate with the Processing language installed on your PC, and to allow PC-side applications to interact with your Arduino sketch.

One of the most underutilized tools in the IDE is the **SERIAL PLOTTER**. Similar to the **SERIAL MONITOR**, the **SERIAL PLOTTER** allows you to generate a real-time graphical plot of a data string that your Arduino project sends to the serial port.

Adding New Boards

Starting with version 1.6.4, the Arduino IDE includes a **BOARD MANAGER** feature that allows you to add support for new Arduino and Arduino-variant boards more easily than in the past. The IDE **BOARD MANAGER** also keeps track of any updates and allows you to update existing board definitions and libraries as they are released.

Adding support for a new board has never been easier, and many Arduino board suppliers are adding support for their products in the IDE **BOARD MANAGER**. To add a new board to your Arduino IDE, all you have to do is add the support URL to the **ADDITIONAL BOARDS MANAGER URLs** section of the IDE. You can edit this preference and have multiple entries, allowing you to add support for any number of new boards quickly and easily. An unofficial list of third-party URLs for the IDE is available at **github.com/arduino/Arduino/wiki/Unofficial-list-of-3rd-party-boards-support-urls**. Your board manufacturer may also provide you with a URL for supporting their boards that you can add to the IDE **BOARD MANAGER** directly.

Arduino Libraries

Including libraries in your sketches is an integral part of programming the Arduino. An Arduino library is a prewritten set of functions that you can add to your sketch to support shields, modules, and devices, and even add additional software functionality, such as trigonometry functions. Through the use of libraries, a large percentage of the programming for a project may already be written for you. Another way to look at Arduino libraries is to think of them as the device drivers needed to interface with all of the various devices you connect to a workstation. By including a library in your sketch, you add the functionality of that library without any additional programming. Then, using function calls from within your sketch, you can interface the shield or device to your sketch quickly and easily. Quite often, your sketch is merely a little bit of programming glue used to tie everything together. The library does all the hard work for you. This is one of the major advantages of the Arduino's open-source model. There are literally thousands of Arduino enthusiasts like you out there, creating and sharing libraries that you can use in your Arduino projects. Arduino libraries can save you hours, even days, of deciphering datasheets and writing test code just to interface to a new device. Many libraries even come with example code that you can integrate into your own sketches, thereby making the development and testing of your own project much faster.

Many libraries are already integrated into the Arduino IDE, and there are a large number of other libraries available on the internet, written and shared by their creators. In many cases, you may find that a library already exists to suit the needs of your project. More recent versions of the IDE include the **LIBRARY MANAGER** feature, which has the built-in ability to search for updates to your existing libraries, as well as allowing you to search for and install new libraries into your IDE.

Installing Libraries

One of the biggest issues I have discovered about using the Arduino is the confusion about what a library is and how to install it. Nearly 90% of the questions I receive about my Arduino books are library-related installation issues. If you have a sketch that uses a library, and that library is not installed or installed incorrectly, the sketch will not compile. To make things even more confusing, starting with version 1.5.7 of the IDE, some of the older libraries no longer work as they used to, and you will need to either use an older version of the IDE that works with that library or update the problem library for your sketch to compile and function correctly. The LCD5110_Basic library is among one of the older libraries that doesn't work with the newer IDE versions. You can download a newer version of this library from **rinkydinkelectronics.com**. Additionally, some of the earlier and more popular libraries are now built into the IDE and no longer need to be installed. There have been numerous continuing issues involving certain library functions, particularly scrolling text on display with the LiquidCrystal_I2C library, since version 1.5.7. Be sure to verify that you're not dealing with a library issue before throwing in the towel. The issue with the scrolling text on the 16 × 2 LCD displays is one of the factors contributing to why I now predominately use OLED or TFT displays in my Arduino projects. If you look at the message area on the IDE, you will see errors relating to the compiler's inability to locate functions contained within a library, along with the name of the .cpp or .h file associated with that library that will indicate a problem with the library itself.

Libraries are not that difficult to install, and newer versions of the IDE allow you to install libraries using several methods. The confusion regarding installing Arduino libraries may simply be the concept of installing a library. On the Arduino, technically, libraries are not installed in the classic sense of software installation. In a normal library installation, no executable file will install a library for use with your IDE. The library installation process is actually far simpler.

The typical Arduino library consists of several files within a single top-level folder. In general, an Arduino library is written in C++ and contains a .h header file and a .cpp file that contains the actual code for the library itself. These files are usually contained in the top level of a folder or series of subfolders, with the top-level folder usually having the same name as the .h and .cpp files it contains. A library may also contain additional files such as keywords.txt, and in the case of newer libraries that can be installed using the IDE Library Manager, a metadata file named library.properties that contains information the IDE can use to install and update the library. A library often has additional subfolders containing example sketches and documentation on how to use the library's features. So, essentially, at the end of the day, an Arduino library is just a group of files in a folder.

A library can be installed manually, as a zip file through the Library Manager, or through the Library Manager itself. When you install a library, you place the library folder and any subfolders it contains into the libraries folder in your Arduino sketchbook folder. In Windows, this is typically in your `Documents>Arduino` folder. This is the same folder that will contain your Arduino sketches. To manually install a library, all you have to do is copy the folder of the library you wish to use into the libraries folder. If the libraries folder does not exist, simply create a folder named libraries in your Arduino sketchbook folder (be sure to use all lowercase for this folder name) and copy your libraries into this folder. For the IDE to see the new library, you may have to exit and restart the IDE. At startup, the IDE will check the contents of the libraries folder, and any libraries in that folder will then be available for use in your sketches. You will see all your available libraries under the `Sketch>Include` Library menu option. Any available Example sketches for your libraries should also be available under the `File>Examples` option of the IDE. All of the libraries for the projects in this book that are on the **kw5gp.com** website were designed to be installed manually, and you will need to download and extract the project compressed .zip file and copy each individual library folder into your sketchbook's libraries folder.

Another way to install an Arduino file is using the Library Manager's .zip file method. While many libraries have been modified to use the new Library Manager, some are still available for download as a compressed .zip file. To install these libraries, all you need to do is download them and use the ADD .ZIP Library option under the `Sketch>Include Library` menu. The Library Manager will extract the zipped library file and place it into the sketchbook libraries folder for you. After you exit and restart the IDE, the library you just installed will be available.

Starting with IDE version 1.5, a Library Manager was added to the Arduino IDE. When you open the Library Manager located under `Sketch>Include Library>Manage Libraries` menu option, you will see a list of libraries available for you to automatically download, install, and update using the Library Manager. All you have to do is select the library you wish to install, and the Library Manager will take it from there. If available, the Library Manager will allow you to choose between the most recent version and older versions of the library. As with the installation of any new library, it's always best to exit and restart the IDE after installing the library to allow the IDE to properly add the new library to the list of available libraries. The new Library Manager feature is very nice, and has really helped with problems related to installing libraries. Additional information on installing Arduino libraries can be found at **docs.arduino.cc/software/ide-v1/tutorials/installing-libraries**.

Using Libraries

To use a library in your Arduino sketch, all you have to do is either select the library using the `Sketch>Include Library` option in the IDE, or you can manually select the library by using a `#include <library.h>` preprocessor definition near the beginning of your sketch. Sometimes, you will see the include statement using double quotes instead of the left and right carat surrounding the library name, i.e., `#include "library.h"` versus `#include <library.h>`. For all intents, these two forms for the include statement are interchangeable; however, the double quote form allows you to have the library files in the same folder as your sketch .ino file instead of the normal libraries folder. If the IDE does not find the library in the sketch folder, it will then look in the libraries folder. The `#include <library.h>` form will only look in the normal libraries folder. In theory, while this does allow you to have multiple versions of a library you have customized, it's generally best to have all of your libraries in the normal libraries folder. When using a library, the `Include` statement must always include the .h file extension as part of the included library's file name.

Some libraries include a keywords.txt file. This is a text file that will list all of the functions available in the library. The IDE also uses this file to highlight keywords of the library's functions while editing your sketch.

One thing to remember about Arduino libraries: many are created by people just like you and me. Sometimes, bugs and glitches slip in and cause things not to work as expected. I have found that these library glitches are rare and are often quickly addressed and resolved. However, you need to be aware that if you are having issues, you should check that the library functions you use in your sketch are working as expected. For example, there was a recent issue with the built-in Liquid Crystal I^2C library, where the library would only print the first letter of a string sent to the LCD. A workaround was quickly found and posted to the various Arduino forums, and the issue was fixed in a library update. Still, you need to be aware that libraries can sometimes not work as intended and give you all sorts of strange problems.

Troubleshooting

Generally, the Arduino IDE is a stable, reliable platform for developing sketches for your Arduino projects. There are a few issues you need to be aware of when using the Arduino IDE, particularly with some of the newer Arduino-compatible boards and variants. With the introduction of the Arduino Leonardo and the Atmel 32U4 chip, the USB functions on the Arduino board were implemented on the 32U4 chip, and the 16U2 USB controller chip on the original Arduino boards is no longer needed. As mentioned earlier, the USB controller in the 32U4 chip functions a little differently than the one in the Uno. When you upload a sketch to a 32U4-based Arduino, the auto-reset that the IDE performs on the board as part of the upload process can cause your workstation to lose connectivity with the Arduino briefly. When it reconnects, the Arduino may be assigned to a different USB port than the one you have set in the IDE. The simple answer is to select the new USB serial port in the IDE and continue normally. Sometimes, the board will not reconnect after uploading a sketch. This is primarily due to the USB driver on your workstation not seeing the Arduino before timing out while trying to reestablish communication. In these cases, you might have to disconnect the USB cable from your workstation to the Arduino a couple of times before everything gets figured out.

Some Arduino-compatible boards use the WCH CH34x series of the USB converter chip in place of the Atmel 16U2. When you connect an Arduino-compatible board that uses the CH34x USB chip, your workstation will not recognize the new board until you install the CH34x USB drivers on your workstation. Locating these drivers can be somewhat troublesome, but there are more and more websites now offering the CH34x drivers for download. I was able to locate the drivers for my CH34x boards on the CH34x manufacturer's website at **wch.cn/download/CH341SER_EXE.html**. The driver for the newer CH341 will also work with the CH340 chip, so all you need is the CH341 driver file for your workstation (PC, MAC, or Linux). If you are unable to locate the drivers on the site listed above, you can find them by doing a web search for Arduino CH340 or CH341 drivers. Once the CH34x driver is installed, your workstation should be able to identify and connect to the Arduino-compatible board.

With some of the newer, more powerful Arduino-variant boards, as well as some of the inexpensive Arduino Uno and Nano-compatible boards, I had issues with a workstation not being able to recognize and communicate with the board, even when the correct drivers had been installed. The board worked fine when I connected it to a different workstation. As it turned out, the board I was using was drawing too much power for the USB port I had it connected to. I had to use a powered USB Hub for the workstation to recognize the board. Also, there are now two versions of the bootloader for the Arduino Nano board. If using an older Nano, you may need to select the old bootloader version when you select the board style.

The message area in the Arduino IDE can provide a great deal of information if you have issues compiling a sketch. In your `File>Preferences` menu, you can select verbose compiler output to get more details on any compiler errors or upload messages, as well as set the type of compiler warnings shown. Compiler warnings alert you of a possible issue. Many compiler warnings can be ignored, but it is beneficial to pay attention to the information provided if and when you see a compiler warning. As with many compilers, a compiler error may not show you the exact statement that failed and why. However, if you carefully read the information, it will usually provide enough to track down the issue.

Memory Issues

The Arduino memory architecture is not the same as in a workstation. The Arduino has three, and in some cases, only two separate types of memory onboard. The Arduino Uno has just 32 K of Flash memory, which is used to store the compiled sketch, and 2 K of Static RAM (SRAM) to store your program variables, data, and memory required by the Arduino sketch to operate properly. The Uno and Nano also have 1 K of EEPROM to store information permanently. Still, few sketches actually use or need the EEPROM, and many Arduino-variant boards do not have EEPROM.

As such, memory in the Arduino Uno and similar boards is at a premium. It is important for you to keep the limited memory capacity of the Arduino in the back of your mind while creating your Arduino sketches. Unlike a regular workstation that will politely give you an "insufficient memory" error and refuse to run the program, the Arduino will do exactly as it is told, keep on running, and then start doing weird things and failing to execute the simplest of commands properly. There are some programs and libraries, such as the Memory Free library, you can incorporate into your sketches to keep track of available memory and let you know how much memory you have remaining. In general, you can fit the vast majority of your sketches in memory just fine. With the larger memory capabilities of the newer Arduino and variant boards, lack of memory is becoming less and less of an issue.

If you need to save precious SRAM, you can instruct the IDE to place static data and constants into Flash memory by including the avr/pgmspace.h library and use the PROGMEM keyword in your sketch. The pgmspace library allows you to store static variables, constants, strings, and arrays in flash memory instead of SRAM. You can also use the FLASH library to store strings, arrays, tables, and string arrays. Finally, you can also use the `F()` syntax for storing string constants in Flash. Just remember that if you put data in Flash memory, it cannot be modified during the execution of the sketch, so you can only use these memory-saving features for data that doesn't change.

Simple Debugging Methods

You will see in many of the sketches for the projects in this book that I have included the preprocessor directive, `#define debug_mode`, or a similar definition to incorporate a form of debugging mode. When `debug_mode` is defined, set to 1, or true, the IDE will compile an additional debug code that has been added throughout the sketch. This debug code is used to output additional information to the **SERIAL MONITOR** that can be used to monitor critical points in the sketch to aid in troubleshooting. Once the sketch is tested and working, the debug mode is undefined, set to 0 or false, and the debug code is ignored and not compiled. This is a quick and easy way to add troubleshooting code to your sketch and then turn it off once troubleshooting is complete. If you use the `#ifdef` and `#endif` preprocessor directives around the block of troubleshooting code, when debug mode is active, the IDE compiler will ignore the troubleshooting code and not compile it, making your finished, compiled sketch smaller and more efficient.

Arduino Create

Arduino now has an integrated online platform known as Arduino Create, which will allow you to write code, save it to the cloud, and even upload your cloud-based sketches to your Arduino boards, all from within your web browser. You can even share this content with others. It has a free version, and all you need to do is create an account at **create.arduino.cc**. The web IDE will automatically detect any Arduino board connected to your PC and configure it accordingly. All of the libraries available in the Arduino Library Manager are automatically detected, so you don't need to install any libraries to get your code to compile. The web IDE even supports the **SERIAL MONITOR**. The Arduino web editor is supported in Windows, Linux, Mac, and Chrome OS.

Arduino Create also gives you access to the Arduino Project Hub, a tutorial platform dedicated to discovering Arduino, electronics, and software. All the content is categorized, searchable, and covers a wide variety of project categories and projects.

Arduino IoT Cloud is another piece of Arduino Create. IoT Cloud is designed to be an easy-to-use Internet of Things applications platform. The IoT Cloud allows you to access your internet-connected objects quickly, simply, and securely. You can connect multiple objects and allow them to exchange real-time data. You can also monitor your objects from anywhere using a simple user interface. Arduino IoT Cloud supports a number of third-party devices and has both free and paid options for all levels of Arduino enthusiasts. You can even use Arduino Create to create Arduino sketches that run on the Raspberry Pi.

Multitasking and Real-Time Arduino Operating Systems

By nature of its design and intent, the Arduino in and of itself is not considered to be a multitasking system. This is changing with the advent of several libraries and multitasking methods designed for microcontrollers such as the Arduino. For simplicity's sake, the firmware used to manage the microcontroller's functions is often referred to as its "operating system." These newer "Real-Time Operating Systems" are more along the lines of a multitasking kernel for embedded and microcontroller systems.

HeliOS is a very small multitasking kernel written in C that runs on most 8-bit microcontrollers like the Arduino. HeliOS is installed as an Arduino library using the Library Manager and integrates easily into your Arduino project. It's easy to learn with an Application Programming Interface (API) that consists of only 21 function calls. HeliOS allows you to implement cooperative and event-driven multitasking operations with task notifications, messaging, timers, and memory management. HeliOS utilizes a run-time balancing feature that ensures tasks with shorter run times are prioritized over tasks with longer run-times. HeliOS is definitely something to consider if your Arduino project needs to deal with multiple conditions and events simultaneously.

Another multitasking option to provide real-time operating system features to your Arduino project is FreeRTOS. While small enough to run on a microcontroller such as the Arduino, FreeRTOS is not limited to just microcontroller applications. It includes a kernel and a growing set of software libraries suitable for use across multiple industry sectors and applications. FreeRTOS is installed as a library to the Arduino IDE using the Library Manager. The FreeRTOS main kernel offers a number of features, including fixed priority preemptive time slice scheduling, inter-process communication, task management features such as task priority, task suspension, deletion, and delay, along with software timer and interrupt management.

While the concept of multitasking with the Arduino is relatively new, the real-time multitasking libraries offer up some interesting possibilities for a variety of Arduino projects that would require the ability to handle multiple conditions, events, and operations simultaneously. They are definitely something to consider experimenting with.

5 Tools, Construction Techniques, and Troubleshooting

Part of the fun of working with the Arduino is creating projects. There's nothing quite like the satisfaction of building something yourself and then getting it to work as intended. As with any electronic project, you will need a few specialized tools. You'll also need some basic test equipment to make sure things are working properly, and to troubleshoot with when they're not. Developing a good troubleshooting method is an important part of getting your Arduino projects to work when they don't do what they're supposed to do.

Specialized Tools

Breadboards

While creating projects on a breadboard is often a good way to start out, eventually you will want a more permanent solution, involving soldering your project's components to some form of prototyping or etched circuit board, and mounting the finished project in an attractive enclosure similar to the one shown in **Figure 5.1**. For this, you will need a soldering iron or soldering station. You'll also want to have a multimeter handy to read voltages, double-check component values, check continuity, look for shorts, etc. As your projects become more and more advanced, particularly if you're working with analog signals, you will probably want to get some more sophisticated test equipment such as a frequency counter, oscilloscope, and/or logic analyzer.

When I begin designing and building a new project, I usually build a prototype version of the project on a breadboard. A breadboard is typically a rectangular plastic block, in my case 7 × 3 inches, with a series of interconnected holes for plugging in components and jumpers to quickly build a project. It allows you to make wiring changes simply by moving the jumpers around.

Figure 5.2 shows my original Arduino development breadboard during the development of the DDS VFO project for my second book. Everything was mounted on a 12 × 15-inch piece of wood. This setup allowed me to easily store the board without taking everything apart when I needed my workbench for other purposes. In addition to having two breadboard units for those really big projects, it had an Arduino UNO, a Digilent Uno32, and an Arduino Due mounted around the edges

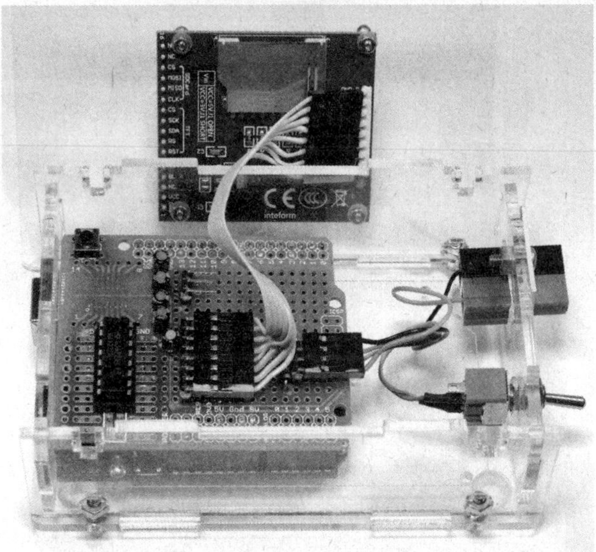

Figure 5.1 — A Completed Project in a Solarbotics Enclosure.

Figure 5.2 — My original homebrew breadboard setup.

Figure 5.3 — The Phase Dock WorkBench with a prototype of the AR-40 Rotator project.

of the board. Also mounted on my development breadboard was a 9 V battery clip, a servo, a 16 × 2 I^2C LCD display, and a terminal block to connect to off-board devices such as antenna rotator motors. This setup allowed me to design and develop many projects without having everything spread out all over the workbench. In between projects, I could quickly and easily store it on a closet shelf until it was time for the next project.

However, as times changed and new Arduino processor boards became popular, others such as the Due and the Digilent Uno32 faded from popularity, and in the Due's case, were discontinued altogether. To modernize my development board, I had to unscrew the boards and mount different, more commonly used processor boards. As this progressed, eventually the wooden board accrued so many screw holes it needed replacing.

The good news is that as the Arduino becomes more popular, so do the commercially available development systems. Rather than upgrade my original breadboard system, I switched over to the Phase Dock WorkBench (see **Figure 5.3**) for my proof of concept and larger prototype development projects. The Phase Dock WorkBench is unique in that each building block-style component is mounted on a "slide," which is then mounted to a "click" platform that then snaps into holes in the work area itself. There are a number of pre-built clicks and slides available, and you can also customize your own. A particularly handy feature is the plastic cover option, which keeps your project safe from cats and other hazards when you're not working on it. A complete review of the Phase Dock WorkBench can be found in the April 2020 issue of *QST*.

Component Testers

When building electronic projects, sometimes it's difficult to determine the value of a component by its color code or the value stamped on the body of the component. When looking for a transistor, how can you easily tell if it's a PNP or an NPN transistor, what the correct pinout is, or if it's even any good at all? Sure, you can test it with a multimeter, or look up the datasheet online, but there is a far easier way. There are now a number of inexpensive multifunction component testers (see **Figure 5.4**) that you can use to test and determine the value of many electronic components such as resistors, capacitors, diodes, transistors, inductors, MOSFETs, and other components with the press of a button. Powered by a 9 V battery, the multifunction tester will try to determine the type of component, test it, and display the value of the component, along with some basic parameters such as pin configuration and gain in the case of a transistor, forward voltage drop in the case of a diode, etc. I don't know about you, but ever since they changed over to the 5-band color codes on resistors and started using 3-digit numbers on capacitors, I can't figure out what the value is without using a cheat sheet or multimeter. Now, I just put the component to be tested into the multifunction tester's onboard socket and verify that the part is good and that it's the correct value with the press of a button. The newer versions of the multifunction testers use a graphic LCD display to

Figure 5.4 — Component Tester.

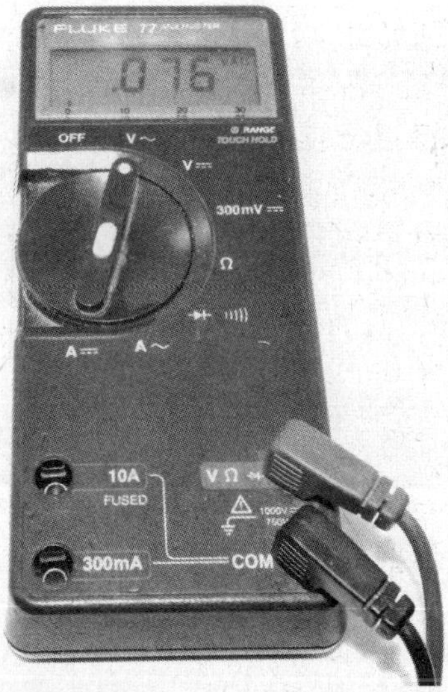

Figure 5.5 — A Multimeter.

actually draw the schematic symbol of the component with the pin configuration of the device overlaid on the schematic symbol. They will also display some basic device parameters. I got my multifunction tester from eBay for about $10, and it is easily the most used piece of test equipment on my workbench.

Multimeter

At the very least, you will also need a multimeter to test your Arduino projects. It doesn't have to be a fancy one (see **Figure 5.5**); just about any multimeter will do. You can get one for as little as $8 (and sometimes even free with a coupon and purchase) from home improvement and hardware stores such as Harbor Freight. A typical multimeter allows you to read voltage, current, and resistance. Some models will even read capacitance, test diodes, and other component values. Generally, you will use your multimeter to read the voltage on the Arduino power and I/O pins, and to check for continuity and short circuits in your project.

Soldering Station

Once you have finished developing and testing your project on a breadboard, the next step is usually to solder everything up on either a piece of copper-clad perfboard or an Arduino prototyping shield. Things sometimes get tight when working with small boards, so you'll want a good soldering iron or maybe even a soldering station similar to the one shown in **Figure 5.6**. For years, I used a standard soldering iron and was quite happy with how things turned out. Then, when I needed to try my hand at soldering some surface mount components, I discovered rather quickly that using a regular soldering iron just wasn't going to get the job done. I borrowed a soldering station from my friend, Tim Billingsley, KD5CKP (now SK), to take a crack at it. For a variety of reasons, I failed miserably at my first surface mount soldering attempt, but I was hooked on the soldering station itself. I liked the fact that the soldering pencil had an adjustable temperature control and used standard interchangeable tips. It also had a heat gun attachment with adjustable airflow and temperature controls. While I may have been less than successful in my dealings with surface mount components (I have gotten better at it since then), the heat gun worked great with heat shrink tubing. I ended up getting a soldering station of my own, with all the bells and whistles, from eBay for about $60.

Another soldering aid you might want to consider is something that I discovered at the Orlando Hamcation hamfest a few years ago. The inexpensive 12 × 12-inch Newtex High-Temperature Resistant Fiberglass Pad is actually a high heat resistant (1,800+ °F) woven cloth-type welding pad to protect against sparks and burning. I use one between my workbench tabletop and the circuit board as I solder my projects together. While I got mine at the hamfest, you can also order one online from sellers such as **Amazon.com**.

Dr.Duino

As I mentioned in **Chapter 3**, the Dr.Duino Explorer shown in **Figure 5.7** is a great way to

Figure 5.6 — An adjustable temperature soldering station with hot air attachment.

Figure 5.7 — The Dr.Duino Explorer.

Figure 5.8 — The Dr.Duino Inventor.

quickly create a proof of concept or prototype project. The Dr.Duino Explorer can act as an I/O shield inserted between either an Arduino UNO or Nano and your prototyping or other type shield, and it allows you to use a multimeter or oscilloscope to view the status of the Arduino's power and I/O pins. You can also do some basic proof of concept and prototyping projects directly on the Explorer board itself and use its onboard components to simulate inputs and outputs to help you with your project design. You can even do a hybrid of both methods and use the Explorer's onboard I/O jumpers to reroute the I/O pins to either your prototyping shield or the Explorer's onboard components. The Explorer also has small breadboard and soldering areas for you to construct projects right on the Explorer board. This is a great compact prototyping solution, especially if you're cramped for space in your work area. I relied heavily on the Explorer to create the working prototypes and sketches for several of the projects in this book. A complete review of the Dr.Duino Explorer can be found in the January 2021 issue of *QST*.

The recently released Dr.Duino Inventor (see **Figure 5.8**) is similar to the Explorer board, but supports the faster and more powerful ESP32 series of microcontrollers, with prototyping features similar to the Explorer. The Inventor board was reviewed in the June 2023 issue of *QST*.

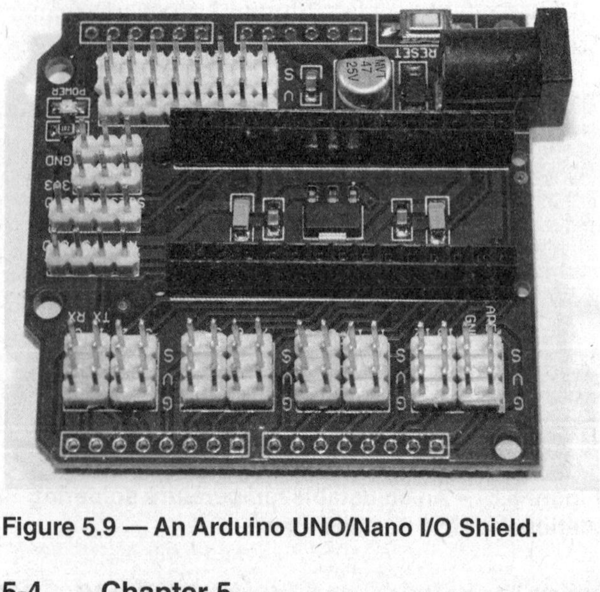

Figure 5.9 — An Arduino UNO/Nano I/O Shield.

I/O Shields

An I/O Shield (see **Figure 5.9**) was one of the first add-ons I bought for my Arduino. An I/O shield brings all the Arduino I/O pins out to 3-pin headers, with the I/O pin, power, and ground in each 3-pin group. You can then use jumper wires from there to a breadboard or access the pins with a multimeter or oscilloscope for testing and troubleshooting. Many of the modern I/O shields for the UNO also have a socket onboard that will allow you to use the shield with a Nano as well. You can also get an I/O shield that will support the Arduino Mega footprint type boards. (see **Figure 5.10**).

Frequency Counter

As you begin working with some of the modules capable of generating audio and RF frequencies, you will most likely want to have a good frequency counter handy. I was able to pick mine up at a hamfest (see **Figure 5.11**) for about $50. All it needed was a new display driver chip, and it was as good as new. Mine goes up to 50 MHz, so it's only usable for audio and HF frequencies, but

Figure 5.10 — Arduino Mega I/O Shield.

that's generally all I need it for. I was able to pick up another perfectly functional frequency counter at the same hamfest for $50 that goes up to 1.3 GHz, but I haven't needed to use it, so it usually sits on the shelf as a backup.

Having a frequency counter is extremely helpful for both audio and RF signals. You may not have a need for a frequency counter in your Arduino project development, but it never hurts to be on the lookout for an inexpensive one at a hamfest. There are also some very inexpensive ones available online from eBay and **Banggood.com** that look like they'd work well for some applications where you need a basic frequency counter.

Oscilloscopes

If I could have only one piece of test equipment in my lab, it would be an oscilloscope. With an oscilloscope, you can view waveforms, measure voltage, and with a little math, you can determine the frequency of the waveform. Some oscilloscopes have add-on units that can also display voltage and frequency, with some of the newer digital storage scopes incorporating that as part of the waveform display. There are basically two types of oscilloscopes: the older, traditional analog oscilloscopes that use a small cathode ray tube (CRT) with a high persistence phosphor to display the signal waveform (similar to how an older CRT-based TV does), and the newer, digital storage scopes. Digital storage scopes actually sample the input signal at a high rate using an A/D converter and graph the signal on either a CRT or TFT display. Because they use digital data, a digital storage scope can save and analyze the waveform and add calculations such as voltage and frequency to the display. Many of the newer digital storage scopes can also interface with a PC via a USB port to capture the waveform on the PC and to control the oscilloscope settings from the PC.

Depending on what kind of Arduino projects you plan on building, you can even build your own mini-oscilloscope using an Arduino. There are several Arduino oscilloscope projects on the Instructables website (**instructables.com**) that work quite well for basic troubleshooting on the Arduino. The main limitation with these Arduino-based oscilloscopes is the speed of the Arduino itself, and you can only use it to view waveforms up to about 200 KHz depending on the oscilloscope project you build. This type of oscilloscope would therefore only be usable at lower audio-type frequencies. If you don't want to build one, there are several pre-built minidigital oscilloscopes that use a TFT display available on eBay for around $10.

PC-Based USB Digital Storage Scopes

Moving up in functionality, you have the PC-based USB digital storage scopes. These are units that connect to your PC via a USB port and are controlled by a software application running on the PC (see **Figure 5.12**). This type of scope generally does not have a standalone display, and all waveforms are displayed on the PC monitor. A good quality 100 MHz USB digital scope costs about $65 on eBay. The one downside of this type of oscilloscope is that you need a PC to use it, so while this type of scope is portable, they tend to take up workspace on your lab bench. I used one for quite a while before I decided to invest in an old analog scope, simply because that's what I was used to using.

Figure 5.11 — A Frequency Counter.

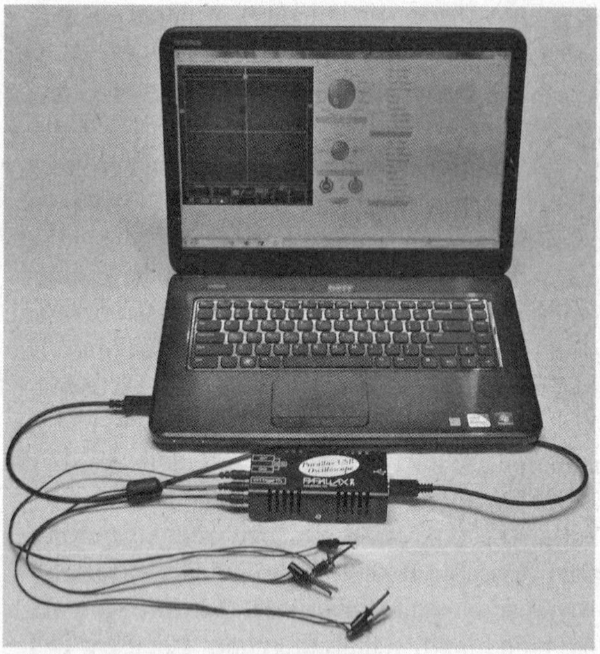

Figure 5.12 — A USB digital oscilloscope. (Photo courtesy Pat Vickers, KD5RCX)

Traditional Analog Oscilloscopes

When you think of an oscilloscope, typically you think of the larger analog scopes such as the Tektronix 475 scope shown in **Figure 5.13**. When I decided to move up from my USB-based scope hooked to a laptop PC, I picked up an old-school Tektronix 465 100 MHz dual-channel oscilloscope with the voltage/frequency add-on at a hamfest for about $150. At the same hamfest, I ran into an old friend, Henry Wingate, K4HAL, who gave me a 200 MHz Tektronix 475 scope that was the exact same scope I used while working with him at Control Data Corporation many years ago. He managed to get it when they closed the doors a number of years back, so there's a sentimental attachment to this particular scope, and it has been a mainstay in my lab for several years now. A typical analog scope can display two channels worth of waveforms on a small CRT, with a high persistence phosphor coating inside the CRT that allows you to see the full screen trace, similar to the way the old-style CRT-based TV sets displayed the picture. While there were some polaroid camera options available that you could use to take pictures of the waveforms on the CRT screen, most analog scopes can't be interfaced to a computer, and they only display the waveform in real-time, meaning a waveform can't be saved and analyzed later.

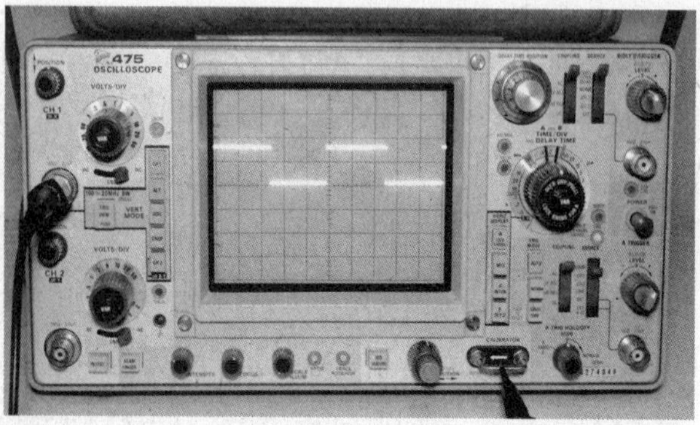

Figure 5.13 — Tek 475 Scope 1.

Digital Storage Oscilloscopes

In the past, the best digital storage oscilloscopes were the big, expensive commercial units that really didn't fit in the average ham's budget, including mine. With the introduction of the larger high-resolution color TFT displays, good quality digital storage oscilloscopes are becoming more affordable. A few years ago, I treated myself to a Hantek DSO5202P 200 MHz dual-channel digital storage scope (see **Figure 5.14**) for around $300. While not top-of-the-line, it's about the size of a lunchbox, with a 7-inch high-resolution color TFT display, and it has most, if not all, the features of the high-priced top-end digital storage scopes. It even has a USB port and a PC application where you can view and control the scope from a PC, allowing you to save and print waveforms like the one shown in **Figure 5.15**. Being about the size of a lunchbox, the Hantek scope easily fits on a shelf above my workbench, and I can hook up the USB cable to save any waveforms to the PC when I need to. While $300 for a piece of test equipment may seem a bit on the expensive side, if you do a lot of building and work with audio, HF, and VHF frequencies, it's hard to beat a scope of this type

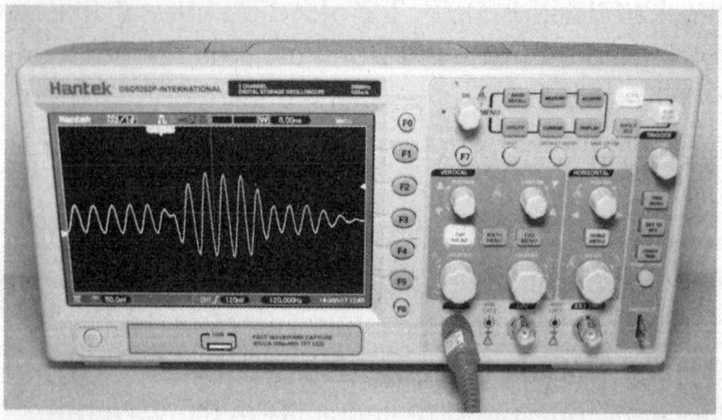

Figure 5.14 — A Digital Storage Oscilloscope.

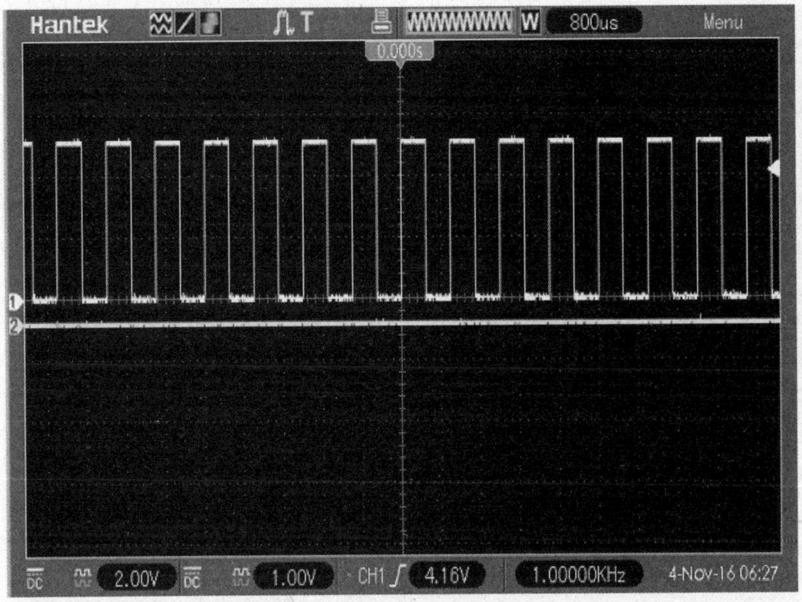

Figure 5.15 — A Digital Oscilloscope's waveform displayed on a workstation.

and price. I'd been worried what I would do if — and when — my nearly 40-year-old analog scope died, but now I have a backup plan. I usually use the new digital storage scope for quick tests and troubleshooting and break out the old analog scope for more involved troubleshooting, mainly because of my comfort and experience with using the older scope.

Logic Analyzer

One of my most recent test equipment acquisitions is a USB-based 16-channel logic analyzer (see **Figure 5.16**). In the past, logic analyzers tended to be a lot like analog oscilloscopes: bulky, unwieldy, and when it came to displaying the actual multitrace analyzer output, just not very good quality unless you bought an expensive, high-end one. As with just about everything else through the years, the logic analyzer has been vastly improved, and it is now available as a USB device that attaches to a workstation with excellent functionality and display quality.

As a major oversimplification, think of a logic analyzer as an oscilloscope for digital-only signals, with a large number of individual display traces, typically 16 or 32 signal traces as shown in **Figure 5.17**. A logic analyzer has a programmable triggering point based on individual or multiple conditions. This allows you to see multiple points of data simultaneously and is ideal for viewing complex timing conditions such as SPI or I²C bus transmissions. This logic analyzer can also decode and display SPI and I²C communications directly.

For my logic analyzer, I went a bit overboard and bought a 16-channel 100 MHz Kingst LA1016 unit for about $80 from eBay. This particular device can safely allow input voltages from –50 to +50 V, so it would take a lot of effort on my part to damage it input-wise. Less expensive and lower bandwidth ones are available with as few as 8 channels and for less than $10. You may never need a logic analyzer, but in my case, there will always be a time when I would give just about anything to have one at hand in the lab.

Bluetooth LE Sniffer

As you start working with Bluetooth projects, having a Bluetooth 4.0 LE sniffer such as the one in **Figure 5.18** will be a great benefit. Unlike your regular Bluetooth devices, a sniffer allows you to see and decode all Bluetooth LE activity within range. You can also use software such as Wireshark to see the data inside each individual packet. This is useful when you're trying to figure out why your Bluetooth LE projects aren't working properly. The sniffers I was able to locate are only for BLE 4.0. I was unable to locate a Classic Bluetooth sniffer, although they may be available as well.

Figure 5.16 — The Kingst Logic Analyzer.

Figure 5.17 — An example of Logic Analyzer output.

Design Tools

Designing and documenting is a major part of every project. While many of us start from a hand-drawn circuit design, at some point in your project, you will want a more professional-looking method of documenting the project, especially if you plan to share it with others. If you plan to have a circuit board fabricated for your project, you will need to use design software that will generate a printed circuit board (PCB) layout in a format that your board fabricator can support. There are a number of design tools to choose from, but we'll focus on a few of the most popular ones here.

Personally, I am a big fan of Autodesk's free version of *Eagle*. *Eagle* allows you to create schematic diagrams as well as the PCB layout based on the schematic diagram and has the capability to automatically route the PCB traces for you. *Eagle* has a huge library of components you can incorporate into your projects, and you can even create new parts as needed. *Eagle* has both a free and paid version. For most of us, the free version will do just about everything you need. It is free for hobbyists and allows you to create two schematic sheets, two signal layers, and an 80 × 80-centimeter (12.4-inch) board area.

Another popular PCB design tool is *KiCad*. *KiCad* is free and open source. Unlike *Eagle*, there are no licensing restrictions with *KiCad*. You can even download the source code and modify it if you choose. If you are an *Eagle* user like me, the transition to *KiCad* is not difficult at all. *KiCad* also has a number of keyboard shortcuts for doing standard tasks such as component rotation or drawing a wire, whereas *Eagle* does not. Compared to *Eagle*, creating components in *KiCad* is a bit more confusing to start out with, but as with all things new, it gets easier in time. *KiCad* has a very nice 3D board viewer to help you with component placement on the board. Both can generate board layouts in the industry-standard Gerber file formats as well as the Numeric Controlled (NC) file for drilling holes in the PCB.

KiCad and *Eagle* are both supported on Windows, Mac, and Linux workstations. Both have their pros and cons, and the choice as to which one to use is up to you. I would recommend working with both to see which one you prefer.

Circuit Simulators

An expanding and generally unexplored area of electronic project design is Circuit Simulation. There are a number of general-purpose simulators, as well as Arduino/Microcontroller-specific simulators. As with the design tools, we'll focus on just a couple of the more popular circuit simulators and recommend that you do further research to determine which one may be best for you.

LTspice is a SPICE-based analog electronic simulator from Linear Technology. Based on the original open-source simulation program with integrated circuit emphasis (SPICE), *LTspice* is free and easy to use. Although it comes with component libraries that contain only Linear Technology products, additional components can be added by the user. The good thing about using

Figure 5.18 — Adafruit's Bluefruit LE Sniffer. (Photo courtesy of Adafruit.com)

a SPICE-based simulator is that once you learn one version, it's relatively easy to switch to other versions that may better suit your needs.

Spectrum Software's *Micro-Cap 12* is a powerful, full-featured, commercial SPICE-based integrated schematic editor and mixed analog/digital simulator that is now free from **spectrum-soft.com**.

Tinkercad is a free, easy-to-use, web-based modeling program from Autodesk, the makers of the *Eagle* PCB design tool discussed earlier. *Tinkercad* can be used for 3D design, electronic simulation, and coding. *Tinkercad* allows you to design your circuit by placing and wiring components, programming your project using either Codeblocks or C++, and then simulating how the circuit performs in real-life. *Tinkercad* supports the Arduino UNO and ATtiny microcontrollers, but you are limited to the devices in the built-in component library, as *Tinkercad* does not support user-created virtual "hardware." However, it does support a broad range of libraries and components. Codeblocks is a very interesting part of *Tinkercad*. Rather than coding each individual line of an Arduino sketch, you can use a "codeblock" to incorporate a standardized block of code into your sketch that can then be expanded into normal Arduino code and edited as you would normally. Once the circuit and sketch are complete, you simply select the **START SIMULATION** button and *Tinkercad* will then virtually "run" your sketch. *Tinkercad* not only supports a virtual version of the Arduino IDE's Serial Monitor, but it also incorporates a virtual multimeter and oscilloscope to allow you view the voltages and signals in your project. Once you have everything working in *Tinkercad*, you can easily export the Arduino sketch, import it into your real Arduino IDE, and upload it to a physical Arduino. *Tinkercad* even includes a debugger that allows you to step through your virtual sketch and look at the variables while the sketch is running.

The Virtronics *Simulator for Arduino* v1.12 (**virtronics.com.au**) is considered one of the better Arduino simulators currently available. This simulator has a free version that incorporates a delay on loading sketches after the initial 45-day trial period expires or the 200-sketch limit is exceeded. It does remain full featured after the trial period ends, with just the initial sketch loading delay. You can also purchase the Pro version for $19.99. *Simulator for Arduino* supports the Arduino UNO, Nano, Mega, Leonardo, and other common Arduino boards. It has a debugging mode that allows you to step through your sketch line-by-line, and it can simulate all the analog and digital I/O pins, as well as a number of devices such as an LCD display, Ethernet, Servo, etc.

Construction Tips

There are several tips and techniques I have found helpful when constructing Arduino projects. First and foremost is the use of connectors between my Arduino project and any external components such as switches, rotary encoders, displays, etc. This allows me easily to remove the Arduino assembly from its enclosure for testing and modification. I have found that using the DuPont-style 2.54-millimeter headers and sockets (see **Figure 5.19**) is ideal for working with Arduino prototyping shields and copper-clad perfboards. The prototyping boards also have a 2.54-millimeter spacing between holes, so no modifications are needed to use these connectors. The DuPont-style headers and sockets have a male header assembly like the standard jumper pins that you find on many electronic devices. The male part of the header typically comes in 40-pin strips that are designed to be separated into shorter strips containing the number of pins that your project requires. The male header is also available in a right-angle configuration, allowing for a lower overall height of the connector assembly for use in tight quarters.

The female portion of the DuPont style header, also known as a "header socket," consists of a plastic shell with holes for inserting a locking female crimped pin. The female connector is available in a wide variety of pin configurations, from a single pin to up to 10 pins and higher. I have found that using these female sockets is perfect for cabling between the Arduino projects and external components, such as switches and displays, using old PC ribbon cables. I was able to buy an entire box of old PC ribbon cables for $5 at a hamfest several years ago, and I still have about a quarter of the box left. More than likely, you'll want the older cables that were used for

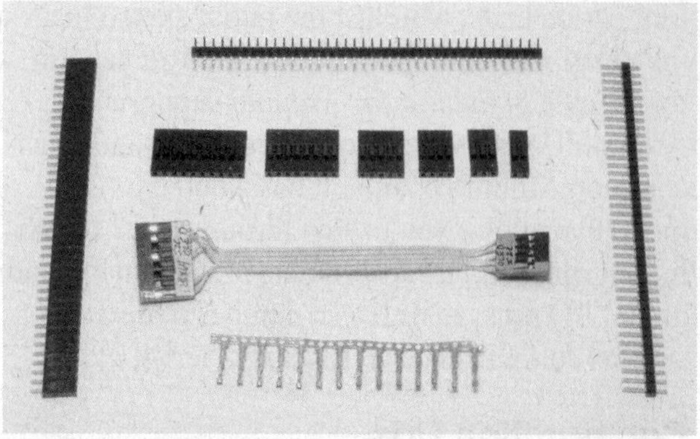

Figure 5.19 — DuPont Style headers, pins, and cables.

Figure 5.20 — Example of two rows of header sockets to create a socket for a module..

floppy drives and the older IDE hard drives, as the conductors in these cables are typically stranded rather than solid. I've found that the stranded wire in these ribbon cables is much easier to work with and more durable compared to the solid conductor versions. To complete the cable, I just crimp and solder the female socket pin to the end of the ribbon cable and insert the socket pin into the appropriate hole in the female header socket. I'll also use a permanent marker to identify pin 1 on the cable and on the board to help prevent wiring errors.

I learned early on in my work with the Arduino that all ICs, including Arduino Nanos and module assemblies that will be mounted on a protoshield or perfboard, should be in sockets. After having to desolder and replace a Nano a couple of times, I started putting everything I could into a socket. While as a general rule, the Arduino is forgiving and pretty hard to burn up, it can be done. When soldering a new project, wiring errors can, and do, occur. Trust me, it's a royal pain when you've just spent five hours wiring up that new project only to discover you were one pin off and just put 12 V into a digital I/O pin on a Nano. So, from here on out, if I can use a connector or a socket, I will. You can get 40-pin strips of the female version of the 2.54-millimeter male headers that can be soldered to a board. For chips and modules such as the Nano, I'll cut these strips down to the number of pins I need and use two rows to create a socket for the device as shown in **Figure 5.20**.

The DuPont 2.54-millimeter headers and sockets, along with the 40-pin SIP (Single In-Line Package) header and socket strips, are very inexpensive. Typically, you can get them in a bulk quantity on eBay for $1 per 100, with free shipping. I generally order each individual type of these connectors separately, so for roughly $10, I have enough various connectors to last me quite a while.

When building projects that use an Arduino UNO-style footprint, I like to use a prototyping shield, also known as a protoshield (see **Figure 5.21**). The protoshield is designed with the same footprint as the Uno, so you can use header pins soldered to the board and connect it to your UNO just as you would with any other shield. This allows me to easily remove the protoshield from the project so I can fix wiring errors or make additional modifications to the board. This technique also allows me to remove the protoshield and

Figure 5.21 — An Arduino UNO Prototyping Shield.

Figure 5.22 — Building a project on a pre-cut perfboard.

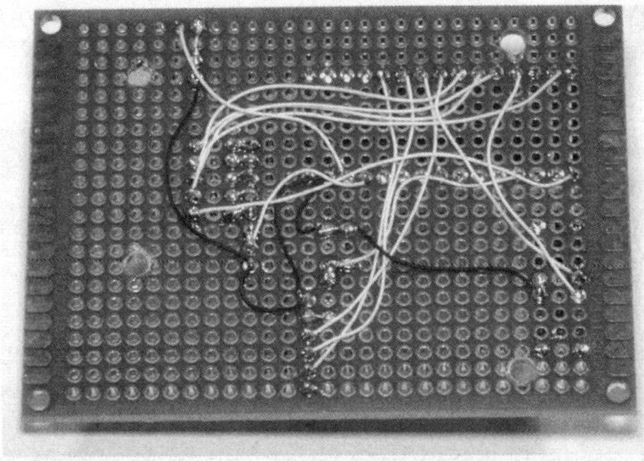

Figure 5.23 — The bottom side of a perfboard project.

reuse the Arduino UNO in another project simply by attaching a different protoshield and uploading a new sketch.

For non-UNO projects, I like to use perfboards cut to fit inside the desired enclosure and solder all the components to the board (not forgetting to use sockets for the chips and modules) as shown in **Figures 5.22** and **5.23**.

When I need a power connector on a project enclosure, I will usually use Anderson Powerpole connectors (see **Figure 5.24**). A Powerpole connector is a plastic housing with a latching crimped contact pin inserted into the housing. Powerpole connector housings are keyed in such a way that multi-conductor cables can be built simply by mating as many Powerpole housings together as necessary. While most often used for providing dc power to ham equipment, larger multi-conductor cables can also be built using Powerpoles. As an example, for portable events such as ARRL Field Day, our club's antenna rotator control cables have a block of Powerpole connectors on each end, with pigtails on the rotator and control unit used to adapt to the long rotator cable. Using a common wiring scheme, we can use just about any club member's rotator and controller without having to modify the cable. All we have to do is plug in the rotator cable to the rotator and controller pigtail cables and we're ready to go. No more messing with those screws on the back of the rotator or controller that always fall into the grass never to be seen again, and no more wiring errors while attaching the rotator cable. You can also get chassis mount adapters and connectors that allow you to easily integrate Powerpole connectors in your Arduino project enclosures, as I have done with several of the projects presented in this book. Powerpole adapters come in a wide variety of current ratings, up to 180 amperes and with different colored housings, allowing you to color-code your connections. Most hams use the 15/30/45 ampere version, as these all use the same size connector shell with just different size contacts.

Troubleshooting

One of the most frustrating things about building your own electronics projects is that they don't always work as they should. While there's no simple tried-and-true method to magically fix everything that doesn't go right, there are a few things you can do to help make the troubleshooting process easier.

The first step in troubleshooting is to sit back and analyze what is happening — or not happening. Quite often, a simple analysis of the problem's symptoms will get you to the problem.

Figure 5.24 — PowerPole connectors.

Troubleshooting basically comes down to logical, deductive reasoning combined with the old divide-and-conquer technique we mentioned earlier. And of course, a little bit of luck never hurt anyone. When something is not working correctly, trust nothing. Every component is suspect, every solder joint is suspect, every wire is suspect, your Arduino sketch is suspect, and the new library you're using is suspect; everything is suspect until you prove to yourself otherwise. Using deductive reasoning, narrow the problem down to its most likely cause, and starting with the most logical suspect, work your way down the list. Test for shorts, test the continuity of each wire, and verify that you're not soldered to the wrong pin.

If you're not making any headway, stop, get up, and walk away. Take a break and work things out in your head. When I was just starting out in my career, I was the assistant to the project's design specialist on a completely new prototype VHF and Microwave data telemetry project. Occasionally, when he'd hit a snag in his design, the circuit didn't do what it was supposed to, or he had to figure out a completely new way of making things work, he'd just get up and take me along for a brief walk. Along the way, he'd throw some thoughts about the problem my way to get my insight and viewpoint. For a 19-year-old kid, it was a tremendous experience to be thought of highly enough to offer some insight into the problem and serve as his sounding board. Often, about halfway through the walk, he'd have that "Eureka" moment, and we'd go back to it and resolve the issue. It's those kind of experiences through the years that have taught me some of the best troubleshooting techniques I have learned.

It is key to evaluate what is and what isn't working, and to come up with possible reasons that match the symptoms you're experiencing. Have a friend double-check your wiring. You would be amazed by how many times you look at something and swear that it's right, while a fresh set of eyes often spots the problem in a matter of seconds.

If that fails, don't hesitate to ask for help or search online for possible answers. There's a good chance that someone somewhere has experienced the same issue you have and can offer up a simple solution or troubleshooting idea. Break your project down into sections and test each piece separately. Again, this will help you narrow the problem down to a specific area. As you can see, troubleshooting is more of a method than using a specific set of tools. But with the Arduino, there are some additional considerations and tools you can use to help troubleshoot your projects.

Arduino has no built-in tool to monitor your memory usage, so it's easy to run out of available memory and not know it. When this happens, things just go totally bonkers and don't make sense. Fortunately, there are some small memory usage monitoring sketches you can integrate into your sketches if you feel that they may be running out of memory. As we mentioned earlier in **Chapter 5**, using the preprocessor `#define` and `#ifdef…#endif` statements to incorporate **SERIAL MONITOR** debug code is another method for peeking inside your sketch. Finally, while the Arduino does not have a debugger built in, there is a sketch debugging tool you can use to troubleshoot your Arduino sketches.

Debuggers

Debuggers are invaluable tools for testing and debugging programs and have been around since the early days of computing. Most often used with the lower-level programming languages such as assembler and C, debuggers allow you to run the target program under controlled conditions that allow you to track the program's operation and monitor changes in variables along with other components of your program. Unfortunately, the Arduino does not have a built-in debugger. However, there is an Arduino library that provides many of the features of a debugger.

The serialDebug library adds multi-level debugging capability to your sketch, displaying the debug messages on your **SERIAL MONITOR** in different colors according to its debug level. You can also implement the *serialDebug* app, which runs on your PC, Mac, or Linux workstation to better format and display the debugger output. While not a full-featured debugger, the serial Debug library does offer several valuable debugging features, including the ability to call a function, show and change global variables, and add or change "watch" conditions for global variables. A debugger watch condition means that the debugger will monitor and notify you if the watched variable changes value. The serialDebug library is compatible with any board supported by the Arduino IDE. There is even a converter that can convert an existing sketch and generate a new sketch with the debugger functions implemented.

The recently released version 2.0 of the Arduino IDE includes a built-in debugger. This feature alone will be a major factor to my moving up to this new IDE as soon as possible. Until that time, I'll be using the serialDebug library to fill the role of an Arduino debugger to help troubleshoot my sketches.

And finally, I can offer up one last — but major — piece of troubleshooting advice. I'm guilty of this more often that I care to admit, but I've found it to be true: don't overthink the problem. Sometimes, we get so wrapped up in the complexity of things that we don't stop and think things out on a simpler level. Instead, we just push on and get more frustrated as we go. The Arduino is a simple microcontroller, not a supercomputer. When installed correctly, the libraries work and can be thought of as "black boxes" you don't need to see inside. More often than not, while trying to figure out some complex solution to a problem, backing up and "keeping it simple" has worked just about every time for me. So, if you find that your design isn't working out, stop and look for a simpler way. You'd be surprised by how many times you'll come across a much simpler and more elegant solution to the complex corner you've painted yourself into.

Reminders

There are a couple of things to keep in mind when building projects. These are some of the "gotchas" that can catch you unaware and cause you all kinds of grief. The I^2C bus is an open-drain bus that requires pull-up resistors to function properly. The internal pull-up resistors on the Arduino SCL and SDA I/O pins (typically analog pins A4 and A5) used by the I^2C bus are not sufficient to serve as pull-up resistors. You will need some form of external pull-up resistor for the I^2C to work right. A 4.7 KΩ resistor is a commonly used value for the pull-up resistor, however, some modules have pull-up resistors built into the module. You want only one set of pull-up resistors on the I^2C bus, so be sure to check whether any module on the bus already has the pull-up resistors onboard before adding external ones.

Some Arduino UNO-style boards have the circuit board copper traces run very close to the mounting hole on the corner of the board near the A4 and A5 pins. It is very easy to chafe through any insulating material on top of the board and short these pins when mounting the board. It's best to use a nylon washer between the screw head and the board to prevent this.

Also, remember that there are now multiple versions of the Nokia 5110 LCD display, as we mentioned in **Chapter 3**. Be sure that you verify the wiring for the type of Nokia LCD module that you are using in your projects. When in doubt, use your breadboard to test the wiring and pin configuration of your Nokia 5110 using one of the Example sketches so you'll know how you need to wire your Nokia display into your project.

Finally, be sure to keep track of your 3.3 V and 5 V devices and power requirements. Don't forget to use some form of level shifters, especially with the 3.3 V modules and a 5 V Arduino, including the Chinese TFT displays even though they are rated for 5 V.

Finishing Touches

As a finishing touch to your project, you'll probably want to have everything in a nice-looking enclosure. I tend to prefer the enclosures (see **Figure 5.25**) available from places such as **Solarbotics.com**, but there are several different types of enclosures available from eBay and other sellers as well. If you have access to a 3D printer or laser cutter, you can also create your own. **Thingiverse.com** has a large number of enclosure designs available for free that you can download. As part of putting your project in a nice enclosure, you'll also most likely want some form of labeling as to what each switch and knob does. Trust me, I don't have the handwriting skills to even think about hand-labelling my projects, so I need some form of printed label. In the past, I've used your typical label maker or I've gone back to my early days and tried my hand at dry transfer labels (yes, you can still get those at a hobby shop or craft store). While at the local hobby store looking for enclosure ideas, I came across the latest in model decaling. They now have decal paper that you can use with an inkjet printer. Using this specially coated paper, you can create and print full-color decals for your projects, complete with text and graphics. All you do is print the design on the paper and spray it with a decal bonding spray to keep the ink from running when you apply the decal. Once the bonding spray is dry, you can apply the decal as you would with a plastic model. Simply dip the decal in water for a few seconds, and the clear decal just slides off and adheres to the enclosure right where you want it. As with a regular decal, you can reposition the decal as needed. When dry, you have a nice-looking, permanently mounted label without the mess. You can also get decal paper that works with laser printers and doesn't require the bonding spray.

Cheat Sheets

There are a number of "Cheat Sheets" and online calculators that really help me when creating my Arduino projects. I have the cheat sheets printed out and next to my work area in my "lab" for quick reference. These are the ones that I have found to be most helpful when creating my Arduino projects:

Arduino UNO Pinout Chart — **itp.nyu.edu/physcomp/wp-content/uploads/Uno_pinouts.png**

SparkFun Arduino Cheat Sheet — **learn.sparkfun.com/resources/90**

Arduino Cheat Sheet — **blog.arduino.cc/2010/04/23/arduino-cheatsheet/**

Teensy Pinout Chart — **pjrc.com/teensy/pinout.html**

Electronics Reference Sheet — **akafugu.jp/images/electronics_reference_sheet.pdf**

Capacitor Codes — **dnatechindia.com/Capacitor-Colour-Code.html**

Online LED Current Limiting Resistor Calculator — **ledcalc.com/**

Online Toroid Calculator — **toroids.info/**

Wire Gauge and Current Limits — **powerstream.com/Wire_Size.htm**

Figure 5.25 — The Solarbotics Mega S.A.F.E. Enclosure.

Final Notes

All the projects in this book were created and compiled using version 1.8.19 of the Arduino IDE. If you are having issues compiling any projects using a later version of the IDE, try using version 1.8.19 to compile and upload the sketch.

All the sketches and libraries for the projects in this book are available for download at **kw5gp.com**. The libraries from this site will have to be installed using the manual library installation process discussed in **Chapter 4**. However, many of them can be directly installed from their original source online using the IDE's built-in Library Manager.

My editor and I have worked hard to make this book as error-free as possible, but as with any project of this scope, sometimes errors do happen. Any updates, corrections, and other relevant information will be listed in the Errata link on the download website and corrected as soon as possible in later printings of this book.

Now it's time to have fun and build some projects. Enjoy!

6 Random Code Practice Generator

Figure 6.1 — The random code practice generator.

Surprisingly, one of the most popular projects from *Arduino for Ham Radio* turned out to be one of the easier ones: the random code practice generator. This time around, I upgraded it to use an Arduino Nano and a 2.2-inch color thin-film-transistor (TFT) display, and I mounted it in a smaller Solarbotics Mega S.A.F.E. enclosure, as shown in **Figure 6.1**. The piezo buzzer used in the original project has been replaced with an LM386 audio amplifier IC, a volume control potentiometer, and a small speaker.

Morse code, or CW, has always been my first love when it comes to amateur radio. My first transceiver back in my Novice days was a CW-only Heathkit HW-16. It didn't have the external VFO, just a handful of crystals for 15 and 40 meters. All I had for an antenna was a two-element beam for 15 meters, so 40 meters was pretty much out of the question. With my three 15-meter crystals, I hammered out tons of CW contacts and had a blast. When I finally got my General-class license, I discovered the joy of 20-meter SSB, 2 meters, and the local 2-meter RTTY gang, so my CW skills faded into near uselessness in favor of the fancy digital modes, and of course, HF SSB.

I've never lost my love for CW though, and I kept promising myself I would dive back in. The first step was to relearn all the Morse code I've forgotten. An Arduino-based code practice generator seemed like just the thing to get me back on track and to familiarize myself with the Arduino at the same time.

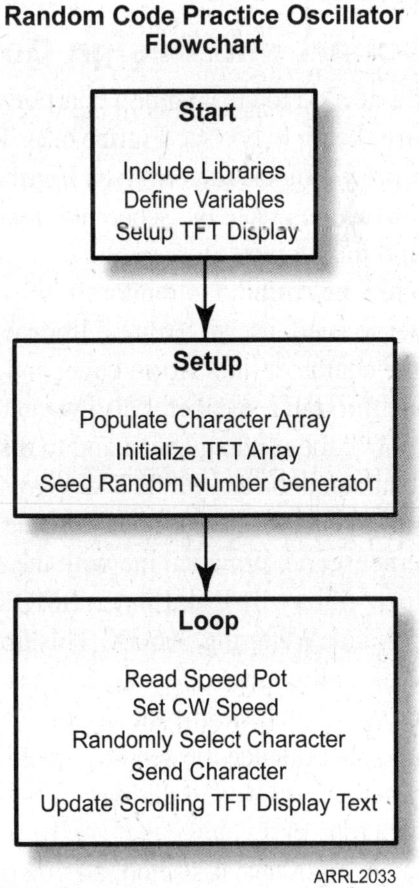

Figure 6.2 — Flowchart for the random code practice generator.

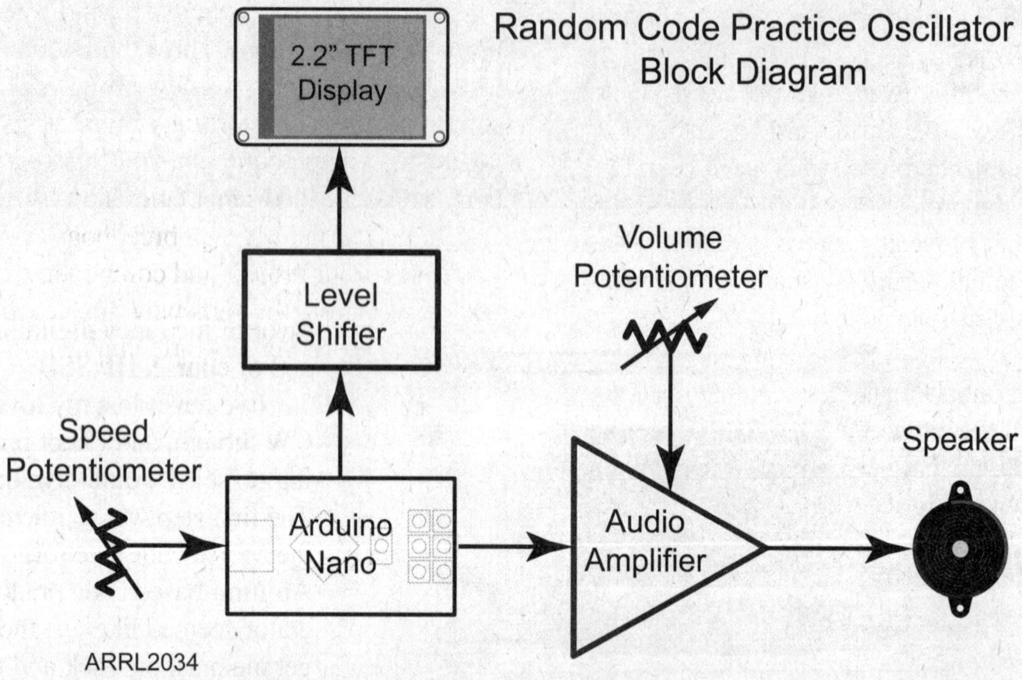

Figure 6.3 — Random code practice generator block diagram.

Flowchart and Design Goals

First, I needed to create a flowchart of what I wanted my sketch to do (see **Figure 6.2**). Then, I quickly drew a block diagram (see **Figure 6.3**) that I could use to create the schematic I would use to build the project.

I wanted the Arduino to randomly select a character to send, use an Arduino library to convert the character into Morse code, and play it as a tone through a speaker. I also wanted to be able to control the sending speed and to display the sent characters on a color TFT display for visual reinforcement of what I was hearing. A brief internet search provided me with the Arduino Morse library by Erik Linder, SMØRVV, and Mark VandeWettering, K6HX. This library is easy to use; all you need to do is include the library in your sketch, define the digital I/O pin you wish to use, include the sending speed, and include whether you want the library to send CW using a tone or a voltage that can be used to key a relay or rig. Once it's setup, all you do is call the library's `send` function with the desired ASCII character, and it will output the Morse code character at the desired speed on the assigned I/O pin.

As an upgrade from the original 16 × 2 LCD display, this project now uses a 2.2-inch color TFT display. Because the ILI9341-type displays don't do well with 5 V on the I/O pins, you'll need to use some form of level shifting, either with 1 K resistors in series with the I/O pins or by using a level shifter. As part of my personal transition to the 3.3 V Arduino world, this project uses a standard 8-channel TXB0108-type level converter module (see **Figure 6.4**) to convert the

Figure 6.4 — TXB0108 level converter module.

Arduino's 5 V signals down to the 3.3 V desired by the TFT display.

Once I had figured out all the pieces, I needed to create the sketch and figure out the electronics I needed to support the sketch. I began laying the project out on the breadboard for testing. What I love about most Arduino projects is that the actual wiring is usually simple. The Arduino and its libraries do most of the heavy lifting for you. The key is to have your concept clearly laid out ahead of time and the rest should go easy. All the parts you'll need for this project are shown in the parts list included with the schematic shown in **Figure 6.5**.

After gathering all the parts, start by assembling the project on your breadboard, based on the schematic. If this is a new project design, be sure to update your working schematic as you go, so that when you finish, you'll have a current copy of the finished schematic. Initially constructing your project on a breadboard will allow you to test your project and components quickly and easily before the somewhat more permanent process of soldering the parts in place.

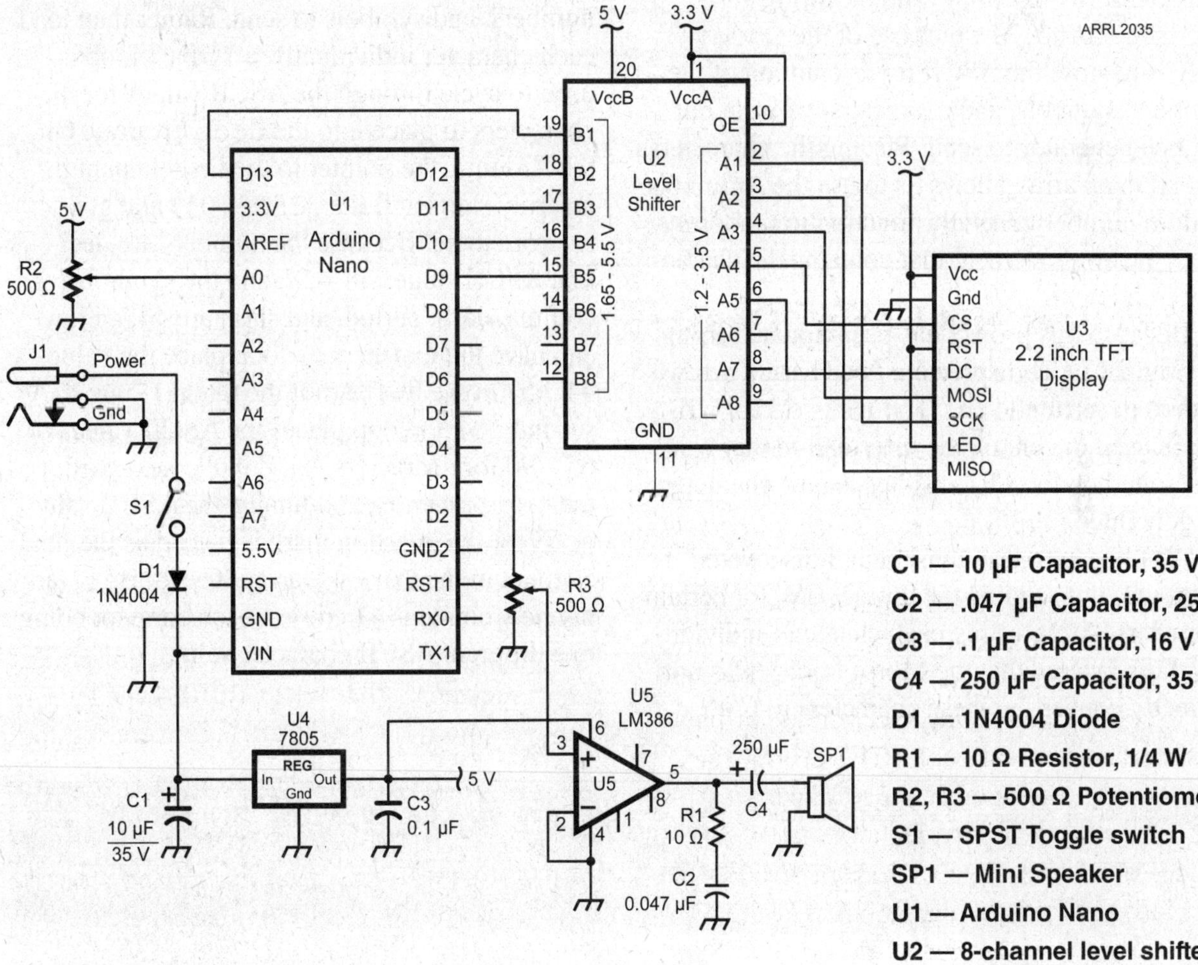

Figure 6.5 — The random code practice generator schematic.

C1 — 10 µF Capacitor, 35 V
C2 — .047 µF Capacitor, 25 V
C3 — .1 µF Capacitor, 16 V
C4 — 250 µF Capacitor, 35 V
D1 — 1N4004 Diode
R1 — 10 Ω Resistor, 1/4 W
R2, R3 — 500 Ω Potentiometer
S1 — SPST Toggle switch
SP1 — Mini Speaker
U1 — Arduino Nano
U2 — 8-channel level shifter module (TXB0108 type)
U3 —2.2-inch ILI9341-type TFT Display
U4 — LM7805 5 V Voltage Regulator
U5 — LM386 audio amplifier IC
Enclosure — Solarbotics Mega S.A.F.E.

The Random Code Practice Generator Sketch

Once you have the circuit wired on the breadboard, you can begin to create your sketch. Rather than printing the entire sketch here, the completed sketch and libraries for this project can be found online at **kw5gp.com**.

Using the flowchart as a guideline, you can break the sketch out into smaller building blocks, testing each block as you go. At the top of the sketch, you will find the usual library `#include` statements, variable definitions, and the object assignments for the TFT display. There are two key variables defined in this section, the character array `Code[]`, and the string variable `tft_data`. The purpose of the `Code[]` array is to store the ASCII representation of the numbers, symbols, and characters we want our practice generator to send. Placing the characters to send in an array allows us to use the Arduino's random number generator function to randomly select the array element that contains the character to send next.

Because this project has been updated from the original LCD display, we need to add a new method to scroll the text. The `tft_data` variable is used to contain the information that will be scrolled on the TFT display as new characters are generated.

As I began to test my sketch, I discovered that it was not sending the correct CW for certain characters. After testing each character individually, I discovered that the Morse library was not correctly converting these characters to CW. Digging through the library, I found that the CW translation for the characters in question was incorrect. (I'm not the best at deciphering C++ code, but once you get used to it, you can usually get yourself in the area you're looking for, so there's no reason to fear libraries if you have to go digging into them for any reason.) After figuring out how the library converted the text to CW and correcting the encoding value for the incorrect characters in the Morse.h file, the library was now sending the correct CW for the desired characters. When you build this project, be sure to use the Morse.h library from **kw5gp.com**.

In the setup portion of the sketch, we need to populate our `Code[]` array with the letters, numbers, and symbols to send. Rather than load each character individually, a `for()` loop is used to cycle through the ASCII values for the characters to place into the `Code[]` array. On the Arduino, the pointer to the first element in an array starts at 0 (i.e., `Code[0]`) and goes up from there. Because the numbers are decimal ASCII values 48 – 57, and the symbols for comma, dash, period, and slash are 44 – 47, we can have the first `for()` loop place the values 44 – 57 in the first part of the `Code[]` array. A second `for()` loop places the ASCII values of 65 – 90 for characters A – Z in the next part of the `Code[]` array, and finally, the ASCII value of 63 for the question mark is placed as the final element in the `Code[]` array. The `Code[]` array now contains 41 entries, each corresponding to a different ASCII character value.

```
// populate the array containing the characters to use (0-9 , . / ? A-Z)

for (int x = 0 ; x<14 ; x++) // 0-9,-./

{

 Code[x] = char(44 + x);

}


for (int x = 14 ; x<40 ; x++) // A-Z

{
```

```
Code[x] = char(51+x);
}

Code[40] = char(63); // add ? character
```

The last step in the setup loop is to randomize, or seed, the Arduino number generator we will be using to randomly select the character to send. The Arduino, like many computers, does not generate a true random number. Instead, the Arduino `random()` function returns a pseudorandom number that will repeat the exact same sequence of random numbers every time the Arduino resets. To add a more realistic randomness to the sequence, you can seed the random number generator using the `randomSeed()` function with a value, thereby altering the starting point in the pseudorandom sequence. A simple way to randomize the Arduino random number generator is to read an unconnected analog input pin and use the value returned as the seed value for the random number generator.

```
// randomize
randomSeed(analogRead(1)); // Seed the Random Number Generator
```

The main loop is where it all comes together. First, read the value of the speed potentiometer. The speed potentiometer varies the voltage on the analog input pin from 0 to 5 V. The Arduino 10-bit analog-to-digital converter will convert this voltage into a value from 0 to 1,023. Rather than trying to figure out what voltage equates to the desired speed, we can use the `map()` function to do the work for us. For this project, I wanted the code speed to be variable between 5 and 35 words per minute (wpm).

```
// Read the potentiometer to determine code speed
key_speed = map(analogRead(0),10,1010,5,35);
```

The `map()` function translates the value of the speed potentiometer analog input to a number between 5 and 35. You may notice that in the `map()` function, I don't use the full range of the possible A/D values. This is done to allow us a little bit of extra range on either end of the analog-to-digital conversion to ensure that the `map()` function can map to the entire 5 to 35 wpm range. This number is then used to set the speed used by the Morse library in words per minute. For this project, the speaker is connected to digital I/O pin 6, so the Morse library is configured to send a tone to the speaker connected to that pin.

```
Morse morse(beep_pin, key_speed, 1); // Set the Code Library to Beep at the
selected Key Speed
```

Once the Morse library is setup for use, we need to randomly select the ASCII value of the character we want to send. Because we know we have 41 elements in the `Code[]` array, we can have the Arduino random number generator select a random number between 0 and 40. This value will be used as the index of the Code[] array, and the ASCII value of the selected character to send is placed in the variable c.

```
index = int(random(41)); // Randomly pick a character from the character

array

tft_char = Code[index]; // Assign the value of the selected character to c
```

Before the character is actually sent, it is displayed on the first line of the TFT display, along with the sending speed displayed on the bottom line. The tft_data variable contains the last 78 characters that have been sent. When sending the next character, the first character in the tft_data string variable is trimmed off using the Arduino substring() function and the current character to send is added to the end of the tft_data string. The string is then sent to the TFT display, which will then appear to be scrolling the text.

```
if (tft_data.length() >= 78) // Scroll bottom line up a line if at end of line 6
 {
  tft_data = tft_data.substring(13); // Remove the first 13 characters (the
top line)
  tft.setCursor(0, 5); // Move the cursor to the top line
  tft.fillRect(0, 0, 319, 209, tft_black); // Clear the tft data area
  tft.print(tft_data); // Print the tft data string variable
  tft.setCursor(0, 165); // Set the cursor to the start of line 6

 }
```

```
  tft_data = tft_data + tft_char; // Add the character to the tft data string

  tft.print(tft_char); // display the received character

}
```

Finally, the character is sent to the Morse library, which will send the character in Morse code on the speaker.

```
morse.send(c); // Send the character in CW
```

When the character is sent, the Arduino will repeat the main loop and endlessly send and display random Morse code.

Once I had the project working on the breadboard, I moved it into an enclosure so it could be put to actual use. For most of my projects that use the color TFT displays, I like to put them in the Mega S.A.F.E. This enclosure is designed for the Arduino Mega footprint, but by drilling just a couple of holes to mount your breadboard, you can put your entire project inside the enclosure. I like to use the Mega S.A.F.E. because it usually allows plenty of room for my project and a 9 V battery to power it, if desired. There's also plenty of room to add the external components, such as the speed potentiometer, volume control, power switch, speaker, and TFT display, as shown in **Figure 6.6**. I like to use 2-mil-

Figure 6.6 — The random code practice generator perfboard mounted in the Mega S.A.F.E.

limeter hardware to mount the TFT display to the top cover of the enclosure.

Before I moved the project from the breadboard to the enclosure, I created the final working version of the project schematic using the free software, Autodesk *EAGLE CAD*. This program allows you to create professional-looking schematics in a matter of minutes. Once your schematic is complete, you can also create a PC board design file that can be used to have your finished design etched by a commercial PC board etching service. The free version limits you to noncommercial use; board size is limited to 100 × 80 millimeters (4 × 3.2 inches) and a maximum of two layers on a single sheet drawing. *EAGLE CAD* runs on Windows, Mac OS X, and Linux. Alternatively, you can use *KiCad*, a free open-source software, without any restrictions.

After updating the schematic from my build notes, I printed a copy of the final schematic as a construction guide. This helps verify that my schematic is documented correctly. Next, I wired the enclosure and the perfboard from the printed schematic, remembering to use the 2.54-millimeter Dupont-style pins, sockets, and headers for connectors, and 2.54-millimeter female headers to create sockets for the Arduino Nano and the level converter module. I then soldered the components and header pins to the prototyping board (see **Figure 6.7**). This design style allows me to easily remove the project assembly to correct the inevitable wiring errors and make any desired modifications to the project. By having the Nano, modules, and integrated circuits in sockets, they can be easily replaced, if necessary.

After wiring the perfboard assembly, I installed it in the enclosure, built cables (old PC floppy drive cables work great for this) using the 2.54-millimeter Dupont-style sockets and female pins to build connectors for the external components, and mounted the external components to the enclosure. The style of the Mega S.A.F.E. that I used for this project makes troubleshooting easier, as you can have your project open

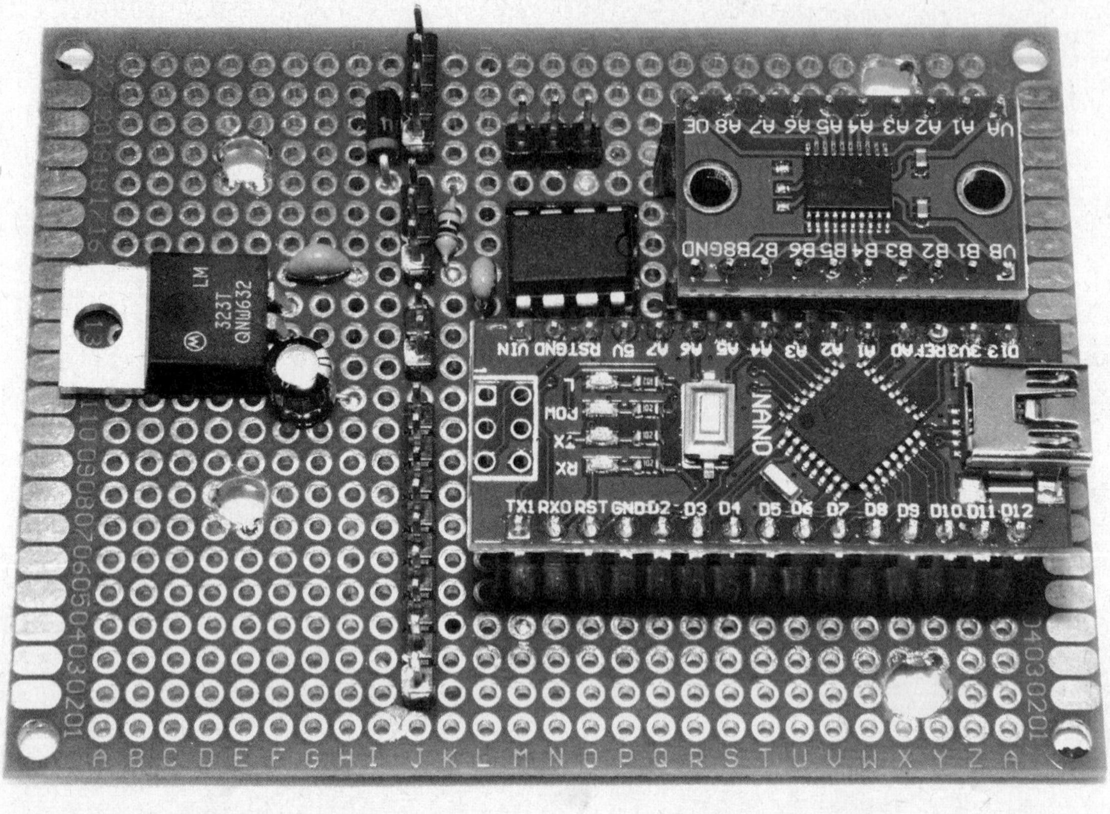

Figure 6.7 — The assembled random code practice generator perfboard assembly.

to access the various pins and connections for troubleshooting while the project is powered up and running.

When you've finished your wiring and testing, button up the enclosure. You now have a fully functional random code practice generator in a nice-looking, rugged case. One of the more interesting side effects of this project came when I was doing some final testing and letting it run on the test bench while I soldered another project. After about an hour of hearing random CW in the background, I found myself copying the code in my head. Leaving the project running (see **Figure 6.8**)., I turned the speed up a notch every now and then. At the end of just a few hours, my code speed had gone from about 10 up to 18 wpm, and I was copying it with 100% accuracy. Continuing the experiment, I let it run in the background for a day or so, gradually increasing the speed. At the end of this experiment in mental osmosis, I was copying about 23 wpm in my head. Because the characters are random, you can't guess what the next letter will be, enhancing the learning process, as you must copy each character as it was sent. Now, if I can get my sending speed to match my receiving speed, I'll be ready in time for the next Field Day. Maybe there's an Arduino project for that too.

Figure 6.8 — The assembled random code practice generator TFT display.

7 Iambic Keyer

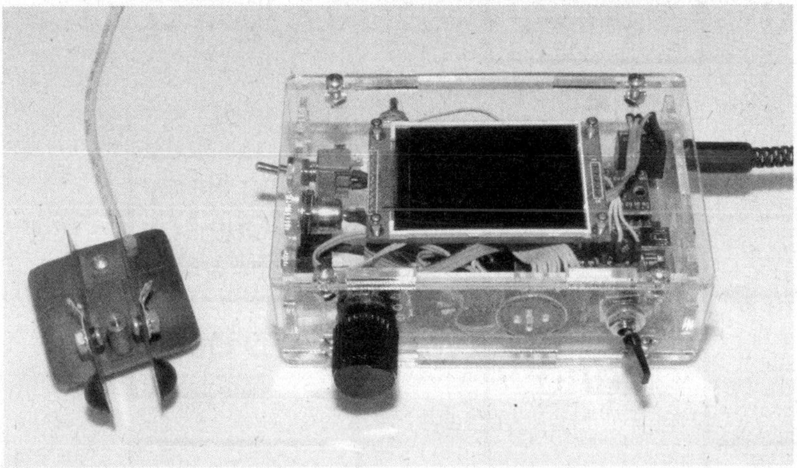

Figure 7.1 — Iambic keyer.

Another popular project from my first book, *Arduino for Ham Radio*, the iambic keyer (see **Figure 7.1**), gets an upgrade from the original Nokia 5110 LCD display to a 2.2-inch color TFT display, as shown in **Figure 7.2**.

It really can't be said enough that the open-source world is a wonderful place. It's like having a whole internet full of developers working side by side with you and sharing everything they do. When you start out working with a new module or project, take some time to look for similar projects and ideas online. While you may not find exactly what you have in mind for your project, there's a good chance you can find someone who has been where you are now and has a similar project you can adapt to help you on your way.

Such is the case with the iambic keyer project. Electronic CW keyers have been around since the early microprocessor days. I remember when the Curtis keyer chips were hot off the press, and I built my first electronic keyer with the Curtis 8044 chip in 1977. Naturally, I had to build an electronic keyer project for the Arduino. Having been away from the CW world for a while, before starting this project, I did some internet sleuthing to see how far keyers had come since my last keyer project. Iambic keying was a new concept to me, and now I've discovered there are not one, but two iambic modes. Naturally, this project would have to support both.

Now that I had a good idea of what I was going to do, the next step was to figure out how to do it. Turning to the open-source community, I did some research and came across a nice little keyer project for the Arduino that, with some minor adaptation, would be perfect for the task. "The Arduino Iambic Keyer," by Steven T. Elliott, K1EL, at **k1elsystems.com**, had almost everything I wanted in my iambic keyer project for the Arduino; it just needed a few extra things to make it exactly what I had in mind. Rather than reinvent the whole wheel, all I had to do was add the features I desired to the foundation Steven had provided. This is the power of the open-source community, and I would like to extend my thanks to Steven for allowing me to adapt his work to the iambic keyer project.

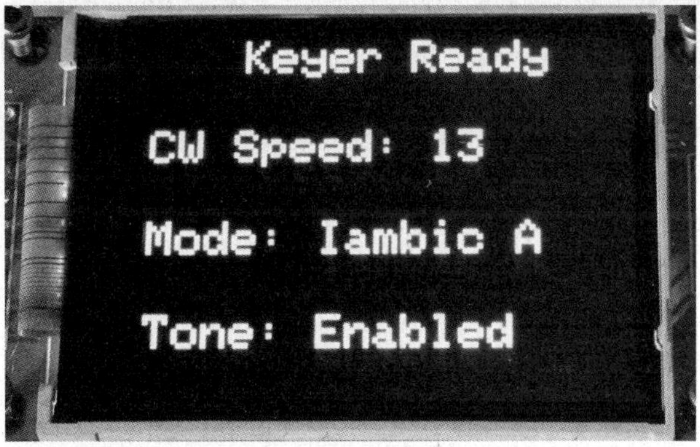

Figure 7.2 — Iambic keyer TFT display.

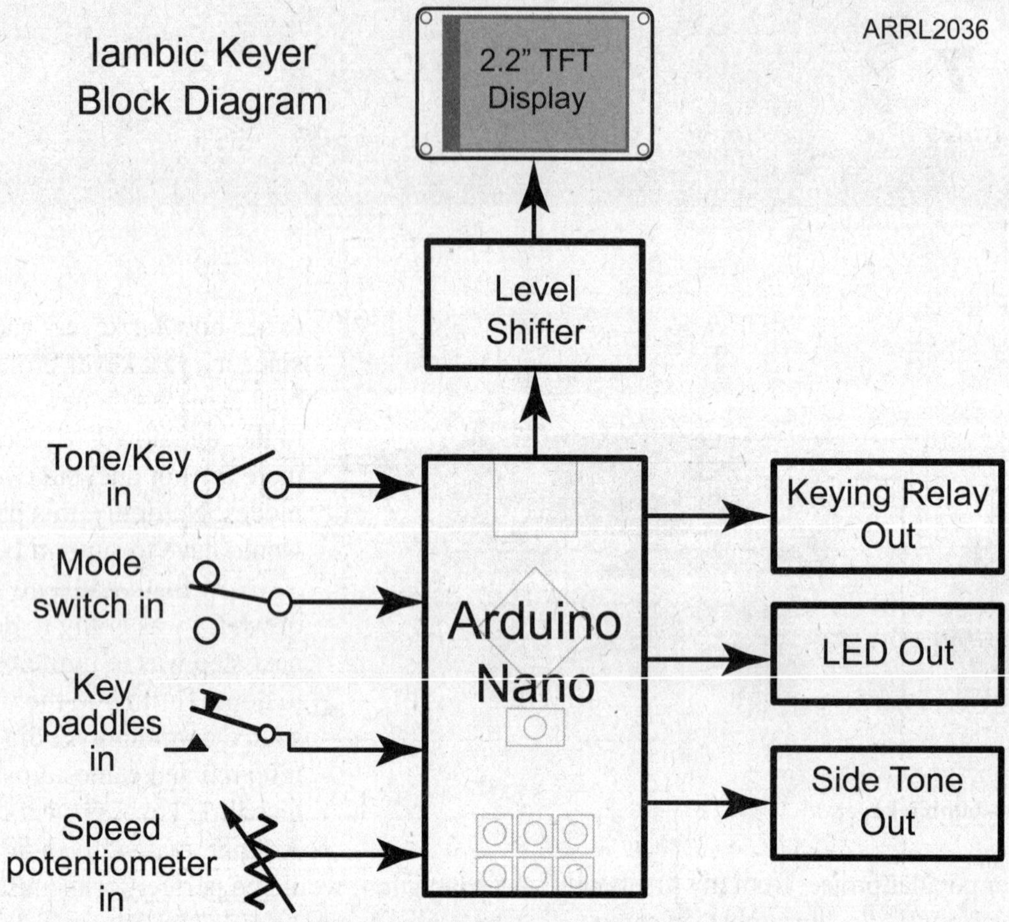

Figure 7.3 — The iambic keyer block diagram.

To start, we'll create our block diagram (see **Figure 7.3**) to determine the basic functions of our keyer. We'll be using an SPDT center-off switch to choose the keyer mode. The switch will select iambic Mode A when switched on in one direction, and the switch will select iambic Mode B when switched on in the other. The center-off position will allow us to use either paddle as a straight key. We'll also add a potentiometer to adjust the keying speed. For this project, we'll have the keying speed variable between 5 and 35 wpm. For our outputs, we'll have our keyer read a switch to either send a tone to the speaker or key the keying relay to drive your radio. We'll also have an LED to indicate when the keyer is sending. Lastly, we'll upgrade the original display from a Nokia 5110 LCD to a 2.2-inch color TFT to display our mode and speed settings.

Iambic Mode A versus Mode B

There are two major modes of iambic keying: Mode A and Mode B. As the electronic keyer evolved, two separate methods of generating the keyed signals when both paddles were squeezed came about. In either mode, when both paddles are squeezed, the keyer sends alternating dits and dahs, starting with whichever paddle was squeezed first. The first iambic mode, Mode A, is derived from the original Curtis Keyer chip design. In iambic Mode A, the keyer sends alternating dits and dahs as long as both paddles are squeezed. When the paddles are released, the keyer completes sending the current dit or dah and then stops, waiting for the next swipe of the paddles. In iambic Mode B, the keyer sends alternating dits and dahs as long as both paddles are squeezed. However, in Mode B, when the paddles are released, the keyer will complete the current dit or dah and then send the opposite

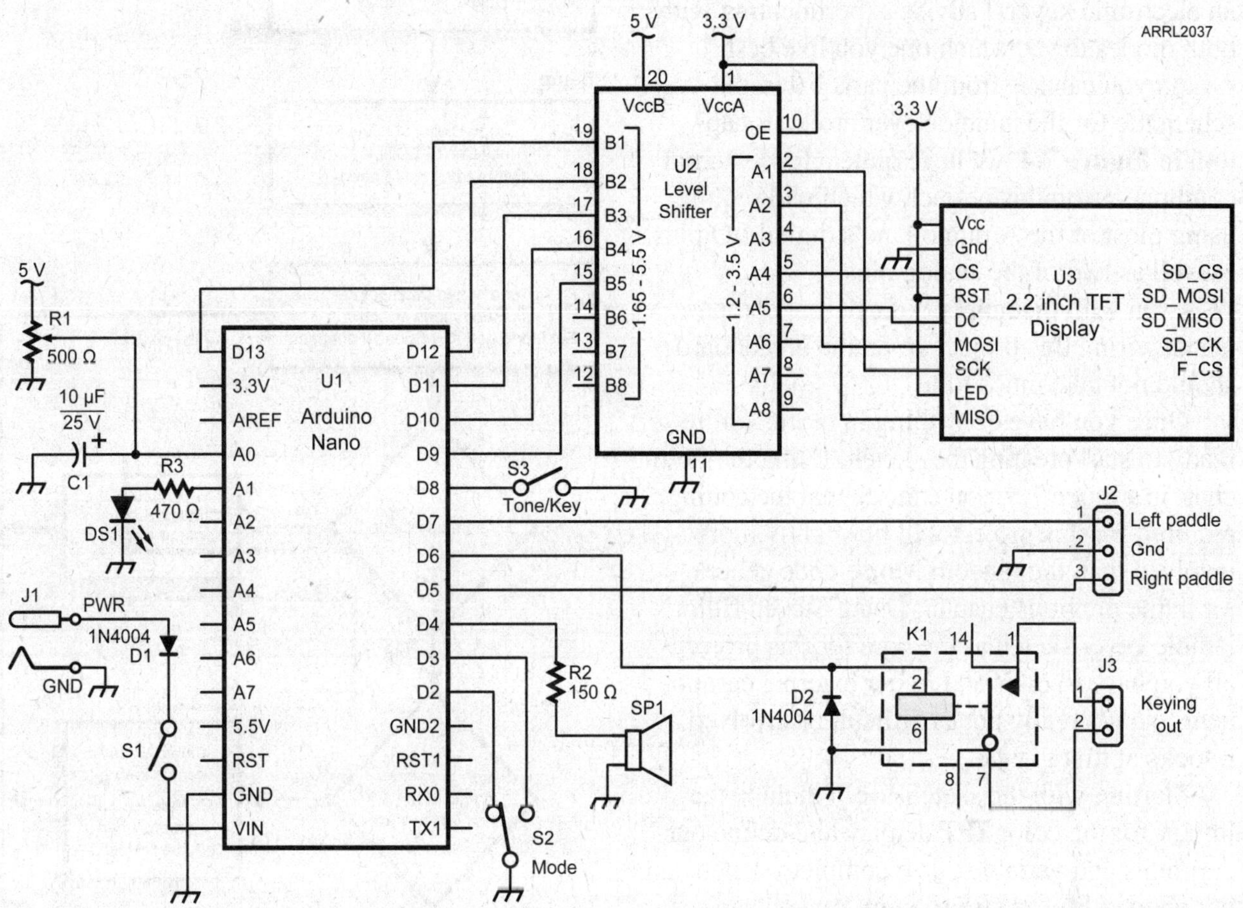

Figure 7.4 — The iambic keyer schematic and parts list.

Part	*Value*	*Part*	*Value*
D1, D2	**1N4004 diode or equivalent**	**U1**	**Arduino Nano**
DS1	**5 mm LED**	**U2**	**8-channel level shifter module (TXB0108 type)**
J1	**dc power connector**	**U3**	**2.2-inch ILI9341-type TFT display**
J2	**3-pin 2.54 mm Dupont-style header connector**	**K1**	**SPST 5 V DIP reed relay**
J3	**2-pin 2.54 mm Dupont-style header connector**	**R1**	**1 KΩ potentiometer**
S1, S3	**SPST toggle switch**	**R2**	**150 Ω resistor, 1/4 W**
S2	**DPST center-off toggle switch**	**R3**	**470 Ω resistor, 1/8 W**
		SP1	**8 Ω mini speaker**

code element before it stops sending.

The choice of iambic modes is purely a matter of personal preference and largely depends on which mode you learned when you started with an electronic keyer. I advise experimenting with both modes to see which one you like best.

As you can see from the parts list in the schematic for the iambic keyer project's caption in **Figure 7.4**, we have quite a few external components on this project, which will require using most of the Arduino Uno's digital I/O pins, as well as two of the analog pins.

Even with the number of external components, wiring this project up on the breadboard should not take much time.

Once you have everything in place, you're ready to start creating the sketch. Using the flowchart in **Figure 7.5**, you can see that the coding required for this project will be slightly more involved than the random Morse code generator in the previous chapter. Using Steven Elliot's iambic keyer sketch as the core for this project, all you have to do is add in our external components, so it's really not as difficult or involved as it looks at first glance.

Starting with the sketch, we'll include the library for the color TFT display and define our constants and variables. The complete sketch can be found online at **kw5gp.com**. As you can see below, we use preprocessor `#define` statements to assign a name to each I/O pin we'll be using. By assigning a name instead of using the pin number, it's much easier to keep track of what each I/O pin does, as well as allowing the flexibility to change pin configurations.

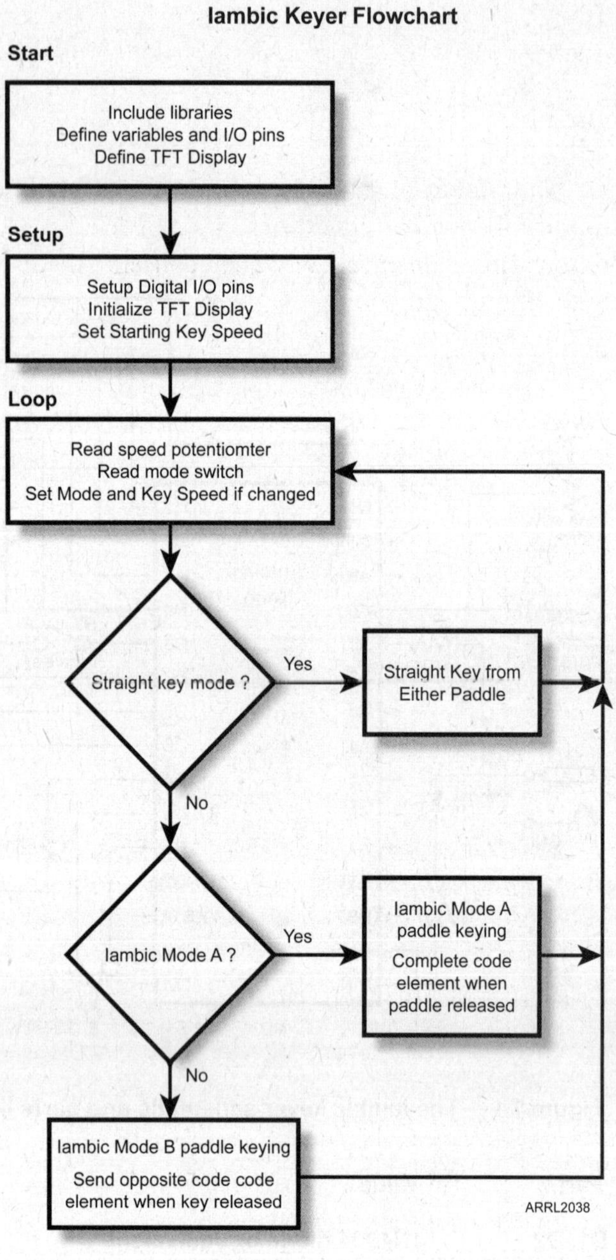

Figure 7.5 — The iambic keyer flowchart.

```
#define TFT_CS      10  // Assign the TFT CS to Pin 10
#define TFT_DC      9  // Assign the TFT DC to Pin 9
#define ST_Pin 4       // Sidetone Output Pin on Pin 4
#define LP_in 7        // Left Paddle Input on Pin 7
#define RP_in 5        // Right Paddle Input on Pin 5
#define led_Pin A1     // LED on Pin A1
#define Mode_A_Pin  3 // Mode Select Switch Side A
#define Mode_B_Pin  2 // Mode Select Switch Side B
#define key_Pin 6      // Transmitter Relay Key Pin
```

```
#define Speed_Pin A0    // Speed Control Pot on A0
#define ST_Freq 600    // Set the Sidetone Frequency to 600 Hz
#define no_beep 8    // Beep Disable Switch Input
```

The iambic keyer's primary functions are all managed and controlled using a single-byte variable to track the various conditions. A single bit is used rather than an entire integer variable for each condition. This is an excellent way to conserve memory, especially with sketches that push memory usage to the limits, and you're looking for every last byte of memory you can possibly conserve.

```
//  keyerControl bit definitions
//
#define      DIT_L       0x01      // Dit latch
#define      DAH_L       0x02      // Dah latch
#define      DIT_PROC    0x04      // Dit is being processed
#define      PDLSWAP     0x08      // 0 for normal, 1 for swap
#define      IAMBICB     0x10      // 0 for Iambic A, 1 for Iambic B
#define      ULTIMATIC   0x20      // 1 for ultimatic
#define      STRAIGHT    0x80      // 1 for straight key mode
```

In his iambic keyer sketch, Steven Elliot introduces the `enum` datatype. This datatype allows you to use enumerated values to represent the various conditions the variable can contain. This datatype allows you to assign an easily understandable alias to a condition that can be used in other Arduino statements, such as within `switch...case()` decision blocks within your sketch. For example, it's a whole lot easier to read and understand `case IDLE`...do this than it is to understand `case 1`...do this.

```
//  State Machine Defines
enum KSTYPE {IDLE, CHK_DIT, CHK_DAH, KEYED_PREP, KEYED, INTER_ELEMENT };
```

In the `setup()` loop, we start by setting up all of the digital I/O pin modes we'll need to use to interface with all the external pieces we have on the breadboard. You will see how much more readable this sketch is when you define the pins with an understandable name. This also allows you to easily switch your pins around if you prefer to connect the external components to different pins or you make a wiring error and would rather change the pin assignments than rewire your circuit.

Next, we'll initialize the color TFT display, output a startup message, and set the keying and iambic mode along with the keying speed.

In the main `loop()`, we first read the current setting on the speed potentiometer and update the keyer sending speed if the speed has changed. We also check to see if the keying mode has been changed and change the keying mode if necessary. We also check to see if we are in straight key mode or iambic keying mode. If we're in straight key mode, we use either paddle as a keying input, turn on the LED, key the keying relay, and play the sidetone as the paddles are pressed. If any of the above conditions have changed, we update the status information on the TFT display as well. Finally, we monitor the keyer paddle inputs for a key press.

Here is where Steven Elliot's keyer code comes into play. Using a `switch...case()` statement, we check the various conditions of the state machine and set the `keyerControl` variable's bits accordingly.

```
// Basic Iambic Keyer
  // keyerControl contains processing flags and keyer mode bits
  // Supports Iambic A and B
  // State machine based, uses calls to millis() for timing.
  Switch (keyerState)
  {
    case IDLE:     // Wait for direct or latched paddle press
    // check to see if a paddle is pressed or keyControl bits DIT_L or
    // DAH_l (dit or dah latch) are set
      if ((digitalRead(LP_in) == LOW) || (digitalRead(RP_in) == LOW) || (keyer-
Control & 0x03))
      {
        update_PaddleLatch(); // set the paddle latch bits DIT_L or DAH_L
        keyerState = CHK_DIT;  // Tell state machine to check for a dit next
      }
      break;      // Exit the case() loop

    case CHK_DIT:        // See if the dit paddle was pressed
      if (keyerControl & DIT_L)
      {
        keyerControl |= DIT_PROC;    // set the DIT_PROC bit in keyerControl
        ktimer = ditTime;      // set the keying timer to the ditTime
        keyerState = KEYED_PREP; // tell the state machine we're ready to key
      } else {
        keyerState = CHK_DAH; // otherwise, do a CHK_DAH next time
      }
      break;

    case CHK_DAH:        // See if dah paddle was pressed
      if (keyerControl & DAH_L)
      {
        ktimer = ditTime*3;  // set the key time to 3 times the ditTime
        keyerState = KEYED_PREP; // tell the state machine we're ready to key
      } else {
        keyerState = IDLE; // no key is pressed, return to IDLE state
      }
        break;
```

Next, we check for the keying portions of the state machine:

```
case KEYED_PREP:
 // Assert key down, start timing, state shared for dit or dah
  digitalWrite(led_Pin, HIGH);      // Turn the LED on
  tone(ST_Pin, ST_Freq);            // Turn the Sidetone on
  digitalWrite(key_Pin, HIGH);      // Key the TX Relay
  ktimer += millis();               // set ktimer to interval end time
  keyerControl &= ~(DIT_L + DAH_L); // clear both paddle latch bits
  keyerState = KEYED;                // next state
  break;
case KEYED:       // Wait for timer to expire
  if (millis() > ktimer)  // are we at end of key down ?
  {
    digitalWrite(led_Pin, LOW);       // Turn the LED off
    noTone(ST_Pin);                   // Turn the Sidetone off
    digitalWrite(key_Pin, LOW);       // Turn the TX Relay off
    ktimer = millis() + ditTime;      // inter-element time
    keyerState = INTER_ELEMENT;       // next state
  }
  else if (key_mode == 1) // Check to see if we're in Iambic B Mode
  {
    update_PaddleLatch();  // early paddle latch in Iambic B mode
  }
  break;
```

The last state of the state machine is to set the inter-element spacing between the code elements:

```
case INTER_ELEMENT:      // Insert time between dits/dahs
      update_PaddleLatch();        // latch paddle state
      if (millis() > ktimer)  // are we at end of inter-space ?
      {
        if (keyerControl & DIT_PROC)       // was it a dit or dah ?
        {
          keyerControl &= ~(DIT_L + DIT_PROC);   // clear two bits
          keyerState = CHK_DAH;                  // dit done, check for dah
        } else {
          keyerControl &= ~(DAH_L);              // clear dah latch
          keyerState = IDLE;                     // go idle
        }
      }
      break;
```

Key Functions

The `update_PaddleLatch()` function is used to update the status of the `keyerControl` byte when a paddle is pressed:

```
void update_PaddleLatch()
{
   if (digitalRead(RP_in) == LOW) {   // set the DIT_L bit if right paddle
      keyerControl |= DIT_L;
   }
   if (digitalRead(LP_in) == LOW) {   // set the DAH_L bit if left paddle
      keyerControl |= DAH_L;
   }
}
```

The `loadWPM()` function calculates the desired dit timing from the selected words per minute:

```
void loadWPM (int wpm)
{
   ditTime = 1200/wpm;
}
```

As you can see, a lot is going on inside the iambic keyer portion of the sketch, but because Steven had already done this part of the sketch for us, we saved a lot of time and effort by utilizing his sketch as a foundation and wrapping our modifications around it. This is part of the fun of the Arduino and open source; you're rarely totally alone, and, as this sketch proves, there's most likely someone out there doing something similar to what you want to do and sharing their work to help you with your project.

Once everything is working on the breadboard, you can build your keyer on a perfboard, as shown in **Figure 7.6**, and mount the finished project into a Solarbotics Mega S.A.F.E. enclosure, as shown in **Figure 7.7**.

Figure 7.6 — The completed iambic keyer perfboard assembly.

Figure 7.7 — The completed iambic keyer.

8 1 – 30 MHz DDS VFO

One of the items on the ever-growing list of things I want to do is to rebuild an older rig from my past, such as my first rig, a Heathkit HW-16, or maybe a Heathkit HW-7, or an HW-8 QRP CW rig. Of course, the rebuild would have to include an Arduino, with an updated display, a built-in keyer, and a VFO, along with whatever other tweaks I can think up. Fortunately, I managed to acquire both an HW-16 and an HW-8 at recent hamfests, so things should get interesting fairly soon.

Back when I was thinking up projects for my second book, Tom Medlin, W5KUB, had introduced me to the very inexpensive Frog Sounds and Frogs Calling QRP CW transceiver kits, and asked if they could be connected to an Arduino-powered VFO.

That discussion led to the creation of this project, shown in **Figure 8.1**. It's actually a derivative of the direct digital synthesis (DDS) VFO circuit I built for the QRP CW and JT65 transceivers that were in my book, *More Arduino for Ham Radio*. The 40-meter QRP CW Transceiver has also been redesigned and included in this book. While designing the VFO project, my thinking was that at some point I would want it to be able to drive a tube-type radio. I designed it with an output amplifier circuit to provide more drive than the AD9850 DDS module could provide. For this book, the project has been upgraded to use a 1.8-inch color TFT display and an Arduino Nano, and it is mounted on a 6 × 8-centimeter (roughly 2 3/8 × 3 1/8-inch) copper-clad perfboard (see **Figure 8.2**).

Design Goals

The design goals for this project were pretty straightforward. We'll build a 1 – 30 MHz VFO using an Arduino Nano and an Analog Devices AD9850 1 – 62.5 MHz DDS module for the VFO. I didn't choose the Si5351 8 kHz to 160 MHz clock generator module for this project primarily because its output is a square wave, and not the smooth sine wave output of the AD9850. When used in oscillator circuits, square waves generate harmonic frequencies, and I didn't want to have to build a circuit to smooth the square wave output of the Si5351 when I had the AD9850 DDS module that already generated nice clean sine waves just waiting in a parts bin.

Figure 8.1 — The 1 – 30 MHz DDS VFO.

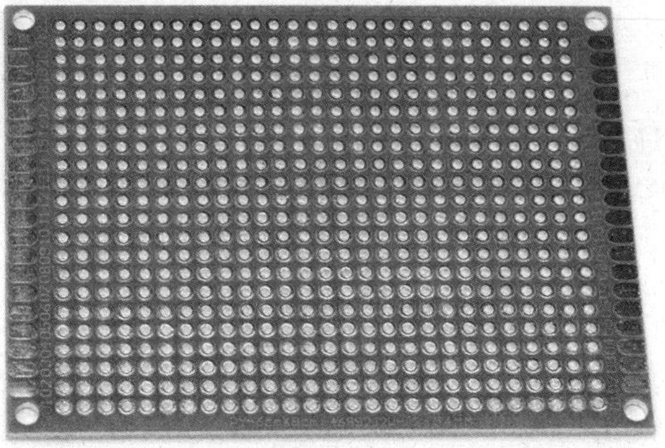

Figure 8.2 — 6 × 8-centimeter copper-clad perfboard.

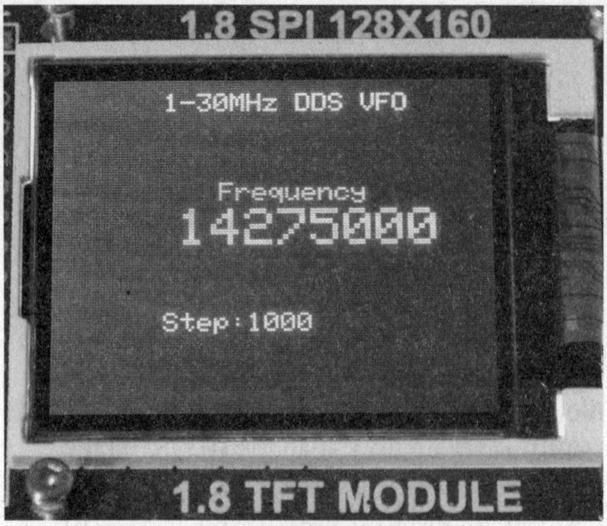

Figure 8.3 — The 1 – 30 MHz DDS VFO updated TFT Display.

Because we are tuning over such a wide frequency range, I wanted the VFO to be able to tune in multiple step ranges. We'll use a rotary encoder with a built-in pushbutton switch to tune the VFO, and we'll use the pushbutton switch to change the frequency step range. We'll upgrade the original Nokia 5110 LCD display to a 1.8-inch ST7735-type color TFT display (see **Figure 8.3**) to show the output frequency and the current tuning step rate.

We'll also provide two levels of output: one just after the AD826 buffer amplifier chip, and a higher power output from a transistor amplifier driven by the AD826 buffer amplifier chip.

Since most of the radios we'll be driving with this VFO are either tube-type station radios or a QRP transceiver board such as the Frog Sounds and Frogs Calling boards, we'll use the readily available 12 V station — or portable battery power — to supply the 12 V needed to power the project.

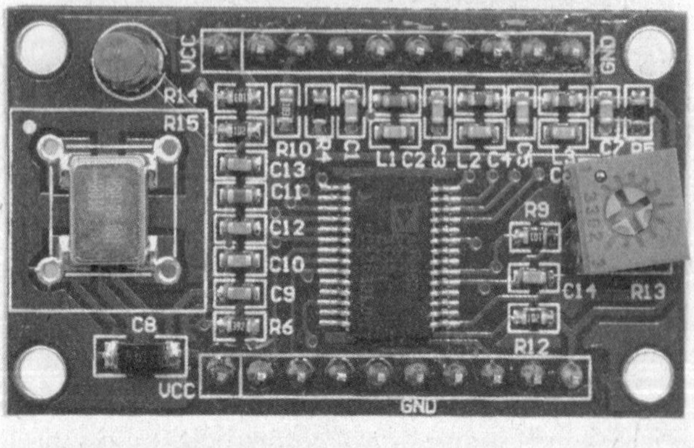

Figure 8.4 — The AD9850 DDS module.

The AD9850 Direct Digital Frequency Synthesis (DDS) Module

The Analog Devices AD9850 DDS module (see **Figure 8.4**) is a direct digital frequency synthesis module that uses a 10-bit internal digital-to-analog converter to generate smooth, clean sine waves from 0 to 62.5 MHz. There is also a square wave output provided on the AD9850, but we won't be using that for most projects. The output frequency is digitally controlled, and with a 32-bit frequency tuning control word, has a tuning resolution of 0.0291 Hz. The output frequency can be changed at a rate of up to 23 million times per second, which opens some very interesting possibilities as you saw with the JT65 QRP transceiver project in *More Arduino for Ham Radio*. The phase of the DDS output can be changed in increments of 11.25°, 22.5°, 45°, 90°, or 180°, which also opens some interesting application possibilities for modes like PSK31.

The 1 – 30 MHz DDS VFO Block Diagram

Figure 8.5 shows the Block diagram for the 1 – 30 MHz DDS VFO project. Note that we've upgraded from the Arduino Uno to a Nano, mounted everything in sockets on a perfboard, and fitted it in a Solarbotics Mega S.A.F.E. enclosure. We'll use Dupont-style 2.54-millimeter header connectors to connect to the ST7735-type color TFT display, the rotary encoder, and the mini stereo phone plug we'll use to provide the two levels of VFO output.

The 1 – 30 MHz DDS VFO Flowchart

Figure 8.6 shows the Flowchart for the 1 – 30 MHz DDS VFO project. We'll start by including the M0XPD DDS library, along with the libraries for the ST7735-type TFT display and the rotary encoder. Next, we'll define the I/O pins used in the project, then declare and initialize the variables we'll need in the sketch. In the last part of the initialization area, we'll create (instantiate) the DDS, TFT Display, and rotary encoder objects.

In the `setup()` loop, we'll set the pin modes for the I/O pins for the rotary encoder, and we'll initialize the TFT display, the DDS,

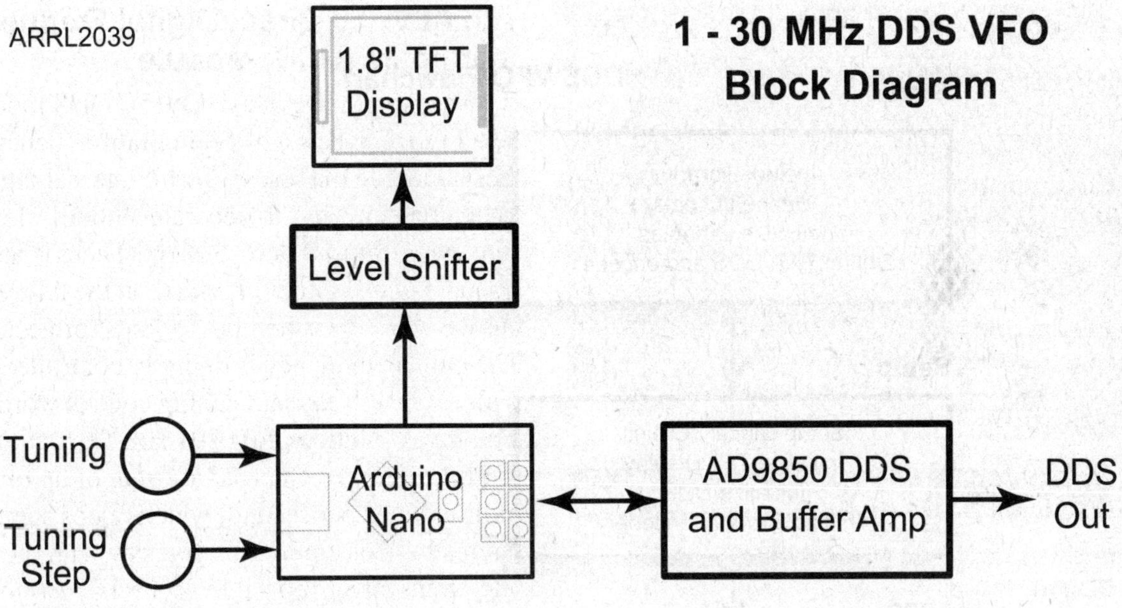

Figure 8.5 — The 1 – 30 MHz DDS VFO Block Diagram.

and the rotary encoder. We'll also display a brief startup screen on the TFT display, and then set up and display the template for the output display. By using a template, all we have to do in the main `loop()` is update the actual displayed values for the frequency and tuning step rate. At the end of the `setup()` loop, we'll display the starting frequency and tuning step values on the TFT display.

In the main `loop()`, we'll read the rotary encoder position, and we'll change the VFO output frequency and update the LCD if the position has changed. If the rotary encoder pushbutton switch has been pressed, we'll change the tuning step rate and update the TFT display with the current step rate. The tuning step rate increments can be set from 1 Hz to 1 MHz, with each press of the pushbutton switch multiplying the step rate by a factor of 10 until it wraps back around to the 1 Hz setting.

Fun with Amplifiers

The hardest part about designing the 1 – 30 MHz DDS VFO project was coming up with a buffer amplifier design that provided a clean, stable sine wave across the entire 1 – 30 MHz spectrum. Remember that I am not the world's best with analog and analog amplifier circuits, especially when using them in RF amplification applications. I experimented with a number of buffer amplifier configurations and ICs, each with varying degrees of success. I ended up with a hybrid design using the Analog Devices AD826 that I had worked with in some of my earlier projects as the first stage of the buffer amplifier. The AD826 is rated to 50 MHz, so it looked to be an ideal candidate for this project. Wanting to add a little more drive for the older tube-type radios, I added a transistor amplifier based on the March 2016 *QST* article by Dr. Jack Purdum, W8TEE, and Dennis Kidder, W6DQ, which is similar to the buffer amplifier used in the application notes for the KN-Q7A transceiver kit, which refers to using the N3ZI DDS kit. I also found a similar design that was created by Gary A. White, W5GW, using the *LTSpice* design application and used in conjunction with the N3ZI DDS kit in his crystal Bank for Vintage Gear.

A lot of my issues with the VFO output level began in the AD9850 DDS module itself. As shown in the graph in **Figure 8.7**, the output of the DDS module goes from 0.96 V down to 0.46 V (more than a 50% decrease) over the range of 1 to 30 MHz. So, that's where the fun began. The first thing I did was try a series of various buffer amplifiers before settling on the

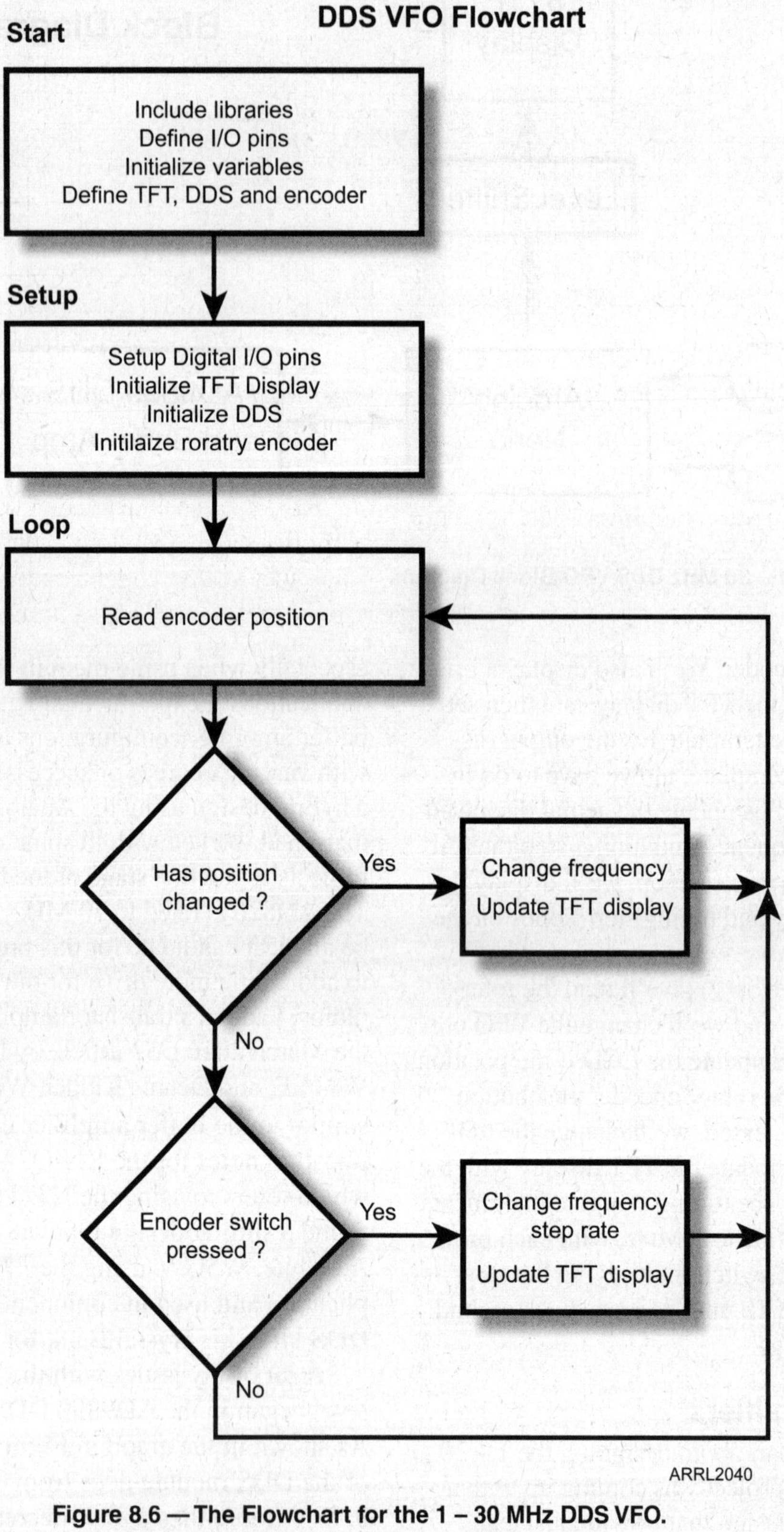

Figure 8.6 — The Flowchart for the 1 – 30 MHz DDS VFO.

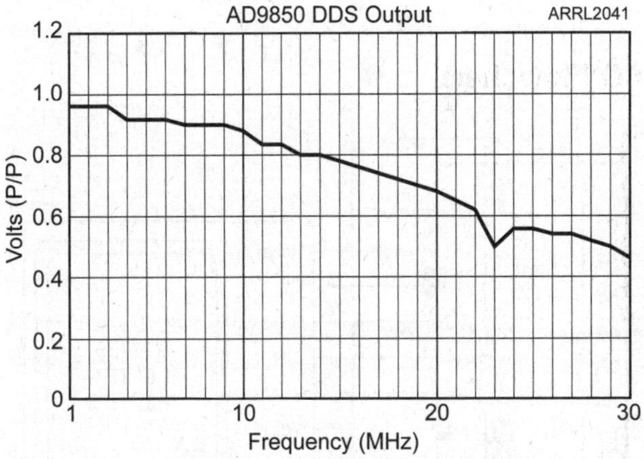

Figure 8.7 — Output graph of the AD9850 DDS module.

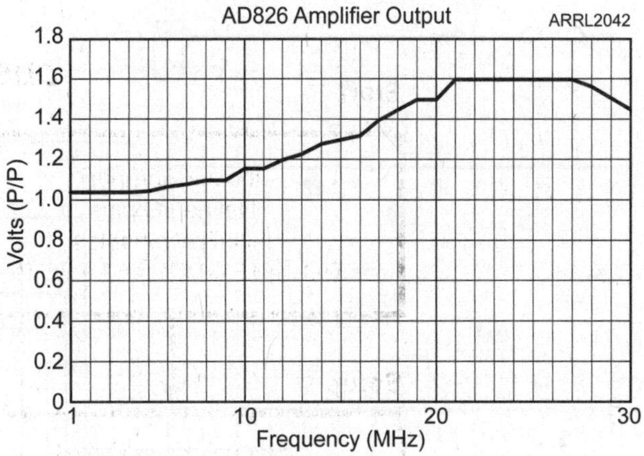

Figure 8.9 — Output graph of the AD826 buffer amplifier.

AD826 instrumentation amplifier. Here is a case where having a breadboard to build and test the various designs was a tremendous help. I was able to quickly build and test a series of different amplifier circuits, including the one based on the AD826 shown in **Figure 8.8**. Of all the circuits I tried, the AD826 was the only one that cleanly reproduced the AD9850's output sine wave, and with a relatively constant output level across the 1 to 30 MHz frequency range. The graph in **Figure 8.9** shows the output of the AD826 buffer amplifier circuit. Interestingly, the output actually increased rather than decreased as it went up in frequency, but only by about 40%. The important thing to me was that the output stayed between 1 and 2 V across the entire range.

Figure 8.8 — The AD826 buffer amplifier design on the breadboard.

Next it came time to add the final output amplifier stage. The design goals stated that this stage should have enough output to drive an older tube-type rig. It's obvious from the output graph (see **Figure 8.10**) that the frequency response is far from optimal, but it should be more than adequate to perform the VFO role for just about any rig we plan to use. At some point, I'll probably revisit this and see if there's a better design out there, but for now, this does what I set out to achieve. One thing all of this fun has taught me is that I need to download the free *LTSpice* design package and the other design tools from Linear Technology and see what I can do to improve my analog and RF design techniques.

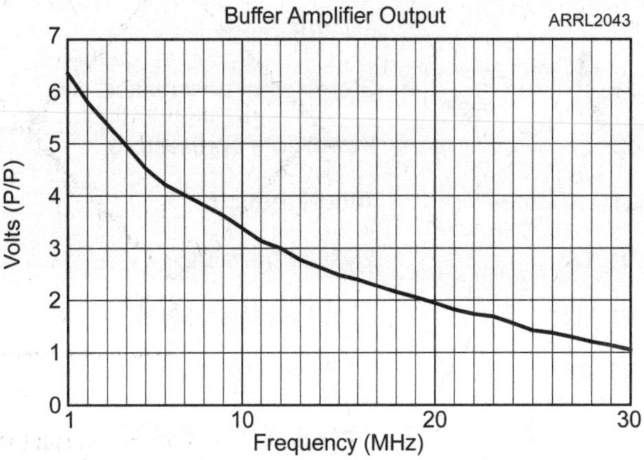

Figure 8.10 — Output graph of the completed buffer amplifier design.

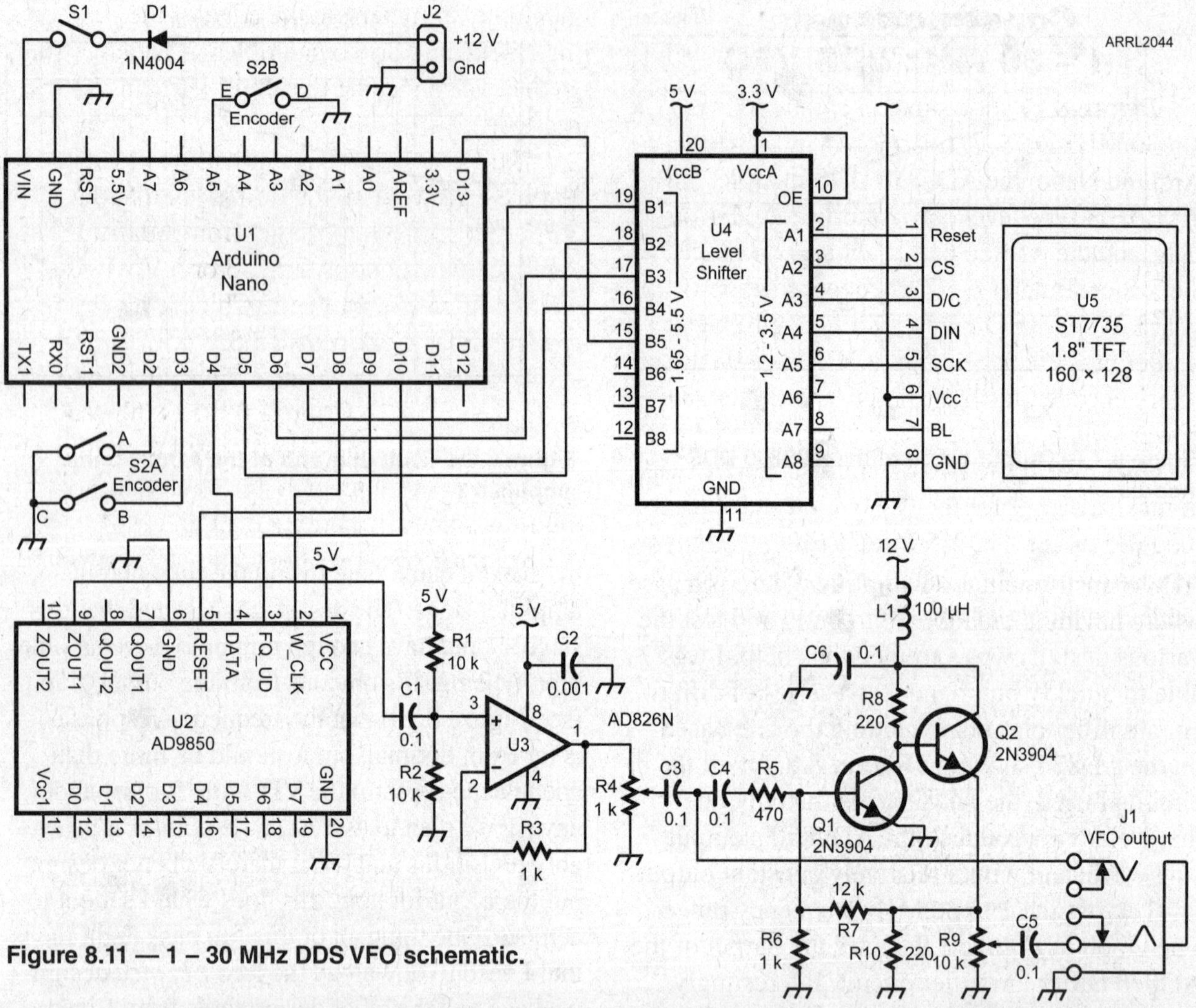

Figure 8.11 — 1 – 30 MHz DDS VFO schematic.

Part	*Value*
C1, C3, C4, C5, C6	.1 µF ceramic capacitor, 50 V
C2	.001 µF ceramic capacitor, 50 V
D1	1N4004 diode
J1	Stereo mini phone jack
J2	Powerpole connectors
L1	100µH inductor
Q1, Q2	2N3904 transistor
R1, R2	10 KΩ resistor, 1/8 W
R3, R6	1 KΩ resistor, 1/8 W
R4	1 KΩ potentiometer
R5	470 Ω resistor, 1/8 W
R7	12 KΩ resistor, 1/8 W
R8, R10	220 Ω resistor, 1/8 W
R9	10 KΩ potentiometer
S1	SPST switch
S2	Rotary encoder with push button switch
U1	Arduino Nano
U2	AD9850 DDS module
U3	AD826N high speed dual operational amplifier
U4	8-channel level shifter module (TXB0108 type)
U5	ST7735 TFT display module
Enclosure	Solarbotics Mega S.A.F.E.

Constructing the 1 – 30 MHz DDS VFO

Figure 8.11 shows the schematic for the 1 – 30 MHz DDS VFO project. We'll use an Arduino Nano, the AD9850 DDS module, the TXB0108-type level converter for the TFT display, and the AD826 buffer amplifier mounted on a 6 × 8-centimeter piece of copper-clad perfboard. We'll use Dupont-style 2.54-millimeter header pins to connect to the TFT display, the rotary encoder, the stereo mini-phone jack, the power switch, and the 12 V Powerpole connector block. We'll use two SIP female header strips to make a socket for the Nano, the level shifter module, and the AD9850, and finally, we'll use a standard 8-pin DIP socket for the AD826 instrumentation amplifier chip. We'll mount the two gain potentiometers on the edge of the prototyping shield, and drill small holes in the enclosure so that we can adjust the amplifier gains as needed.

The Chinese ST-7735 modules shown in **Figure 8.12** work slightly different that the ST7735 module you can get from Adafruit. The Adafruit unit supports either 3 or 5 V without any modification, while the Chinese ST7735-type modules work best at 3.3 V with a level shifter to convert the Nano's 5 V signals to 3.3 V. To select the 3.3 V mode for the Chinese ST7735-type modules, you'll need to bridge the on-board solder jumper J1 to enable the 3.3 V mode as shown in **Figures 8.13** and **8.14**.

Figure 8.15 shows the completed perfboard assembly.

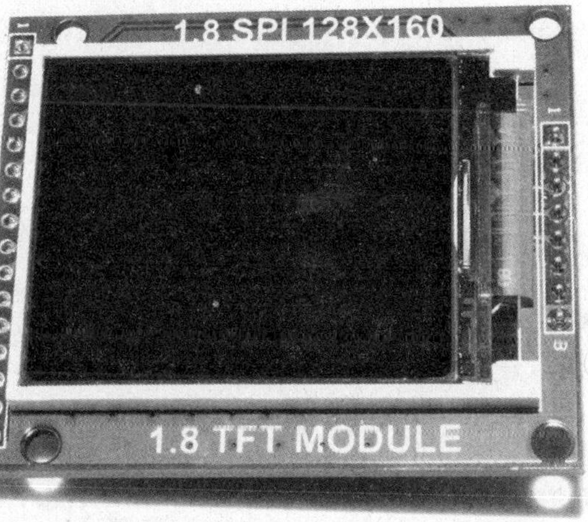

Figure 8.12 — The Chinese ST7735-type 1.8-inch color TFT display.

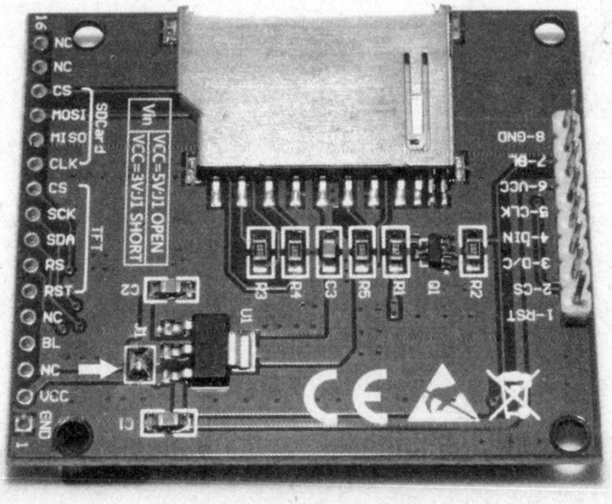

Figure 8.14 — TFT Module with jumper set for 3.3 V.

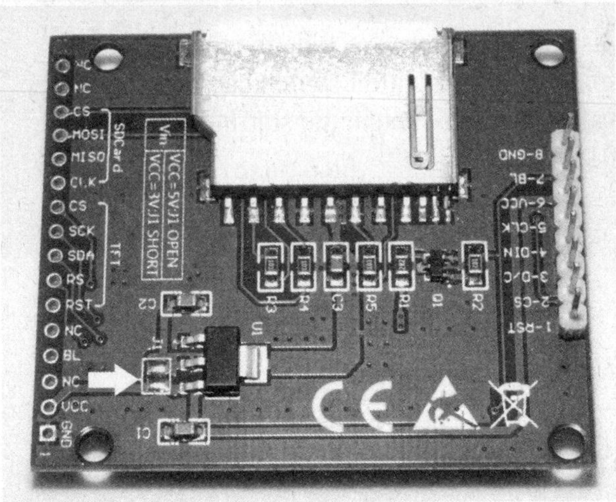

Figure 8.13 — TFT Module with no jumper set for 5 V.

Figure 8.15 — The assembled 1 – 30 MHz DDS VFO board.

The 1 – 30 MHz DDS VFO Sketch

Compared to the sketches for some of my previous projects, the sketch for the 1 – 30 MHz DDS VFO project is much simpler. The M0XPD DDS library created by Paul Darlington, MØXPD, is a simple and easy-to-use library to control the AD9850 DDS. All you have to do is send it the frequency you want, and it does the rest for you. To conserve space, only key portions of the sketch will be shown here. The 1 – 30 MHz DDS VFO sketch and libraries can be downloaded from **kw5gp.com**.

Starting out in the sketch, we'll include the DDS, TFT display, and rotary encoder libraries, and we'll define the I/O pins that we'll need for the AD9850 DDS, TFT display, and rotary encoder.

```
#include <DDS.h>  // include the M0XPD DDS Library
#include <Adafruit_GFX.h>     // Core graphics library
#include <Adafruit_ST7735.h> // Hardware-specific library
#include <SPI.h>  // SPI library so we can communicate with the TFT display
#include <Encoder.h>  // Include the Rotary Encoder Library

//========================================
// AD9850 Module....
// set pin numbers:
const int W_CLK = 8;
const int FQ_UD = 10;
const int DATA = 4;
const int RESET = 9;

#define TFT_CS     5  // Assign the TFT CS to pin 10
#define TFT_RST    7  // Assign the TFT RST to pin 7
#define TFT_DC     6  // Assign the TFT DC to pin 6

#define tft_delay 10  // set the TFT command delay to 10ms

const int encoder_PinA = 3;  // Pin A of the Rotary Encoder
const int encoder_PinB = 2;  // Pin B of the Rotary Encoder
const int encoder_Switch_Pin = A4;  // Encoder Pushbutton Switch input pin
```

Next, we'll declare and initialize the variables we use in the sketch. We'll define the frequency step value in an array, so that each press of the rotary encoder's pushbutton switch can be used to step through the array to select the correct frequency step value. For this sketch, the starting frequency is set to 7.0 MHz, with an initial frequency step value of 1 Hz.

```
double step_value[] = {1, 10, 100, 1000, 10000, 100000, 1000000};

const int debounceInterval = 500;  //disable step change for 500 ms between en-
coder pushbutton press
boolean encoder_Switch;
int current_step = 0;  // Default Tuning Step is 1 Hz
int encoder_Pos = 0;
int old_encoder_Pos = 0;
```

```
const double range_low = 1000000; // Low end of range (1 MHz)
const double range_high = 30000000; // High end of range (30 MHz)
unsigned long int delay_time = 2000;

double start_freq = 7000000;  // Starting freq

double dds_freq = start_freq;
```

In the final section of the initialization area, we'll create (instantiate) the DDS, TFT display, and rotary encoder objects.

```
// Instantiate the DDS
DDS dds(W_CLK, FQ_UD, DATA, RESET);

Adafruit_ST7735 tft = Adafruit_ST7735(TFT_CS,  TFT_DC, TFT_RST);  // Initialize
the TFT display instance

extern uint8_t SmallFont[];  // define the Nokia Font

// Instantiate the Rotary Encoder
Encoder Enc(encoder_PinA, encoder_PinB);
```

In the `setup()` loop, we'll first initialize the TFT display, display a brief startup screen, then we'll initialize the DDS. After a 3-second delay, we'll clear the display and display a template for the frequency and frequency step rate. By using a template in this manner, all we have to do is update the values for the frequency and step rate.

Next, we'll set the pin modes for the rotary encoder, and enable the internal pullup resistors on the encoder input pins.

```
void setup()
{
  tft.initR(INITR_18BLACKTAB);   // initialize a 1.8" TFT with ST7735S chip,
black tab
  delay(tft_delay);
  tft.fillScreen(ST7735_BLACK); // Clear the display - fill with BLACK background
  delay(tft_delay);
  tft.setRotation(1); // Set the screen rotation
  delay(tft_delay);
  tft.setTextWrap(false); // Turn off Text Wrap
  delay(tft_delay);
  tft.setTextSize(3); // Set the Font Size
  delay(tft_delay);
  tft.setTextColor(ST7735_BLUE); //Set the Text Color
  delay(tft_delay);
  tft.setCursor(40, 20);  //Set the Cursor and display the startup screen
```

```
delay(tft_delay);
tft.print("KW5GP");
delay(tft_delay);
tft.setTextSize(2);
delay(tft_delay);
tft.setCursor(35, 60);
delay(tft_delay);
tft.print("1-30 MHz");
delay(tft_delay);
tft.setCursor(40, 90);
delay(tft_delay);
tft.print("DDS VFO");
delay(tft_delay);

// start up the DDS...
dds.init();
// (Optional) trim if your xtal is not at 125MHz...
dds.trim(125000322); // enter actual osc freq

delay(3000);
tft.fillScreen(ST7735_BLACK); // Clear the display
delay(tft_delay);

update_display(); // update the display
first_pass = false; // turn off the update display first pass flag

// Set up the Rotary Encoder
pinMode (encoder_PinA, INPUT);
pinMode (encoder_PinB, INPUT);
pinMode (encoder_Switch_Pin, INPUT);
// Enable the Internal Pull-up resistor on the Encoder Inputs
digitalWrite(encoder_PinA, HIGH);
digitalWrite(encoder_PinB, HIGH);
digitalWrite(encoder_Switch_Pin, HIGH);
```

In the final part of the `setup()` loop, we'll start the DDS, set the starting frequency by calling the `set_DDS()` function, and then display the frequency and frequency step rate on the display.

```
// start the oscillator...
set_DDS(dds_freq);

update_display();

} // End Setup Loop
```

The `set_DDS()` function will check to see if the desired frequency is in the allowable range, and then call the M0XPD DDS library to set the DDS to the new frequency.

```
void set_DDS(double set_freq)
{
  if (set_freq >= range_high)
  {
    set_freq = range_high;
  }
  if (set_freq <= range_low)
  {
    set_freq = range_low;
  }
  dds.setFrequency(set_freq);  // Set the DDS to the freq
}
```

Because we have moved most of the operations into functions, the main `loop()` of the sketch consists of just two lines.

```
void loop()
{

  read_encoder(); // Read the encoder and set the Tx/Rx frequencies
  read_pushbutton();  // Read the Encoder Pushbutton switch and set Fast/Slow
tuning mode

}  // End Main Loop
```

Starting out in the main `loop()`, we'll read the rotary encoder by calling the `read_encoder()` function.

```
void read_encoder()
{
  // Read the Encoder
  encoder_Pos = Enc.read() / 4; // divide by 4 to match encoder detent
  if (encoder_Pos != old_encoder_Pos) // If the Encoder has changed update freq
  {
    if (encoder_Pos < old_encoder_Pos) // If we're increasing frequency
    {
      if (dds_freq >= range_high) // Limit to top end of band
      {
        dds_freq = range_high;
      } else {
        dds_freq = dds_freq + step_value[current_step];
      }

    } else {
      if (dds_freq <= range_low)// We're decreasing frequency, limit to low end
 of band
```

```
        {
          dds_freq = range_low;
        } else {
          dds_freq = dds_freq - step_value[current_step];
        }
      }

    //    dds.setFrequency(tx_freq); // Set the TX frequency
    old_encoder_Pos = encoder_Pos;
    set_DDS(dds_freq);
    update_display();

  }
}
```

The read_encoder() function checks to see if the position of the rotary encoder has changed and will increase or decrease the VFO frequency by the frequency step rate, as long as we are not at the upper or lower limit of the frequency range. We'll then update the TFT display with the current frequency.

The only other statement in the main loop() calls the read_pushbutton() function to change the frequency step rate when the rotary encoder pushbutton switch is pressed.

```
void read_pushbutton()
{
  // Read the Encoder Pushbutton Switch
  encoder_Switch = digitalRead(encoder_Switch_Pin);
  if (encoder_Switch == LOW && millis() > delay_time) // Check to see if pressed
  {
    // if it's changed, toggle the step size but don't allow again for debounce time
    delay_time = millis() + debounceInterval; // if it's changed, toggle the step
size but don't allow again for debounce time
    current_step = current_step + 1; // Change the Step pointer
    if (current_step > 6)
    {
      current_step = 0;
    }

    update_display();
  }
}
```

The read_pushbutton() function will check to see if the rotary encoder's pushbutton switch has been pressed. After a 500-millisecond delay to debounce the switch, the frequency step range is increased to the next step value in the current_step array, and the display is updated with the new frequency step value. If we're at the maximum step value, we'll wrap back around to the 1 Hz frequency step value.

The sketch and libraries for this project can be downloaded from **kw5gp.com**.

9 CW Decoder

The CW decoder (see **Figure 9.1**) is another one of my personal favorites from my first book, *Arduino for Ham Radio*. This time around, we'll upgrade the Arduino Uno to a Nano, and replace the 16 × 2 LCD with a 2.2-inch color thin-film-transistor (TFT) display. Notice that the library for the ILI9341-type display allows us to rotate the display image to present eight lines with 14 characters per line (see **Figure 9.2**). You could also modify the sketch to use a smaller font size to display 17 lines of 29 characters.

How many times have you been working CW, gotten distracted, and asked yourself, "What did they just send?" It's times like that I wish that I had a CW decoder handy to double-check myself. Sure, I can fire up the PC with a sound card interface and have the PC do the decoding for me, but that can be cumbersome, especially if you're operating a portable station in the middle of nowhere without any ac power.

By now you've seen some projects that can send CW, but what about receiving and decoding CW? If they can write a computer program for the PC that uses a sound card interface to decode CW, then surely we can do it with an Arduino.

Figure 9.1 — The CW decoder.

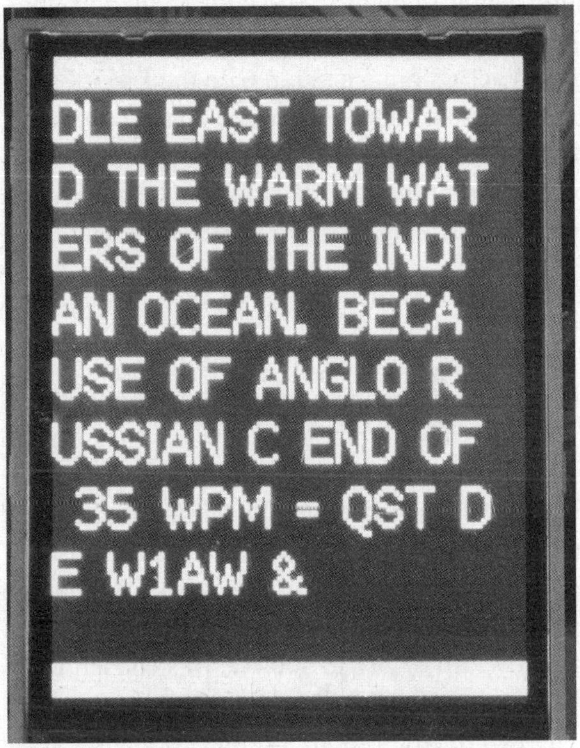

Figure 9.2 — The TFT display set for eight lines of 14 characters per line.

All we have to do is convert the CW signal to a digital pulse and measure the pulse widths to decode the CW. Now, for the really cool part — someone has already created an Arduino library to do just that. The MorseEnDecoder.h library will decode a digital CW signal into the actual letters and numbers; all we have to do is convert the CW audio tone into a digital signal that the MorseEnDecoder.h library can use.

To do this, we'll use an LM358 operational amplifier to buffer the CW audio, and an LM567 tone decoder chip to convert the CW tone into a digital signal. By using the LM567 tone decoder, we also have the ability to limit the decoder bandwidth to help prevent interference from adjacent CW signals. This also gives us the option to add a potentiometer to vary the bandwidth in future versions. The LM567 can be used to decode a tone, and its output will go high when the desired frequency is present on the input. The center frequency and bandwidth of the decoded tone are controlled by the value of a resistor and two capacitors. To calculate the desired center frequency, you can use the formula $f_o = 1/(1.1 \times RC)$, where R is the resistance (in ohms) across pin 5 and pin 6 on the LM567, and C is the capacitance (in farads) on the capacitor between pin 6 on the LM567 and ground. The bandwidth of the tone decoder can be determined with the formula

$$BW(in\ Hz) = f_o \times ((1070 \times (V_{in} / (f_o \times C))^\wedge \left(\frac{1}{2}\right))/100)$$

where V_{in} is the input signal in V_{rms}, f_o is the center frequency, and C is the value of the capacitor (in microfarads) between pin 2 on LM567 and ground. Fortunately, there is an online LM567 tone decoder calculator at **vk2zay.net/calculators/** to make these calculations much easier for us. Using the online calculator, a variable center frequency of 455 to 909 Hz, with a bandwidth of 217 to 300 Hz, was chosen for the CW decoder project. With the potentiometer (R6) for the LM567's center frequency set at a mid-range value of 5 kΩ, the center frequency will be approximately 600 Hz with a bandwidth of 270 Hz.

Now that we know how we can convert the CW audio into a digital signal that the Arduino can use, we can plan our project. The CW decoder block diagram (see **Figure 9.3**) shows that we will use the LM358/LM567 circuit to decode the CW audio signal, and we'll scroll the decoded CW on the TFT display. We will also use an LED on the output of the LM567 as a tuning indicator. Finally, we'll use a potentiometer (R10) to adjust the speed range of the CW decode. The MorseEnDecoder.h library doesn't handle automatic CW decode speed adjustments, so we'll need to use the speed potentiometer to manually adjust it. This speed adjustment is primarily used to determine when to add a space between words.

Now that we have our block diagram, we can start putting all the pieces together on our breadboard using the parts list and the schematic diagram shown in **Figure 9.4**. The majority of this project involves the LM386/LM567 circuit. Potentiometer R9 will be used to adjust the audio input to the LM358 operational amplifier that we're using to amplify and buffer the audio input to the LM567. Potentiometer R5 is used to adjust the audio input level to the LM567. The LM567 requires an input signal level of 200 mV or less to decode properly, so two 1N5711 Schottky diodes in a back-to-back configuration are used as clamping diodes to limit the LM567 input voltage to a maximum of 200 mV. Potentiometer R6 is used to adjust the center frequency of the CW tone to be decoded. As a starting point, adjust this potentiometer to its mid-range point of 5 kΩ.

Once you have the CW audio-decoding circuit on the breadboard, you can test and adjust the center frequency. Connect the audio from your receiver to the audio input of the CW decoder circuit and tune to a CW signal. You can also use an audio signal generator to precisely tune the center frequency of the LM567 tone decoder. There are also sketches available on the internet that allow you to use an Arduino as a precision audio signal generator. Using an oscilloscope or voltmeter, adjust potentiometer R9 until the audio input on pin 2 of the LM358 operational amplifier is about 50 mV ac. The gain on the LM358 is controlled by the values of resistors R1 and R2. For the CW decoder project, the gain was selected to be 10, which is calculated using the simple formula R2/R1. This will provide an output of up to 500 mV ac on pin 1 of the LM358. Once you have the input on pin 2 of the LM358 adjusted to 50 mV ac, you can adjust R5 to control the audio level into the LM567 tone decoder. You'll want to adjust this level to be close to, but not above, 200 mV ac on your voltmeter.

Next, tune your receiver to output the pitch of the CW tone that you prefer to listen to. Alternatively, you can use a waveform generator and adjust its output to the frequency your ears like best, usually a frequency of around 600 to 700 Hz. Adjust potentiometer R6 to find the center of

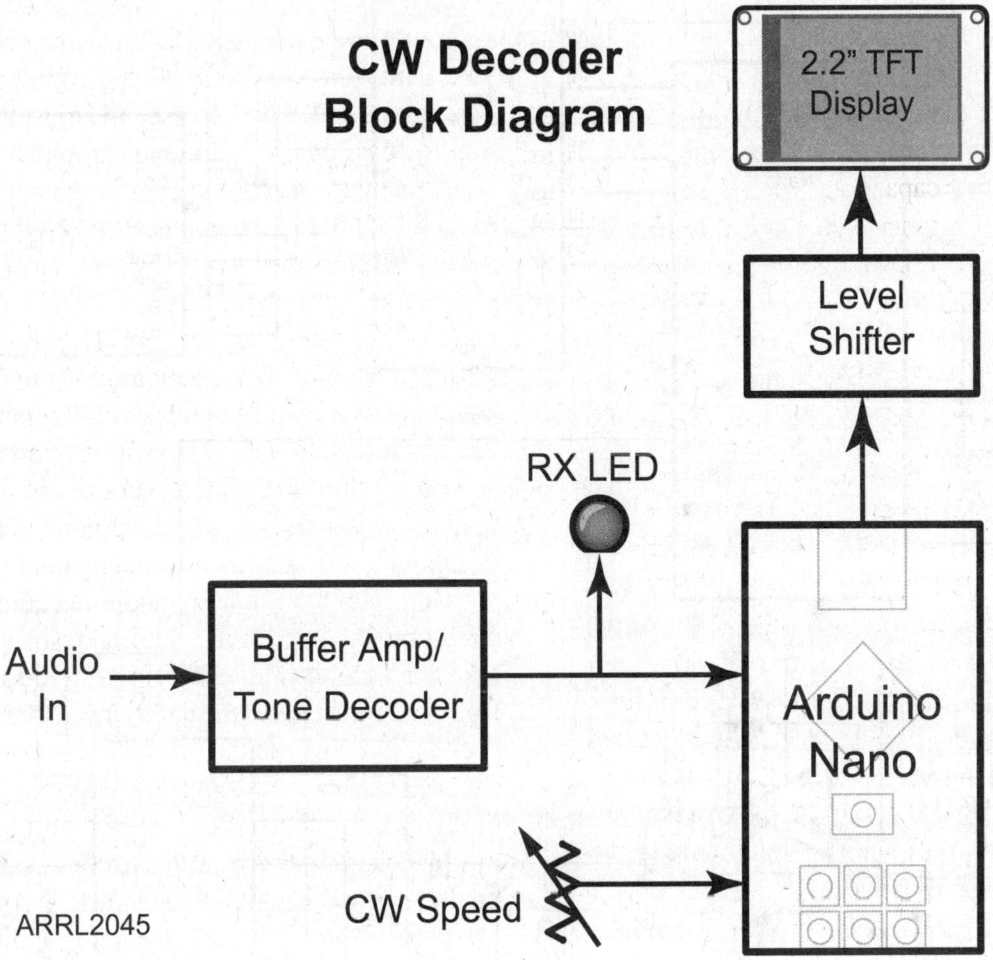

Figure 9.3 — The CW decoder block diagram.

the point where the LED on pin 8 of the LM567 comes on. Your CW tone decoder circuit is now tested and tuned to the desired CW pitch.

Finish hooking the rest of the circuit on your breadboard and we're ready to write the sketch for the CW decoder. Using the flowchart in **Figure 9.5**, you can see that the LM358/LM567 does most of the audio work for us, while the MorseEnDecoder.h library handles the rest of the decoding on the Arduino side of things.

Starting out in the sketch, we'll define the libraries we'll need for the CW decoder project. You'll note that the sketch defines the avr/pgmspace.h library, but you don't see where it is used in the sketch. The MorseEnDecoder.h library stores its CW conversion tables in the Arduino's flash memory, so this library uses the avr/pgmspace.h library itself. So, you effectively have the case of a library using a library to get the job done — it doesn't get much more efficient than that. To display the decoded CW characters, we'll use a color TFT display and include the TFT_ILI9341.h library. You may notice that is not the usual TFT display library used in other projects. The Adafruit library does not scroll the data on the display fast enough at the 35 wpm level. The TFT_ILI9341 library is derived from the Adafruit libraries but has been updated and optimized for better performance. The complete sketch can be found at **kw5gp.com**.

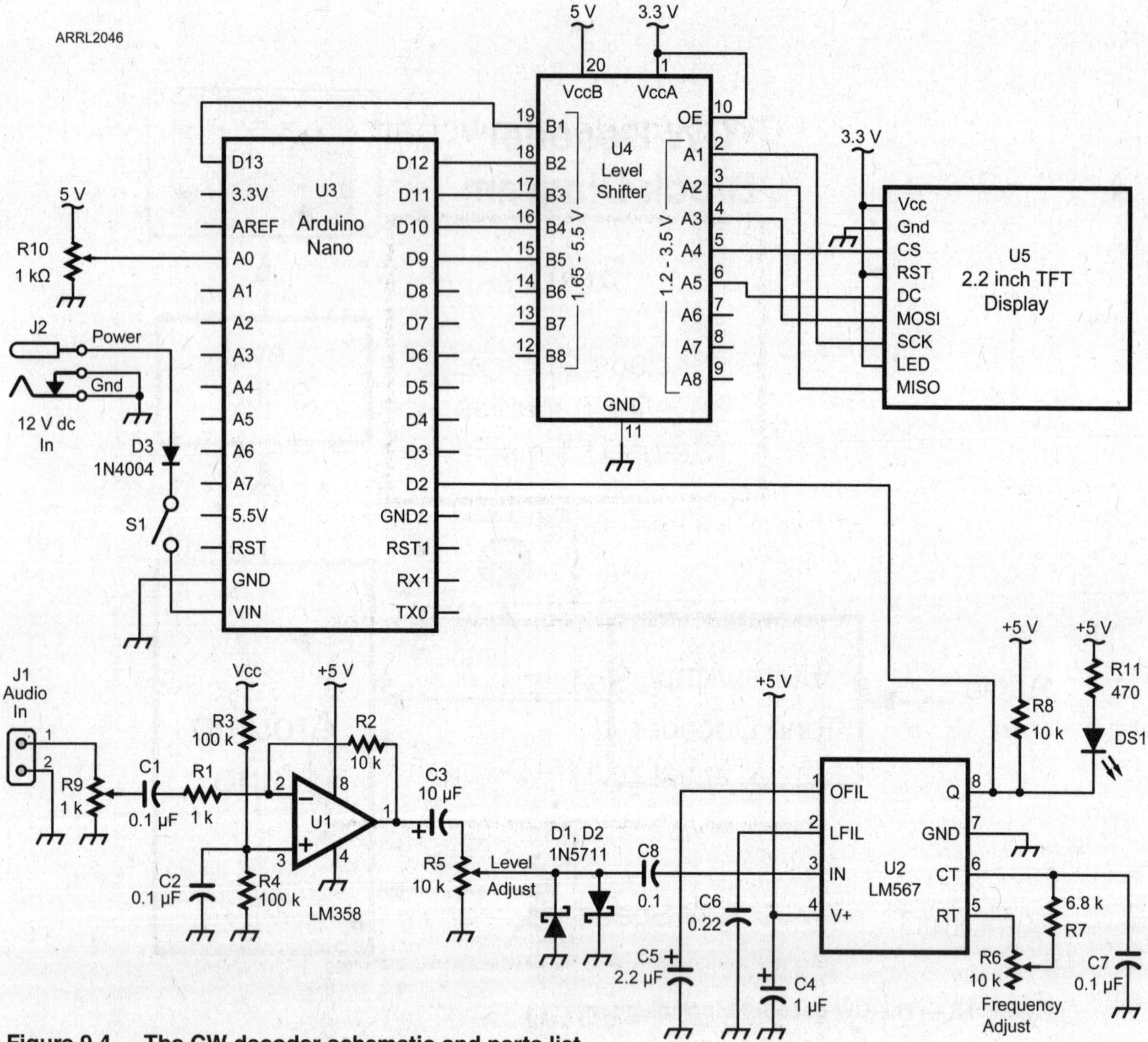

Figure 9.4 — The CW decoder schematic and parts list.

Part	*Value*
C1, C2, C7	0.1 µF capacitor, 25 V
C3	10 µF capacitor, 25 V
C4	1 µF capacitor, 25 V
C5	2.2 µF capacitor, 25 V
C6	0.22 µF capacitor, 25 V
D1, D2	1N5711 Schottky diode
D3	1N4004 diode
DS1	5-millimeter LED
J1	two-pin 2.54-millimeter Dupont male header
J2	dc power jack
R1	1 kΩ resistor, 1/8 W
R2, R8	10 kΩ resistor, 1/8 W

Part	*Value*
R3, R4	100 kΩ resistor, 1/8 W
R5, R6	10 kΩ 10-turn trim potentiometer
R7	6.8 kΩ resistor, 1/8 W
R9, R10	1 kΩ potentiometer
S1	SPST switch
U1	LM358 operational amplifier
U2	LM567 tone decoder
U3	Arduino Nano
U4	eight-channel level-shifter module (TXB0108 type)
U5	2.2-inch ILI9341-type color TFT LCD display

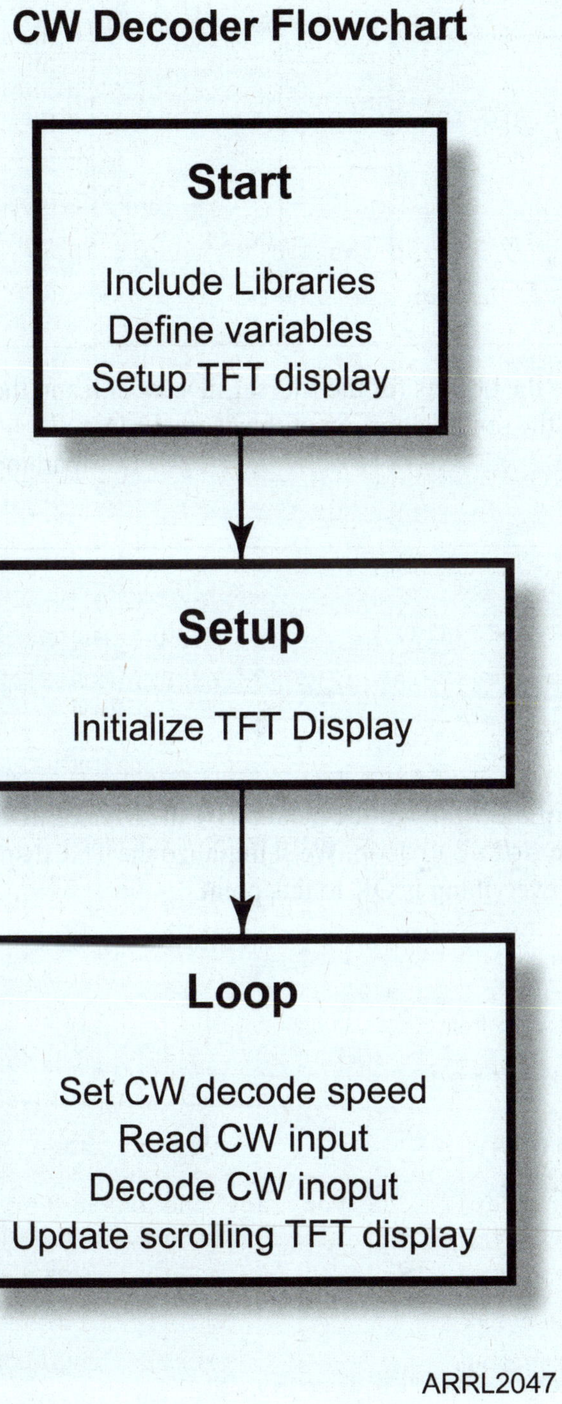

Figure 9.5 — The CW decoder flowchart.

```
#include <avr/pgmspace.h>
#include <MorseEnDecoder.h>  // Morse EnDecoder Library
#include <TFT_ILI9341.h> // Hardware-specific library
#include <SPI.h>
Next, we'll define the Arduino I/O pins and the variables used in the sketch:
#define morseInPin 2
#define Speed_Pin A0
#define TFT_CS      10  // Assign the TFT CS to pin 10
#define TFT_DC      9  // Assign the TFT DC to pin 9
```

Finally, we'll initialize the objects for the MorseEnDecoder.h and the TFT_ILI9341.h libraries. Because we are not using the sending portion of the MorseEnDecoder.h library, we don't need to be concerned about the `MORSE_KEYER` or `MORSE_ACTIVE_LOW` parameters when we initialize the library.

```
// Use hardware SPI for TFT
TFT_ILI9341 tft = TFT_ILI9341();      // Invoke custom library
morseDecoder morseInput(morseInPin, MORSE_KEYER, MORSE_ACTIVE_LOW);  // Define
the Morse objects
```

Because the majority of the work will be handled by the MorseEnDecoder.h library, there's really not much to be done in the `setup()` loop. We'll initialize the TFT display and display a brief start-up message so that we know everything is OK to this point.

```
void setup()
{
  pinMode(morseInPin, INPUT);

#ifdef debug1
  populate_character_array();  // Populate the character array if simulating re-
ceive
#endif

  start_tft();  // Start the TFT display

  //#ifdef debug
  Serial.begin(BAUD_RATE); // initialize serial communication on the USB port
  //#endif

}  // End Setup Loop
```

In the main `loop()`, we'll first read the value of the CW receive speed potentiometer. If the receive speed potentiometer value has changed, the new receive speed will be displayed on the display and the receive decode speed will be updated in the MorseEndecoder.h library.

```
void loop()
{
  read_speed = map(analogRead(Speed_Pin), 10, 1000, 5, 35); // Read the potenti-
ometer to determine code speed

  if (current_speed != read_speed)  // If the Speed Pot has changed, update the
speed and LCD
  {
    current_speed = read_speed;  // Set the current speed to the desired speed
    morseInput.setspeed(read_speed);  // Call the set speed function
    text = String(current_speed) + " wpm";  // set up the LCD display with the
current speed

#ifdef debug
    Serial.print("CW speed set to: ");
    Serial.println(read_speed);
#endif

    update_status();
    // Put the cursor back
    tft.setTextSize(cw_font_size);  // Set the tft text font size
    tft.setCursor(0, 5); // Move the cursor to the top line

    tft.print(cw_data); // Print the tft data string variable
  }
```

Because the MorseEnDecoder.h library doesn't auto-detect the speed of the received CW, the receive speed potentiometer must be manually adjusted to the receive speed. This adjustment does not have to be precise, as it's primarily used to determine when to add a space between words. Once you start listening to the incoming CW, you can estimate the speed and adjust the receive speed accordingly. Because this is a coarse adjustment, you'll find that it's not as cumbersome as it may seem to be at first. The MorseEndecoder.h library can handle some variation in the receiving speed internally, so you'll just need to set the receiving speed close to the actual speed, and the library will take it from there.

```
morseInput.decode();  // Decode incoming CW
if (morseInput.available())  // If there is a character available
{
  cw_rx = morseInput.read();  // Read the CW character
```

The actual CW decoding is done by the MorseEnDecoder.h library. All you have to do is call the `morseInput.decode()` function to decode the incoming character. Using the `morseInput.available()` function to see if there is a new character received, the `morseInput.read()` function will return the decoded CW character. The received characters are then scrolled from right to left on the top line of the TFT display.

Once you have the circuit and sketch working on the breadboard, you can then build the finished project on a copper-clad perfboard and mount it in an enclosure. For this project, I used a Solarbotics Mega S.A.F.E. (see **Figure 9.6**). Even though there are a lot of external components in this project, everything fits easily on a 6 × 8-centimeter copper-clad perfboard with room to spare, as shown in **Figure 9.7**.

The sketch and libraries for the project are available at **kw5gp.com**.

Figure 9.6 — The CW decoder in the Solarbotics Mega S.A.F.E. enclosure.

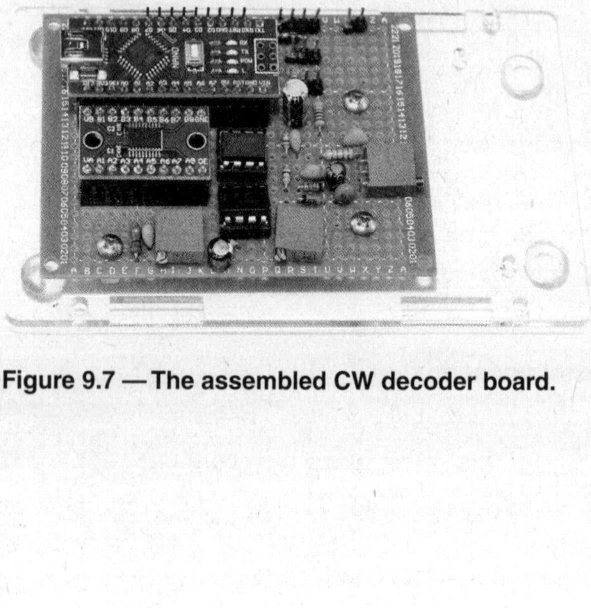

Figure 9.7 — The assembled CW decoder board.

10 Talking GPS/UTC Time/ Grid Square Indicator

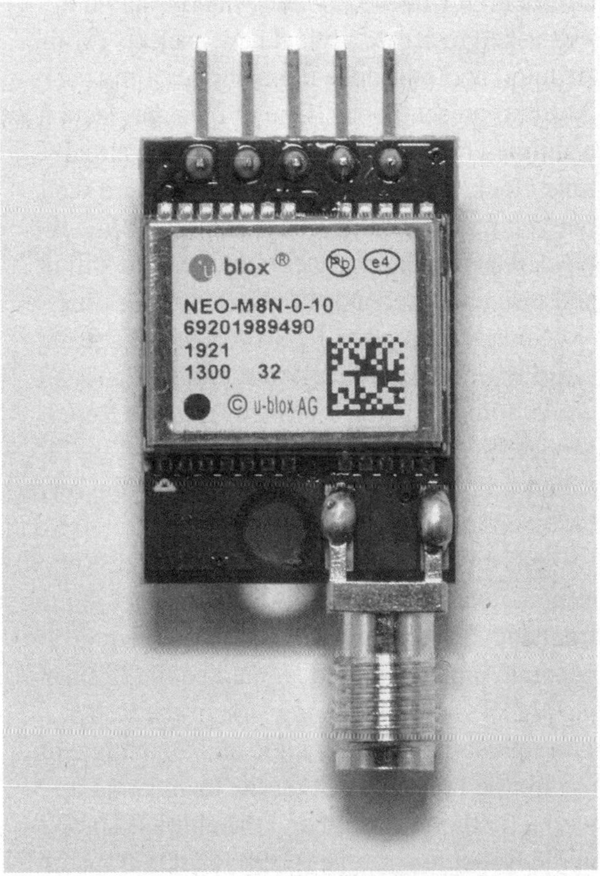

Figure 10.2 — The u-blox NEO M8N GPS module.

Figure 10.1 — The talking GPS/UTC time/grid square indicator.

This is another one of my favorites from my first book. This project was one of several that used the Emic 2 text-to-speech module. Sadly, the Emic 2 has been discontinued, and it took a bit of searching to find a suitable replacement. Be very cautious when experimenting with some of the generic text-to-speech modules, such as the XFS5152CE, SYN6288, and SYN6658. These modules don't do English language text-to-speech, only Chinese. Fortunately, the DFRobot Gravity Speech Synthesis Module V2.0 does excellent text-to-speech conversion and is now my module of choice for text-to-speech projects. The only real downside to the DFRobot module is that the speaker is mounted to the board, and there is no provision for an external speaker. From all appearances, you could probably desolder the speaker from the module and add an external speaker. For this new version of the Talking GPS/UTC Time/Grid Square Indicator, we'll upgrade the discontinued Emic 2 with the DFRobot module, replace the Arduino UNO with a Nano, and the Nokia 5110 LCD display with a 2.2-inch color TFT display. We'll also upgrade the GPS module to a u-blox NEO-M8N GPS module. We'll assemble this on a 6 × 8-centimeter piece of copper-clad perfboard and mount it in a Solarbotics Mega S.A.F.E. enclosure (see **Figure 10.1**).

A Global Positioning System (GPS) module like the one shown in **Figure 10.2** can do more than tell you latitude, longitude, and how to get where you're going. Since a GPS uses highly accurate timing signals to determine your location, it can also provide very accurate time information. In fact, there is a whole series of National Maritime Electronics Association (NMEA) messages, or "sentences," that use the timing information provided by the GPS satellites, and that information is processed by a GPS receiver

MAIDENHEAD GRID SQUARES

What are these grid squares you keep hearing about? The Maidenhead Locator System was devised in 1980 at a meeting of European VHF managers to create a global geographic coordinate system that amateur radio operators could use to denote their location. Named after the town in England where the meeting was held, Maidenhead Grid Squares divide the world into small areas, measuring 1° of latitude by 2° of longitude (approximately 70 × 100 miles in the continental US). A grid square comprises of 4 to 6 characters and numbers. It measures northward starting from the South Pole and eastward starting from the anti-meridian of the Prime Meridian in Greenwich, England (in the middle of the Pacific Ocean).

The first two letters of a grid square, also known as the field, denote 18 separate 20° blocks of longitude and latitude, denoted by the letters A – R. The first letter is the longitude, and the second letter is the latitude. The field is followed by a pair of numbers, known as the square. These numbers range from 0 – 9 and divide the field into "squares" of 2° longitude × 1° latitude, with the first number being longitude and the second number being latitude. Optionally, the grid square can be further divided into sub-squares of 2.5 minutes of latitude × 5 minutes of longitude (approximately 3 × 4 miles in the continental US), denoted by a pair of letters a – x, with the first letter representing longitude and the second letter representing latitude.

As an example, my longitude is 90.0223° W. This places me within the fifth block of 20° of longitude from the anti-meridian, so the first letter of my grid square is "E". My latitude is 34.9799° N, which places me within the thirteenth block of 20° from the South Pole, so the second letter of my grid square is "M". Within the field, the calculations show that I am in the fifth 2° block of longitude. Since the numbering system starts at zero, the first number in my square is "4". Applying the same calculation to my latitude, the calculation shows that I am also in the fifth 1° block of latitude. Therefore, my grid square is EM44. You can continue the calculations down to the sub-square and determine that my exact grid square is EM44xx. Fortunately, the formula for calculating grid squares has already been developed for us, so there is rarely a need to calculate a grid square manually. Some hams have expanded the grid square designation by a few additional characters. While not formally part of the Maidenhead grid square naming convention, these additional characters provide the ability to define a location of just a few hundred feet. This is often used in the case of ham radio pico balloon GPS telemetry to track a downed balloon to its precise location.

into usable information. By using these NMEA sentences, we can extract all kinds of useful information from our GPS receiver.

What can we do with all this GPS information? Up until recently, I was asking myself the same thing. I've had a GPS unit in my car for years, and aside from that one time it sent me down a dirt road out into the middle of a cornfield, it's been very handy and reliable. I was asked to create several new projects for the Arduino and introduce them at one of my early Arduino presentations. One of these projects was a simple UTC/Local Time clock using a real-time clock/calendar module. This worked well, but you had to set the time manually. While the clock was reasonably accurate, it would still drift and lose a few seconds of accuracy over time.

Shortly thereafter, I got involved in a team effort to get the JT65 digital communications mode running on the then-new Ten-Tec Rebel Model 506 QRP CW Transceiver. The Ten-Tec Rebel was an Open Source QRP CW transceiver based on the Digilent chipKit Uno32, an 80 MHz variant of the Arduino Uno. The Ten-Tec Rebel used a Texas Instruments AD9834 direct digital frequency synthesis (DDS) integrated circuit to generate the transmitted signal and the IF mixing signal for the receiving side of the Rebel. Through some serious magic and amazing coding by Joe Large, W6CQZ, the team came up with a method to command the chipKit Uno32 in the Rebel to shift the AD9834 DDS transmit frequency on the fly to generate the 64 tones necessary for JT65. Since the Ten-Tec Rebel was an ideal portable QRP transceiver that could go with you just about anywhere, we wanted a way to acquire accurate time anywhere for the JT65 mode to function correctly. The JT65 protocol requires that transmission to start at precisely 1 second into a new minute. If you're out in the middle of nowhere with no internet access to internet time servers, your only option is to manually set the time on the workstation running the JT65 software. We wanted the JT65 operations on the Rebel to be totally self-sufficient from external time sources for time synchronization. The light bulb came on, and a GPS interface was quickly built on a modified Arduino Protoshield and mounted on the chipKit Uno32's shield expansion pins inside the Rebel. Our Rebel prototype could now provide accurate time, location, and as an added treat, it could also calculate and display the current Maidenhead Grid Square.

What if you don't use JT65, FT8, or similar digital modes or don't have a Rebel? Well, many of the contests, particularly the VHF contests, use your Maidenhead Grid Square as part of the contest exchange. If you're into mobile operating, it's difficult to keep track of your grid square as you travel down the road. It's also convenient to have this project around for ARRL Field Day and other portable events because it provides accurate time, altitude, and grid square information.

The NEOM8N GPS is a fully self-contained GPS module that supports the standard NMEA-0183 messaging format. With a -167 dBm tracking sensitivity, the NEO M8N can provide highly accurate locations, times, and altitude information. The NEO M8N also provides heading and speed information, allowing it to be used for navigation in addition to providing fixed location information. The NEO M8N communicates with the Arduino using a standard TTL Serial port running at a default speed of 9,600 baud.

The NMEA-183 Protocol

The NMEA-183 is an ASCII text-based protocol. All messages start with a `$`, end with a carriage return/line feed (CR/LF), and contain a checksum to detect corrupted messages. The NEO M8N GPS module supports the NMEA-183 standard messages. Of these messages, the GGA message contains the majority of the information we'll be using in our projects. A standard GGA message will look like this:

```
$GPGGA, 150028.000,3458.8015,N,09001.3496,W,1,9,0.90,93.7,M,-30.0,M,,*55
```

This message breaks down into:

`$GPGGA` – GGA Message Type

`150028.000` – UTC Time in hhmmss.sss

`3458.8015` – Latitude in degrees and minutes ddmm.mmmm

`N` – North Latitude

`09001.3496` – Longitude in degrees and minutes ddmm.mmmm

`W` – West Longitude

`1` – Position Fix Indicator. A "1" indicates the reading is in GPS SPS mode and the fix is valid

`9` – The number of GPS satellites used to determine the position fix

`0.90` – The Horizontal Dilution of Precision (HDOP) of the fix

`93.7` – Altitude above mean sea level

`M` – Altitude reading is in meters

`-30.0` – The Geoids Separation value (correction for tidal influences)

`M` – Geoids reading is in meters

`*55` – Message Checksum

And the answer to your next question is yes, if you plug these numbers into your GPS, you will find yourself on top of the worktable in my lab, give or take a few meters. As you can see, there is a lot of valuable information packed into an NMEA sentence. Fortunately, there are some excellent libraries for the Arduino that handle decoding the GPS messages without us having to parse out all that data

Figure 10.3 — The DFRobot Gravity speech synthesis module V2.0.

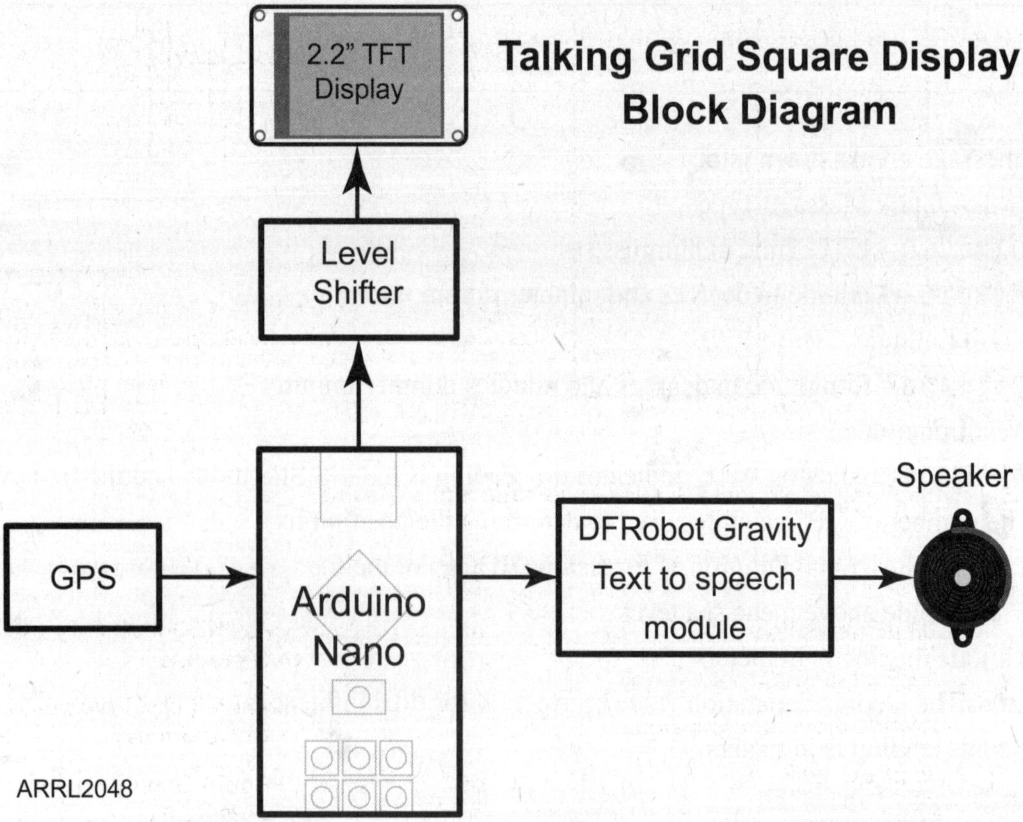

Figure 10.4 — The talking grid square block diagram.

manually. The Arduino TinyGPSPlus library from **arduiniana.org** is outstanding for any sketches needing to extract data from your GPS module.

To kick things up a notch, we'll also use the DFRobot Gravity speech synthesis module V2.0, as shown in **Figure 10.3,** to have our GPS project speak the time and current Maidenhead grid square. This can come in handy when you're operating mobile and don't want to take your eyes off the road. One nice thing about the DFRobot speech synthesis module is that you can interface to it using either standard TTL Serial I/O or via the I²C bus, selected by an onboard switch. For this project, we'll talk to it using the I²C bus.

Figure 10.4 shows the block diagram for the talking GPS/UTC clock and grid square indicator. For this project, we'll have the Arduino read the GPS and extract the longitude, latitude, date, time, and altitude information. The Arduino will calculate and display our grid square on a color TFT display. The DFRobot speech synthesis module will speak the time and grid square when the audio is enabled by a switch mounted to the enclosure.

Now that we know what pieces we'll be using in this project, we'll create the parts list and schematic (see **Figure 10.5**), which shows us how to wire everything on our breadboard. In this project, we'll use a software serial port on the Arduino to communicate with the GPS module.

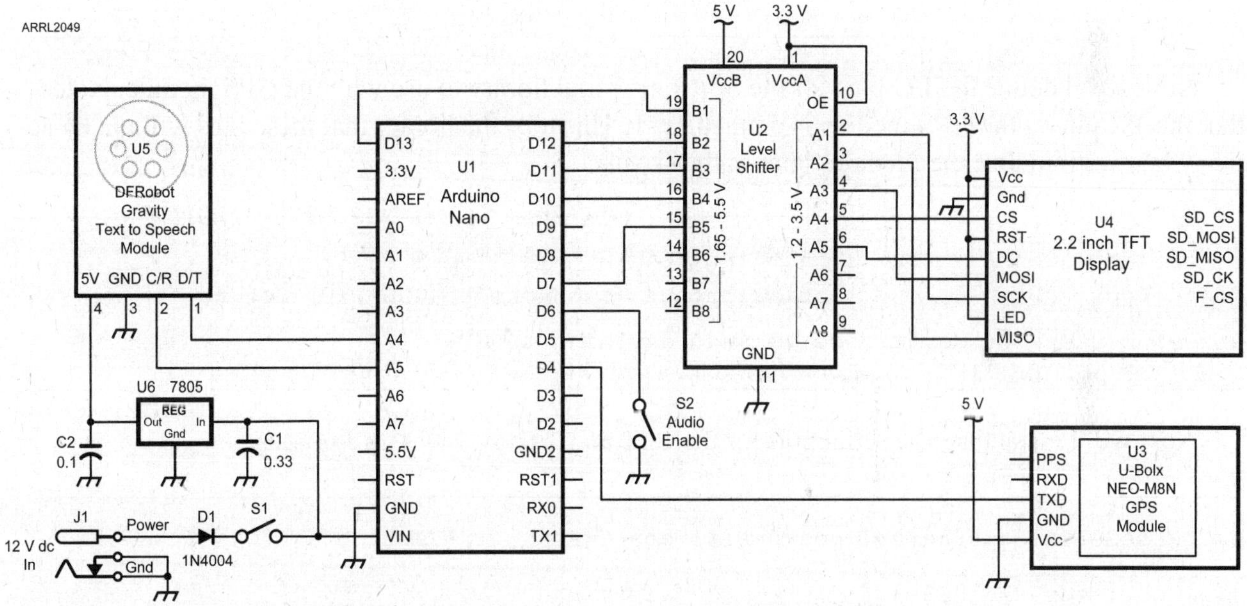

Figure 10.5 — The talking grid square display schematic and parts list.

Part	***Value***
C1	**.33 µF capacitor, 35 V**
C2	**.1 µF capacitor, 35 V**
D1	**1N4004 diode or equivalent**
J1	**dc power jack**
S1, S2	**SPST toggle switch**
U1	**Arduino Nano**
U2	**8-channel level shifter module (TXB0108 type)**

Part	***Value***
U3	**u-blox NEO-M8N (or equivalent) GPS module**
U4	**2.2-inch ILI9341-type color TFT LCD display**
U5	**DFRobot Gravity speech synthesis module V2.0**
U6	**LM7805 5 V regulator**

With everything wired and ready to go, we can begin to write the sketch. Using the flowchart in **Figure 10.6**, we have all the steps laid out in the order we need. You will see as we get into the sketch that it really is as easy as the flowchart makes it look. The libraries do most of the work for us; all we have to do is glue the various pieces together. The sketch and libraries can be found at **kw5gp.com**.

Starting off with the sketch, you will see that we include the libraries we'll need:

```
#include "DFRobot_SpeechSynthesis_V2.h"
#include <SPI.h>
#include <Adafruit_GFX.h> // The Adafruit Graphics Library
#include <Adafruit_ILI9341.h> // The ILI9341 TFT Display Library
#include <SoftwareSerial.h>  // include the SoftwareSerial library so we can use
it to talk to GPS
#include <TinyGPSPlus.h>  // The TinyGPSPlus GPS Library from Arduiniana
#include <maidenhead.h> // The Maindenhead Grid Square Library
```

Now, we'll define the I/O pins for the Software Serial library to use with the GPS module. Note that the TX pin on the GPS module is defined as -1. This tells the library that these pins will not be active when we initialize the modules later in the sketch:

```
#define gps_rxPin 4     // Serial input (connects to GPS TxD)
#define gps_txPin -1    // Serial output (connects to GPS RxD) Set to -1 (dis-
abled) because we don't need to send anything to GPS
```

Next we'll instantiate the serial port for the `GPS` and `TinyGPSPlus` libraries.

```
SoftwareSerial gps_serial(gps_rxPin, gps_txPin);  // set up a new software se-
rial port for the GPS
```

Then we define all the variables needed in the sketch:

```
float LAT, LON;  // floating integer for latitude and longitude values
double gps_altitude, prev_gps_altitude = -9999;   // Variable to hold the alti-
tude value
int gps_tick = 0, gps_timeout = 240;  // GPS startup timer variables
unsigned long gps_time, gps_date; // GPS date and time variables
String s_date, s_time, s_month, s_year, s_day, s_hour, s_minute, s_second, grid_
text; // String variables for date, time and grid square
boolean gps_ok = false;  // gps data ok flag
```

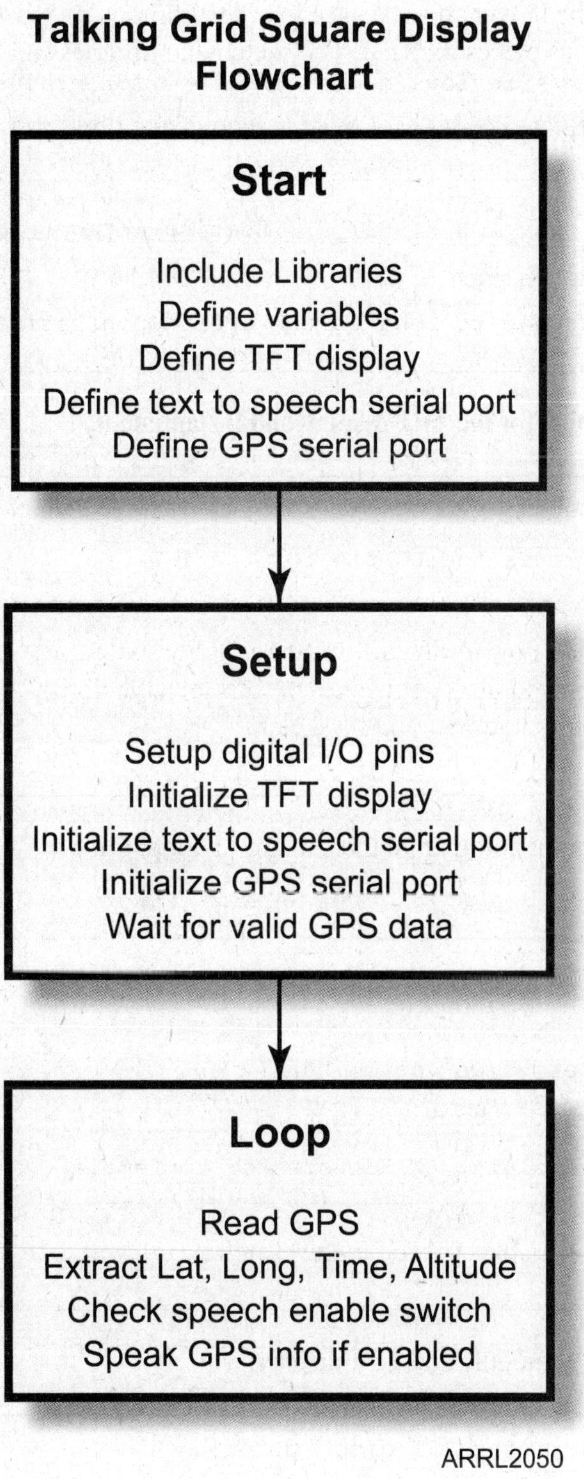

Figure 10.6 — The talking GPS/UTC time/grid square display flowchart.

```
bool audio_on = false; // Speech enabled flag
bool prev_audio;  // Previous state of speech enabled flag
int num_sats = 0; // Number of satellites being received
int prev_sats = -1;  // Previous number received for status update
String prev_date = "", prev_time = " ", prev_grid = ""; // variables to hold the
previous info
float prev_LAT = 999, prev_LON = 999;  // floating integer for previous latitude
and previous longitude values
String msg; // The message to send to the Speech Synthesizer
```

Then we'll define the pins for the TFT display and instantiate it:

```
#define TFT_CS 10 // Defines the CS pin for the TFT
#define TFT_DC 7  // Defines the DC pin for the TFT

// Use hardware SPI for TFT
Adafruit_ILI9341 tft = Adafruit_ILI9341(TFT_CS, TFT_DC);  // Create the TFT in-
stance
```

In this sketch we have a lot going on in the `setup()` loop. First, we'll setup the DFRobot speech synthesis module:

```
ss.begin();  //Init speech synthesis sensor


ss.setVolume(3);  //Set voice volume to 3
ss.setSpeed(5);  //Set playback speed to 5
ss.setSoundType(ss.eFemale1);  //Set speaker to female
ss.setTone(4);  //Set tone to 4
ss.setEnglishPron(ss.eWord);  //For English, speak word
```

Then we'll start the GPS module and TFT display:

```
  start_tft();  // Start the TFT display
  gps_serial.begin(gps_baud_rate);  // set the data rate for the GPS SoftwareSe-
rial port
```

Now, we'll wait for the GPS to acquire the satellites.

```
  // Check for valid GPS data
  clear_display();  // Clear the TFT display
  tft.setCursor(70, 50);  // Display the aquiring satellites message
  tft.print("Acquiring");
  tft.setCursor(60, 90);
  tft.print("Satellites");

  while ((gps.satellites.value() == 0) && (gps_tick < gps_timeout)) // Loop un-
til we start getting satellites
  {
    num_sats = gps.satellites.value();

#ifdef debug
    Serial.print("Sats: "); Serial.print(num_sats); Serial.print("   Location
Valid: "); Serial.println(gps.location.isValid());
#endif

    tft.setCursor(95, 150);
    tft.print("  " + String(gps_timeout - gps_tick) + "  "); // Display the
timeout timer
    smartDelay(1000); // Read and update GPS data
    gps_tick = gps_tick + 1;  // Wait a second and decrement the timeout timer
  }
  if (gps_tick < gps_timeout)  // Check to see if valid message
  {
    gps_ok = true;  // We got valid data before timeout, flag the GPS data as re-
ceiving satellites
    num_sats = gps.satellites.value();

#ifdef debug
    Serial.println("Receiving Satellites");
    Serial.print("Sats: "); Serial.print(num_sats); ; Serial.print("   Location
Valid: "); Serial.println(gps.location.isValid());
#endif
    clear_display();  // Clear the TFT display
    tft.setCursor(60, 50);  // Display the waiting for enough satellites to
determine position
```

```
    tft.print("Waiting For");
    tft.setCursor(80, 90);
    tft.print("Satellite");
    tft.setCursor(130, 130);
    tft.print("Lock");
    smartDelay(30000);
    clear_display();
    update_status();

    // Set the previous audio enable flag to the opposite of what the enable
switch reads
    prev_audio = (digitalRead(audio_enable)); // Read the state of the audio en-
able switch
  } else
  {
    // GPS Failed to lock
    clear_display();
    tft.setCursor(130, 50); // If we get here, the GPS failed to get valid sat-
ellite data after the timeout period
    tft.print("GPS");
    tft.setCursor(100, 90);
    tft.print("Failed");
    tft.setCursor(95, 130);
    tft.print("To Lock");

    while (true);  // Loop here until reset
    {
    }
  }
  // Set the speech enable to the opposite of what it is now so it will display
correctly
  prev_audio = digitalRead(audio_enable); // Read the state of the audio enable
switch

} // End Setup Loop
```

The sketch will allow up to two minutes to acquire a location fix. The NEO M8N GPS data sheet says that the maximum time for a satellite fix from a cold start takes 26 seconds. In my indoor lab, I have seen the satellite fix take up to a minute and a half, so I set the timeout in the sketch for two minutes. From a warm restart, the NEO M8N GPS will usually acquire the GPS satellites in just a few seconds. The TFT display will show a message indicating that we are acquiring the satellites and show the timeout timer as it counts down.

```
void loop()     // Start the Main Loop
{
  audio_on = !(digitalRead(audio_enable)); // Read the state of the audio enable
switch

#ifdef debug
  Serial.print("Audio Enable: ");
  Serial.print(audio_on);
  Serial.print("  Prev: ");
  Serial.println(prev_audio);
#endif

  smartDelay(0);  // smartDelay is used by GPS library to update satellite info
while delating x milliseconds
  num_sats = gps.satellites.value();   // Get the number of satellites
  LAT = gps.location.lat();   // Get the latitude
  LON = gps.location.lng();   // Get the longitude

#ifdef debug
  Serial.print("Sats: "); Serial.print(num_sats);
  Serial.print("   LAT: "); Serial.print(LAT, 9);
  Serial.print("   LON: "); Serial.print(LON, 9);
  Serial.print("   Location Valid: "); Serial.println(gps.location.isValid());
#endif
  smartDelay (3000);

  gps_date = gps.date.value(); // Read the GPS Date info
  s_date = gps_date;  // Put the date in a separate string

  gps_time = gps.time.value(); // Read the GPS Time info
```

```
  s_time = gps_time;  // Put the time in a separate string

#ifdef debug
  Serial.print("Date: "); Serial.print(gps.date.value()); Serial.print("   Time: 
"); Serial.println(gps.time.value());
#endif

  if (s_date.length() < 6)  // If the date string length is less than 6, pad it 
until it is
  {
    s_date = "0" + s_date;
  }

  while (s_time.length() < 8)  // If the time string length is less than 8, pad 
it until it is
  {
    s_time = "0" + s_time;
  }

  s_year = s_date.substring(4, 6); // Break out the date string into Day/Month/
Year
  s_month = s_date.substring(2, 4);
  s_day = s_date.substring(0, 2);
  s_hour = s_time.substring(0, 2); // Break out the time string into Hour/Min-
ute/Second
  s_minute = s_time.substring(2, 4);
  s_second = s_time.substring(4, 6);
  gps_altitude = gps.altitude.feet();  // Get the altitude

#ifdef debug
  Serial.print("Date: "); Serial.print(gps_date); Serial.print("  Date String: 
"); Serial.print(s_date); Serial.print("   Length: "); Serial.print(s_date.
length());
  Serial.print("   Time: "); Serial.print(gps_time); Serial.print("   Time 
String: "); Serial.print(s_time); Serial.print(" Altitude: "); Serial.
println(gps.altitude.feet());
```

```
#endif

  grid_text = get_mh(LAT, LON, 6);  // Get the Maidenhead Grid Square

#ifdef debug
  Serial.print("Grid Square: ");
  Serial.print(grid_text);
  Serial.print("  Length: ");
  Serial.println(grid_text.length());
#endif

  update_status();  // Update the TFT display

  // The Text to Speech part of the sketch

  if (audio_on)  // Only speak if Audio is enabled
  {
    ss.speak("[g2][h2]U T C Time is");  // Say the Time
    msg = ("\"[n1]" + s_hour + s_minute + "\"");  // Prepare the message
    ss.speak(msg);  // Speak the Time

    ss.speak("Grid Square");  // Say the Grid Square
    msg = ("\"" + grid_text + "\"");  // Prepare the message
    ss.speak(msg);  // Speak the grid square
  }

  smartDelay(10000);  // Delay 10 seconds and update the GPS info

}  // End Main Loop
```

Figure 10.7 — The assembled talking GPS/UTC time/grid square indicator board.

Figure 10.8 — The assembled talking GPS/UTC time/grid square board.

In the main `loop()`, we start by checking if the `Audio Enabled` switch is on and update the display if necessary. Next, we'll read and display the GPS information. The NEO M8N GPS can determine altitudes up to 50,000 meters, so even if you do your next Summit on the Air (SOTA) operation from low Earth orbit, you'll still have an accurate altitude reading.

Unlike the original version of this project, where we had to calculate the Maidenhead grid square manually, now all we need to do is call the maidenhead.h library and it will do all that work for us. Then, we'll display the calculated grid square. Finally, if the `Audio Enable` switch is turned on, we'll speak the time and grid square. The GPS information and grid square are shown on the TFT display and updated approximately every 30 seconds, as shown in **Figure 10.7**.

Once the project is fully debugged and operational on the breadboard, the copper-clad perfboard can be assembled (see **Figure 10.8**), consisting mainly of pin headers as connectors for the TFT display, speech synthesis module, GPS module, and the `Audio Enable` switch. The finished assembly is mounted into a Solarbotics Mega S.A.F.E. enclosure. As you can see in **Figure 10.9**, the GPS module and DFRobot speech synthesis module are mounted to the side of the enclosure.

The sketch and libraries for this project can be downloaded from **kw5gp.com**.

Figure 10.9 — The GPS and speech synthesis module mounted to the side of the enclosure.

11 Talking SWR Meter

Figure 11.1 — The talking SWR meter.

Shortly after my first book was published, I was contacted by Dr. Ronald E. Milliman, K8HSY (SK), then-president of the South Central Kentucky Council of the Blind and retired Professor of Marketing at Western Kentucky University. He mentioned that he and his fellow blind hams found the projects that used the Emic 2 text-to-speech module very helpful, and that he would love to see more speech applications with the Arduino. When I was writing the first Arduino book, I thought adding speech capability to Arduino projects was cool, and I knew that it could be of benefit to the visually impaired, but when I found out that I had actually reached such an audience with a useful Arduino project, it really made my day.

I have to stop for a moment and say that I don't write to get rich. Trust me, I'll never get rich writing books of any kind. I write because I enjoy working with the Arduino and sharing my experiences with others. It's rewarding hearing from fellow hams like Ron, who can really put some of these projects to good use.

With the EMIC 2 text-to-speech module discontinued and no longer available, it's time to upgrade the original talking SWR project (**see Figure 11.1**) to use the DFRobot Gravity Speech Synthesis Module V2.0 that we used in the previous chapter. While we're at it, we'll upgrade the Arduino UNO to a Nano and replace the Nokia 5110 LCD display with a 2.2-inch ILI9341-type color TFT display and its associated TXB0108-type level converter module. We'll assemble the project on a 6 × 8-centimeter copper-clad perfboard and mount it inside a Solarbotics Mega S.A.F.E. enclosure.

One of the fun things about the Arduino is that there are more and more new devices and modules available for it all the time. This project was originally designed to be just a standard digital SWR meter, and the discovery of a text-to-speech module took things to a whole new level. With the Emic 2 text-to-speech module no longer available, I began a search for a new text-to-speech module that I could use. Luckily, I came across the DFRobot Gravity Speech Synthesis Module V2.0. Now, we're back in business with talking projects.

When this project was originally created, my first thought was my typical, "How cool

Figure 11.2 — The DFRobot Gravity Speech Synthesizer Module V2.0.

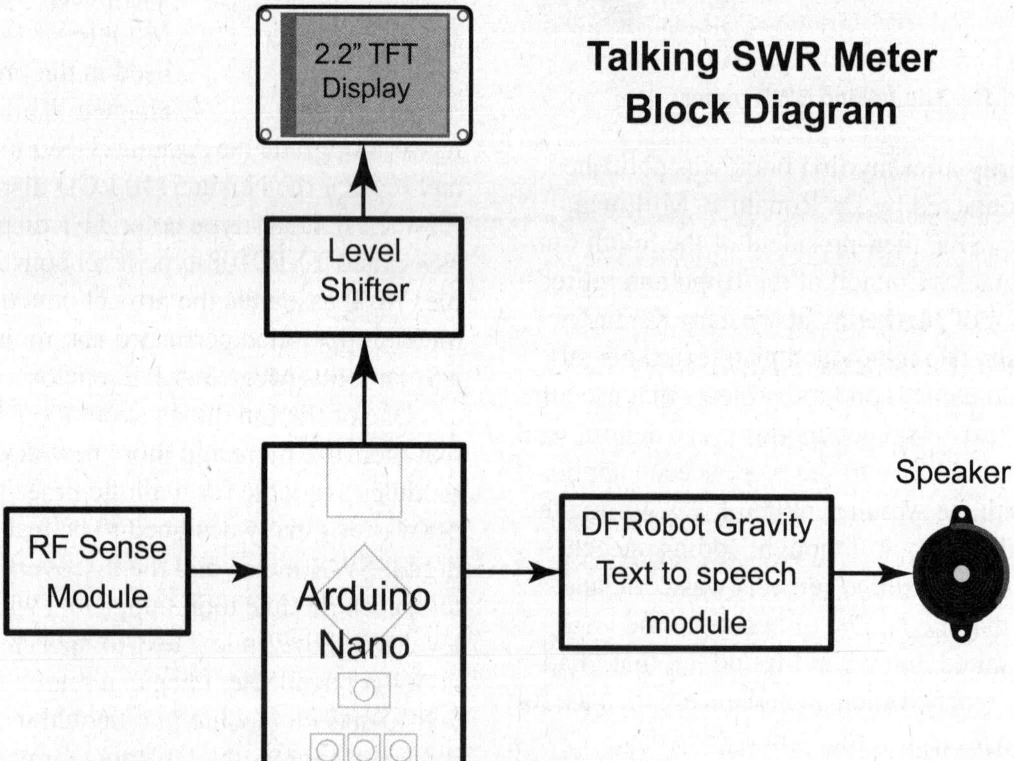

Figure 11.3 — Talking SWR meter block diagram.

would it be to have an SWR meter that could speak and tell you what the SWR is?" This was quickly followed by the realization that a text-to-speech module would open a whole world of new projects for operating in low-light conditions, and for visually impaired hams. Several of my ham friends are visually impaired, and I had ideas for projects they could build (or have someone build for them) while getting some serious usefulness out of them. Also, because many of our club's operating events are at night with the usual dim-to-nonexistent lighting, the idea of an SWR meter that could speak the SWR had my attention. By upgrading the original project to a TFT display, we can now change the background color based on the SWR values, with green meaning the SWR is good, yellow meaning it's not optimal but not excessive, and red signifying the SWR is above the preset limit, meaning the problem should be addressed. Also, you can incorporate speech into these warnings.

My friends and I had hours of fun coming up with phrases we would have the SWR meter speak when it encountered alarm conditions, most of which cannot be repeated here, but the theme was generally, "Hey man, your stuff is broke." Who says you can't have fun with test equipment?

The DFRobot Gravity Speech Synthesizer Module V2.0

The DFRobot Gravity Speech Synthesis Module V2.0 text to speech synthesizer (see **Figure 11.2**) converts standard ASCII text into speech. Interestingly enough, the module supports both English and Chinese, so it actually could be used to build an Arduino-powered English/Chinese translator. It can also speak the current time and environmental data. The Gravity Speech Synthesis Module communicates with the Arduino using either standard TTL Serial data or I^{2}C, selectable by a switch on the module. The Gravity Speech Synthesis Module V2.0 is fully self-contained, including an onboard audio amplifier and speaker. The output volume and other speech parameters can be software-controlled from within your sketches.

The module supports a number of commands that allow you to select different voices, speed, and other features. Unfortunately, for some reason these are not well documented, so you'll need to look at the example sketches that come with the DFRobot_SpeechSynthesis_V2 library, or dig through the library keywords, .h., and .cpp files. At some point, I'll try to compile the complete list, but for now, we'll just stick to the documented commands in our sketches.

Now that we know what we're going to do with the SWR data when we get it, we have to figure out a way to read the SWR and pass that information on to the Arduino (see **Figure 11.3**). Remember in the Introduction when I told you how handy *The ARRL Handbook* can be? As it turns out, others have passed this way ahead of us and left some very useful breadcrumbs. The 2013 *ARRL Handbook for Radio Communications* has a section on building a Microprocessor Controlled SWR Monitor. Unfortunately, this is not in *Handbook 100*. While the circuit design in the 2013 *Handbook* is for a Parallax Propeller microprocessor and not an Arduino, the SWR sensing head used in the handbook project is perfect for our needs. The SWR Sense Head described in *The ARRL Handbook* is based on "The Tandem Match – An Accurate Directional Wattmeter" from the January 1987 issue of *QST* by J. Grebenkemper, KA3BLO. Using a simple and easy-to-construct design, the SWR sense head will output an analog voltage representation of both the forward and reflected SWR simultaneously. We can supply these voltages to the Arduino's analog input and directly read the voltages provided by the SWR sense head. I built mine using the exact circuit design in the 2013 *ARRL Handbook* (see **Figure 11.4**) on a small piece of perfboard mounted inside an aluminum project enclosure to provide RF shielding as shown in **Figures 11.5** and **11.6**. In the *ARRL Handbook* design, you have the option of building the SWR sense head for sensing 10 W or 100 W by simply changing the number of turns of wire on the two toroids used to transfer a small amount of RF energy into the SWR sensing circuitry. For this project, I chose to build the 10 W version, but

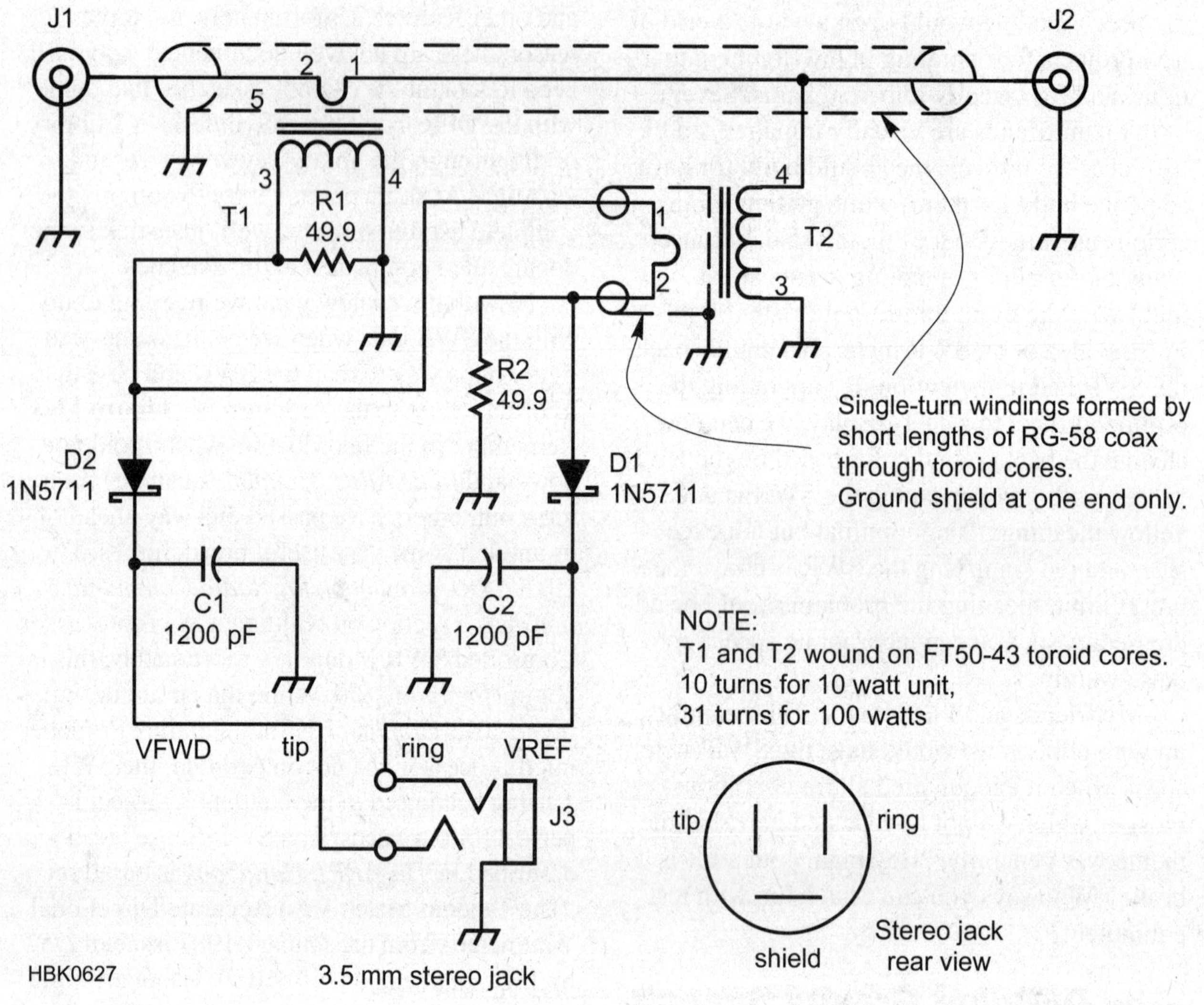

Figure 11.4 — The SWR sense head from the 2013 *ARRL Handbook* and parts list.

Part	*Value*
R1, R2	**49.9 Ω resistor**
D1, D2	**1N5711 Schottky diode**
C1, C2	**1200 pf capacitor**
L1, L2	**FT50-43 toroid core**
Aluminum Project Enclosure	

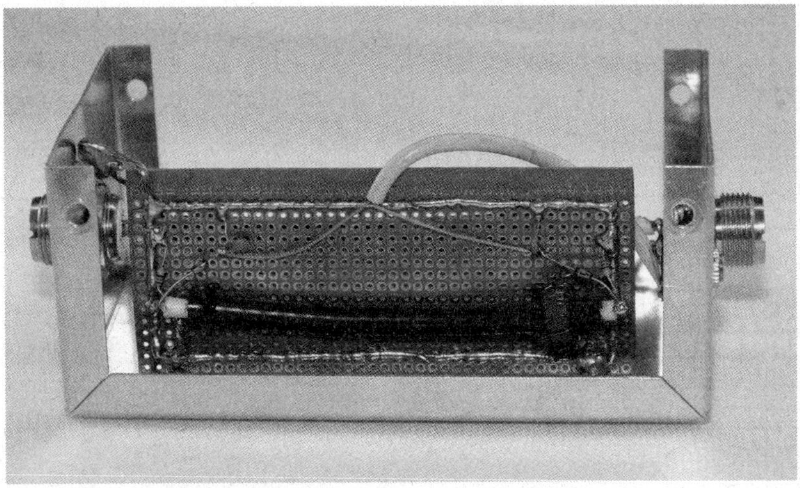

Figure 11.5 — The SWR Sense Head Front View.

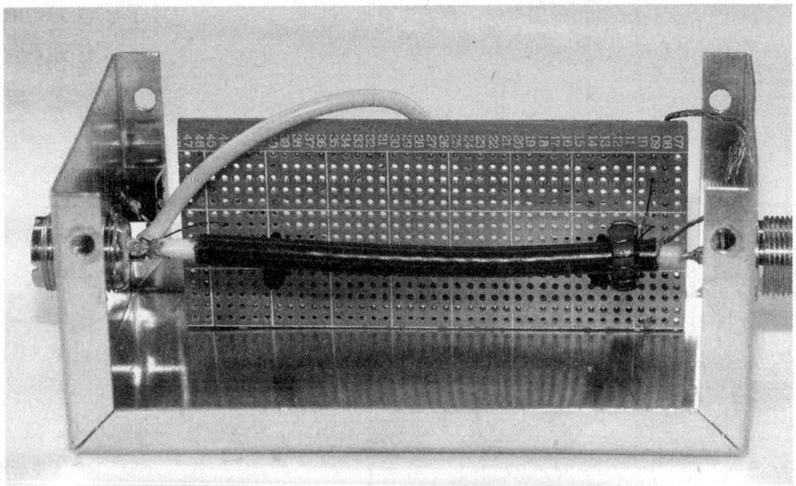

Figure 11.6 — The SWR Sense Head Rear View.

you could as easily build the 100 W version of the sense head for your project.

Now that we have all the parts and pieces ready, we can use the talking SWR meter parts list and the schematic diagram (see **Figure 11.7**) to start breadboarding the Arduino side of things. Whatever you do, don't forget to include the 5.1 V Zener diodes on the SWR sense head inputs to the Arduino. These diodes are there to prevent any SWR signal input voltages from exceeding 5 V and potentially damaging the Arduino's analog inputs.

With the circuit ready on the breadboard, we can start putting together the sketch. While this may seem like a somewhat complicated project, using the flowchart in **Figure 11.8**, we can break it down into bite-size pieces. For me, that is the key to writing Arduino sketches. When you look at some sketches as wholes, they may seem difficult and complex. When you break the sketch down into smaller pieces, you soon realize that it really isn't that difficult. The key is to plan things out ahead of time, using your flowchart like a map to stay on course. You can find the complete sketch and libraries for this project online at **kw5gp.com**.

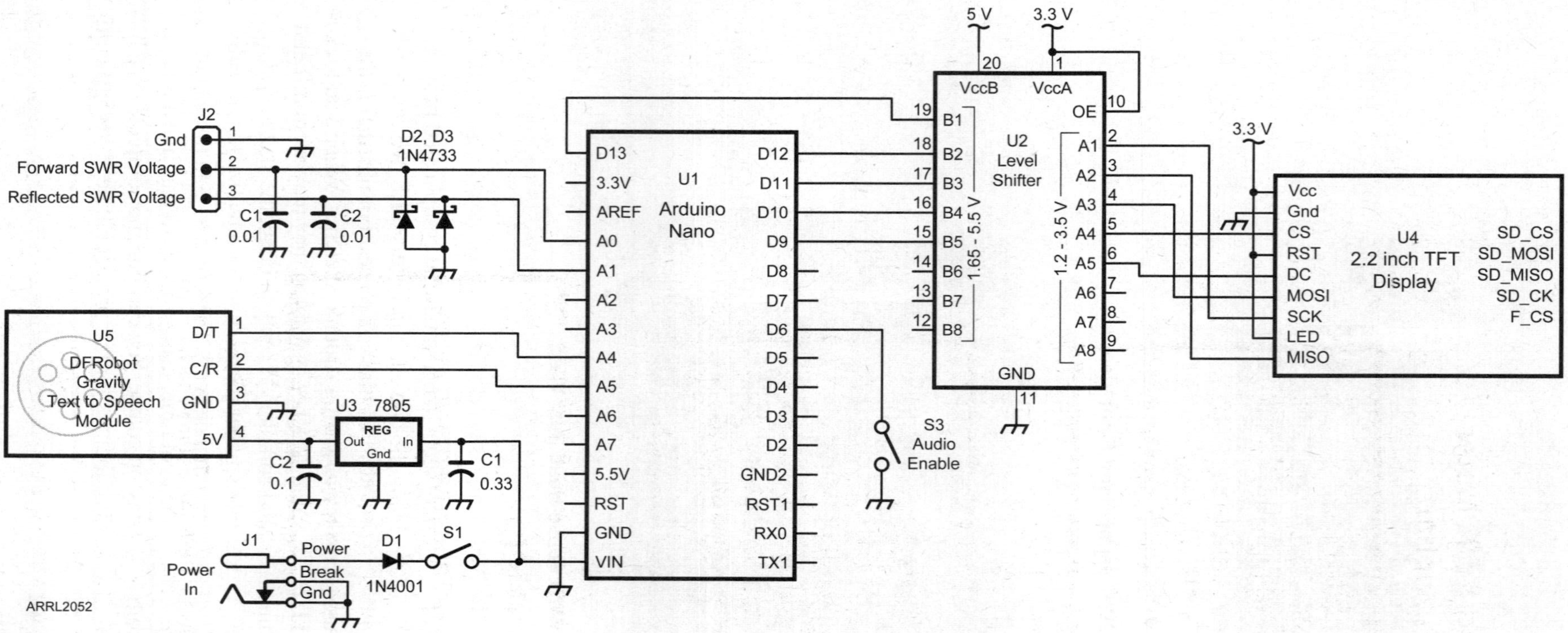

Figure 11.7 — The talking SWR meter schematic.

Part	*Value*
C1, C2	.01 µF capacitor, 35 V
C3	.33 µF capacitor, 35 V
C4	.1 µF capacitor, 35 V
D1	1N4004 diode or equivalent
D2, D3	1N4733 5.1 V Zener diode
J1	dc power jack
J2	3-Pin 2.54 mm Dupont male header
S1, S2	SPST toggle switch

Part	*Value*
U1	Arduino Nano
U2	8-channel level shifter module (TXB0108 type)
U3	LM7805 5 V voltage regulator
U4	2.2-inch ILI9341-type TFT LCD color display
U5	DFRobot Gravity text-to-speech module

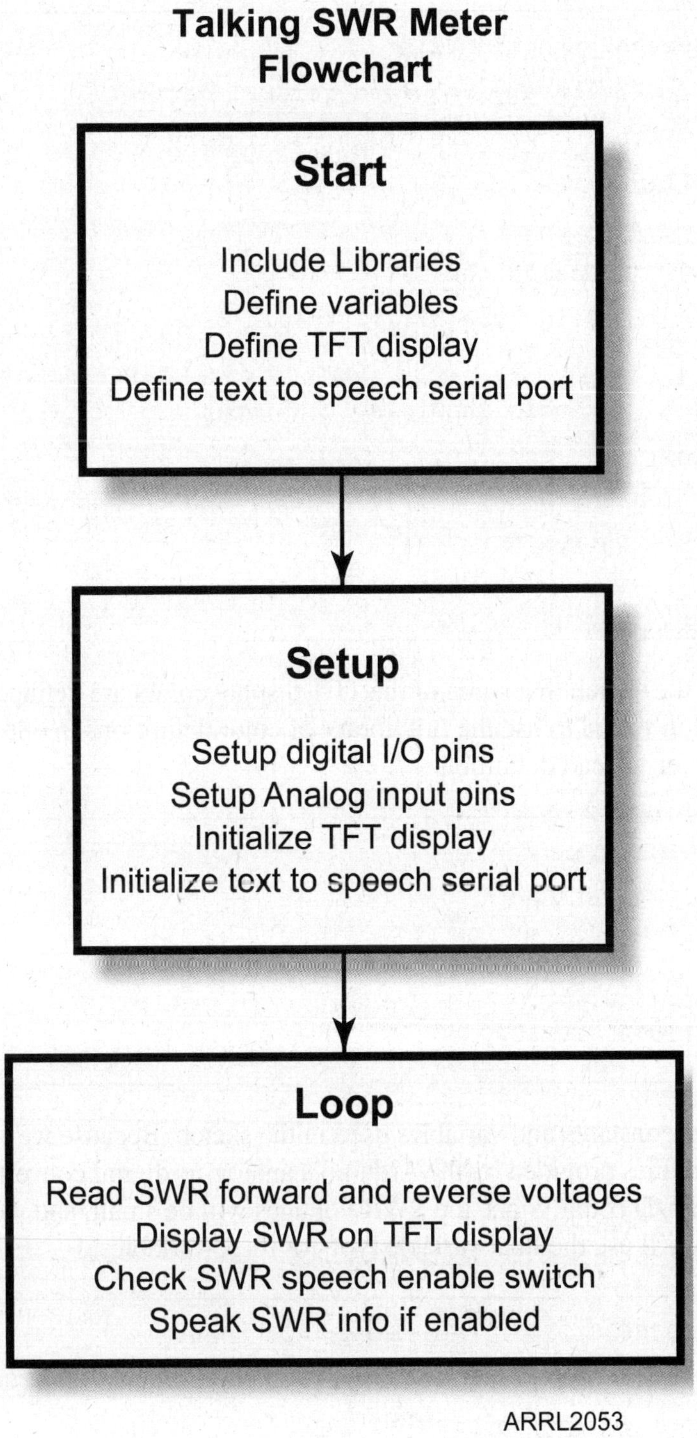

Figure 11.8 — Talking SWR Meter Flowchart.

In the initialization part of the sketch, we start by defining the libraries we'll need to include, assigning the I/O pins, and initializing the library objects:

```
#include "DFRobot_SpeechSynthesis_V2.h"
#include <SPI.h>  // SPI library required for SPI devices
#include <Adafruit_GFX.h>    // Core graphics library
#include <Adafruit_ILI9341.h>

#define TFT_CS      10  // Assign the TFT CS to pin 10
#define TFT_DC      7  // Assign the TFT DC to pin 8

#define audio_enable 6  // Audio Enable switch input
DFRobot_SpeechSynthesis_I2C ss;
// Use hardware SPI (on Uno/Nano, #13, #12, #11) and the above for CS/DC
Adafruit_ILI9341 tft = Adafruit_ILI9341(TFT_CS, TFT_DC);  // Create the TFT in-
stance
```

You'll also notice in the sketch that some of the TFT display colors are defined. This is done as a shortcut method so we don't need to use the full library of color definitions in our sketches. Instead, we use a simpler and easier-to-read definition.

```
#define tft_cyan ILI9341_CYAN
#define tft_black ILI9341_BLACK
#define tft_red ILI9341_RED
#define tft_green ILI9341_GREEN
#define tft_yellow ILI9341_YELLOW
```

Next, we'll define the constants and variables used in the sketch. Because we want to calculate voltage from the digital values provided by the Arduino's analog-to-digital converter, we define the conversion value for one A/D count. Since the SWR voltages will be small, and we'd like them to be as accurate as possible, we'll use the float variable type for these variables.

```
#define adc_count 0.0048828125  // Volts per ADC count
float V_Fwd, V_Ref = 0, V_SWR ; // The variables to contain the SWR values
bool audio_on = false; // Speech enabled flag
bool prev_audio;  // Previous state of speech enabled flag
String msg; // The message to send to the Speech Synthesizer
String s_SWR; // Holds the string representation of V_SWR
int x_SWR;
int (update_time) = 10000;
```

In the **setup()** loop, we start the DFRobot text-to-speech module and set the speech parameters for this sketch

```
ss.begin();  //Init speech synthesis sensor
ss.setVolume(3);  //Set voice volume to 3
ss.setSpeed(7);  //Set playback speed to 7
ss.setSoundType(ss.eFemale1);  //Set speaker to female
ss.setTone(4);  //Set tone to 4
ss.setEnglishPron(ss.eWord);  //For English, speak word
```

Notice that we are also adding the INPUT_PULLUP option to the audio enable switch pin while it is configured as an input. This enables the internal 20 K pull-up resistor on that I/O pin, saving us from using an external pull-up resistor on the audio enable switch pin. Finally, we'll start the TFT display and show a brief startup message so we can tell that things are running.

```
  pinMode(audio_enable, INPUT_PULLUP);  // Setup the Audio enable pin as an in-
put

  // Set the previous audio enable flag to the opposite of what the enable switch
reads
  prev_audio = (digitalRead(audio_enable)); // Read the state of the audio en-
able switch

  start_tft();  // Start the TFT display
```

Now we're ready to start the main **loop()**. First, we'll read the status of the audio enable switch and read the forward and reverse voltages coming in on the Arduino analog input pins.

```
void loop()    // Start the Main Loop
{
  // Read the audio enable switch
  audio_on = !(digitalRead(audio_enable)); // Read the state of the audio enable
switch

#ifdef debug
  Serial.print("Audio Enable: ");
  Serial.print(audio_on);
  Serial.print("  Prev: ");
  Serial.println(prev_audio);
```

```
#endif

  // Read the A/D converters
  V_Fwd = analogRead(0) * adc_count;
  V_Ref = analogRead(1) * adc_count;
```

The values are then converted to their actual voltages and displayed on the TFT display.

```
// Calculate VSWR
  if (V_Fwd > V_Ref)
  {
    V_SWR = (V_Fwd + V_Ref) / (V_Fwd - V_Ref);
  }
  else if (V_Fwd == 0 && V_Ref == 0)
  {
    V_SWR = 1;
  }
  else
  {
    V_SWR = 10;
  }
  if (V_SWR > 10)
  {
    V_SWR = 10;
  }

  background = tft_green;
  if (V_SWR > 2)
  {
    // Turn the background yellow
    background = tft_yellow;
  }
  if (V_SWR > 3)
  {
    // Turn the background yellow
    background = tft_red;
  }

#ifdef debug1
```

```
  Serial.print("V_SWR = ");
  Serial.print(V_SWR);
#endif

  x_SWR = (V_SWR * 10); // Trim the value to 1 decimal place
  V_SWR = float(x_SWR) / 10;

  tft.setTextColor(tft_black, background);
  clear_display();
  tft.setCursor(120, 5); //
  tft.setTextSize(4);  // large font
  tft.print("SWR");
  tft.setTextSize(6);  // large font
  tft.setCursor(70, 60);
  tft.print(V_SWR, 1);
  tft.print(":1");
  tft.setTextSize(3);
  tft.setCursor(20, 140);
  tft.print("Fwd V: ");
  tft.print(V_Fwd, 6);
  tft.setCursor(20, 170); //Set the cursor to the top
  tft.print("Ref V: ");
  tft.print(V_Ref, 6);
```

Then, we'll calculate and display the SWR. As an error check, if the reflected voltage is higher than the forward voltage, the SWR is reported as 0. To protect your equipment, don't transmit without either an antenna or dummy load connected to your radio. Also, you should not transmit when the SWR at the transmitter exceeds 3:1 unless your transmitter has SWR protection and/or an internal tuner and can adjust the SWR match to a safe level. Many modern radios have SWR protection circuits to limit the transmitter power output when high SWR is detected, but that is not a good reason to tempt fate. If your SWR exceeds 3:1, troubleshoot and repair the problem, and do not transmit at full power until you have the SWR within a safe range for your transmitter. For my testing and calibration, I used a Ten-Tec Rebel Model 506 QRP CW transceiver with the output set to 3 W and compared the SWR readings into a 50 Ω dummy load and my GAP Challenger HF vertical antenna with a Diamond SX-600 Dual-Band Power Meter. The SWR readings between the Diamond SX-600 and my homebrew SWR sense head were within 5% of each other on both 20 and 40 meters, and no adjustments or modifications were needed on the SWR sense head.

Now for the fun stuff. We'll check to see if the audio enable switch on digital I/O pin 6 is turned on. If it is, we'll have the DFRobot speech synthesizer speak the SWR for us.

```
    if (audio_on)  // Only speak if Audio is enabled
  {
     if (V_SWR > 3)
     {
       ss.speak("[v3][s7]Warnning"); // SWR is over 3.0 - speak warning (have to
tweak the phonetics)
     }

     ss.speak("[v3][g2][h2][s7]S W R is ");  // Send the desired string to con-
vert to speech
     msg = ("\"[v3][n2][s7]" + s_SWR + "to:1" + "\""); // Prepare the message
     ss.speak(msg);  // Speak the SWR

  }

  delay(update_time);
}
```

Once the circuit and sketch were tested and debugged, the finished project was moved onto an Arduino Protoshield using the schematic diagram in **Figure 11.9**. The completed project was mounted inside a clear Solarbotics Mega S.A.F.E. enclosure to protect the Arduino and internal components from the elements.

You can find the completed sketch and libraries for this project online at **kw5gp.com**.

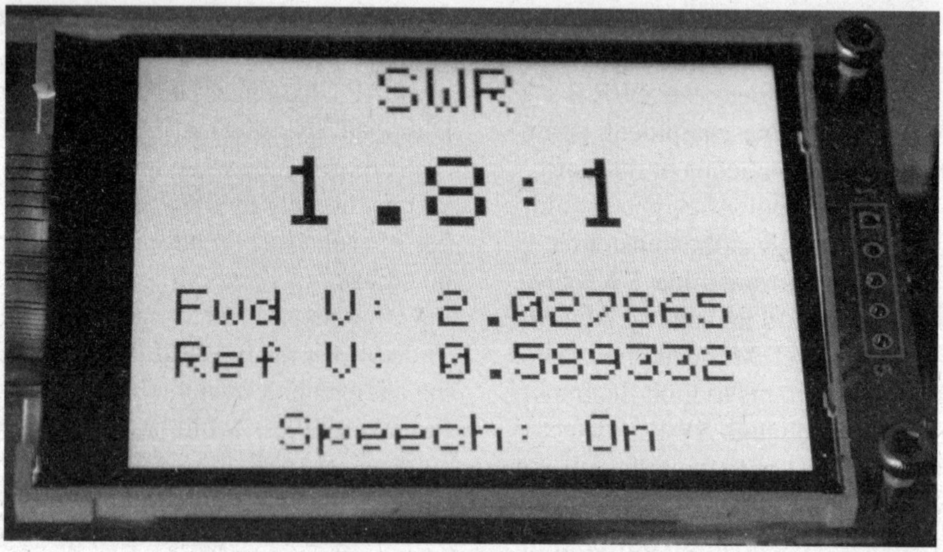

Figure 11.9 — The talking SWR meter TFT display.

12 Antenna SWR Analyzer

Figure 12.1 — The Antenna SWR Analyzer.

Every ham should have an antenna SWR analyzer, shown in **Figure 12.1**, or at least have access to one. Before today's relatively low-cost antenna analyzers existed, measuring the SWR of an antenna required transmitting into it at a low power, watching the needles on an analog SWR meter bounce up and down, and adjusting the antenna over and over until it was tuned. Without access to costly test equipment, that was the best we could do. Needless to say, that was a slow and cumbersome process, especially if the only transmitter you had available was in your shack and the antenna was on the roof.

Today, you can just hook up an antenna analyzer to the antenna and get it tuned in no time at all. There are a wide variety of antenna analyzers available today, from manual analyzers all the way up to computer-driven vector network analyzers (VNAs). VNAs can be used to graph the entire frequency range in question and provide an output graph and detailed data about the antenna under test. Personally, I love my MFJ-269 SWR Analyzer (see **Figure 12.2**), and I don't know how I ever lived without it.

Several of my ham friends have antenna analyzers with a color LCD that can graph the SWR over a selected frequency range, such as the W4RT Electronics miniVNA Pro and the MFJ-223 Color Graphic VNA Antenna Analyzer. The only potential downside to owning an antenna analyzer is the price, with most costing in the $275 – $350 and up range. Much of ham culture relies on the club community's willingness to share and borrow materials like antenna analyzers due to some prices being beyond the average operator's budget.

So, let's stop for a minute and think about what we really need and how we can get it inexpensively and without stealing or borrowing our friend's toys to get the job done. While we're at it, because this is a book about Arduino projects for ham radio, let's do it with an Arduino. First, let's look at what SWR is and why we're so concerned about it.

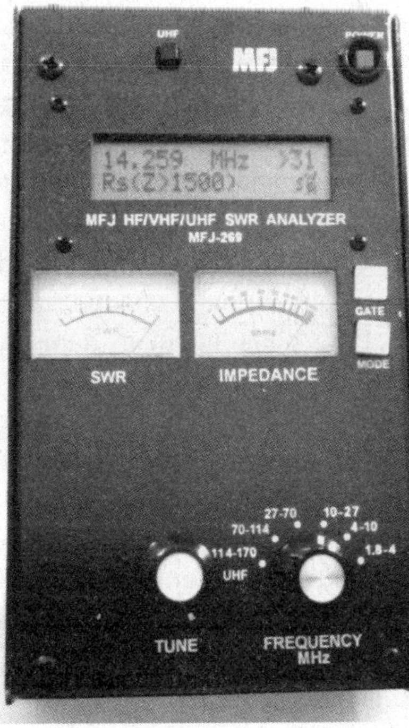

Figure 12.2 — The MFJ-269 Antenna Analyzer.

Standing Wave Ratio (SWR)

Simply put, standing wave ratio (SWR), also known as voltage standing wave ratio (VSWR), is the ratio of forward to reflected power. It is also an indication of an impedance mismatch in your antenna and feed-line system. Most antennas and feed lines use coaxial cable (coax) with an impedance of 50 Ω, although some ladder-line and other feed-line systems have an impedance of 300 – 600 Ω. For the purpose of this discussion, we'll just talk about 50 Ω coax-based antenna systems.

In a 50 Ω antenna system, the goal is to have the impedance of the coaxial cable, antenna, and transceiver all match at 50 Ω. In an ideal world where everything matched perfectly, the entire path would appear to the transmitter as a 50 Ω purely resistive load with an SWR of 1:1, and the transmitted power would be radiated by the antenna. Unfortunately, when we're dealing with long lengths of wire, the antenna, and radio frequencies, the inductance and capacitance in the path create their own form of resistance, which is known as impedance. As the frequency changes in the path, so does the impedance, so we're basically shooting at a moving target. Whereas a 50 Ω resistor dummy load would have the same value regardless of frequency, and the SWR would remain flat at 1:1.

When the impedances of the transmitter, coax, and antenna don't all match, and the transmitted energy hits the point of an impedance mismatch, a portion of that power is reflected to the transmitter. While a high SWR is not a good thing, and we should do all we can to eliminate impedance mismatches and high SWR in our antenna systems, the performance differences between a 1:1 SWR and a 3:1 SWR are virtually unnoticeable over the air. It's at this point that many hams might get confused or start obsessing over SWR for the wrong reasons.

According to the November 2006 *QST* article, "Understanding SWR by Example," by Darrin Walraven, K5DVW, the actual loss from a 3:1 SWR only results in an additional loss of 0.9 dB, resulting in a measurable, but not very significant, loss of radiated power.

A much bigger concern is the issue of the reflected power from an impedance mismatch going back into your transmitter. Impedance mismatch can cause a high SWR, which can then cause high currents or voltages in your transmitter output stage. It's not necessarily the SWR that causes the damage, it's the impedance mismatch resulting from your transmitter transmitting into an improperly tuned load. This can result in heat damage to the transmitter, which is why most modern transceivers will not transmit if the SWR at the transmitter is over 3:1.

Measuring SWR

We should do as much as we can, within reason, to tune our antenna system for the best possible impedance match that also gives us the lowest SWR. There are several ways to read the SWR of your coax and antenna system. Some use an inductive circuit to sense the forward and reverse voltages at the measurement point. This is how many SWR meters operate, which is why they can be used with just about any antenna and feed-line system, because they primarily only show the ratio between the forward and reverse voltages at the measurement point. It is important to note the SWR varies at the transmitter and the antenna, especially when you use an antenna tuner. An antenna tuner does nothing more than adjust the impedance that your transmitter sees; it won't improve an impedance mismatch up at the antenna, which is why we want to take all of our antenna tuning measurements out at the antenna itself.

Another method of measuring SWR is using a resistive SWR bridge, as described in the *ARRL Antenna Book*. Technically speaking, a resistive bridge does not measure SWR, it measures the impedance of the antenna system as compared to a resistive load, in this case, 50 Ω.

Antenna tuning and impedance matching are two different things. With antenna tuning, we're trying to get the actual SWR as low as we can and not necessarily matching the impedance of the antenna system to 50 Ω. With a resistive bridge, we're measuring how close the antenna system impedance is to 50 Ω. While yielding similar results, they actually measure two different, but closely related, things.

The good news is that we can use a resistive bridge to measure the impedance of our antenna system and graph the measured impedance across a selected frequency range, which will give us an output we can use to tune our antenna system.

Design Goals

We want to design a portable resistive antenna impedance bridge that will measure the impedance of our antenna system to allow us to tune it as close to 50 Ω as possible. We'll use a DDS module that will allow us to measure a range of frequencies from 1 to 21 MHz, and we'll graph the output as SWR on a 2.2-inch color graphic TFT display. I originally planned for this project to go up to 30 MHz, but during testing I found that the buffer amplifier circuit was not strong enough to reliably drive a long length of coax out to the antenna and back and still produce usable data, so we'll limit this project to cover 160 to 15 meters. In order to keep things small, we'll use an Arduino Nano and build the project on a piece of copper-clad perfboard and fit it all inside a 124 × 88 × 38-millimeter (4.9 × 3.5 × 1.5-inch) aluminum enclosure. We'll power the project with an external 12 V power source, such as a battery or battery pack, using Powerpole connectors.

I also originally planned to use a larger, 3.2-inch 320 × 240-pixel TFT display, however, that would have required a larger enclosure, so I went with the Adafruit 2.2-inch ILI9340-type TFT display instead. Using the Schottky 1N5711 diodes to detect the forward and reverse SWR is a bit old school. I had fully intended to use the newer Analog Devices AD8307 logarithmic amplifier that I have seen used in other SWR meter designs. The AD8307 does not have the internal 0.4 V forward voltage drop of the 1N5711 Schottky diodes used to sense the forward and reverse voltages, and because it is an operational amplifier, it can't provide the amplification needed for the lower DDS output at higher frequencies. Unfortunately, this is an actual case of "the dog (or in this case, cat) ate my homework." Just as I began to construct the AD8307 version of the project, one of my two parts-thieving cats hid my only pair of AD8307 chips that I had set out on my workbench the previous evening. Only when I was cleaning up a few weeks later did I find the chips, carefully stuffed inside a group of file folders lying on the floor. Of course, by then it was too late to go back and incorporate them into the project, so that version will have to be put aside for a later redesign.

Note that this project is only for 50 Ω antenna systems. You would need to change the resistive bridge resistors for impedances other than 50 Ω.

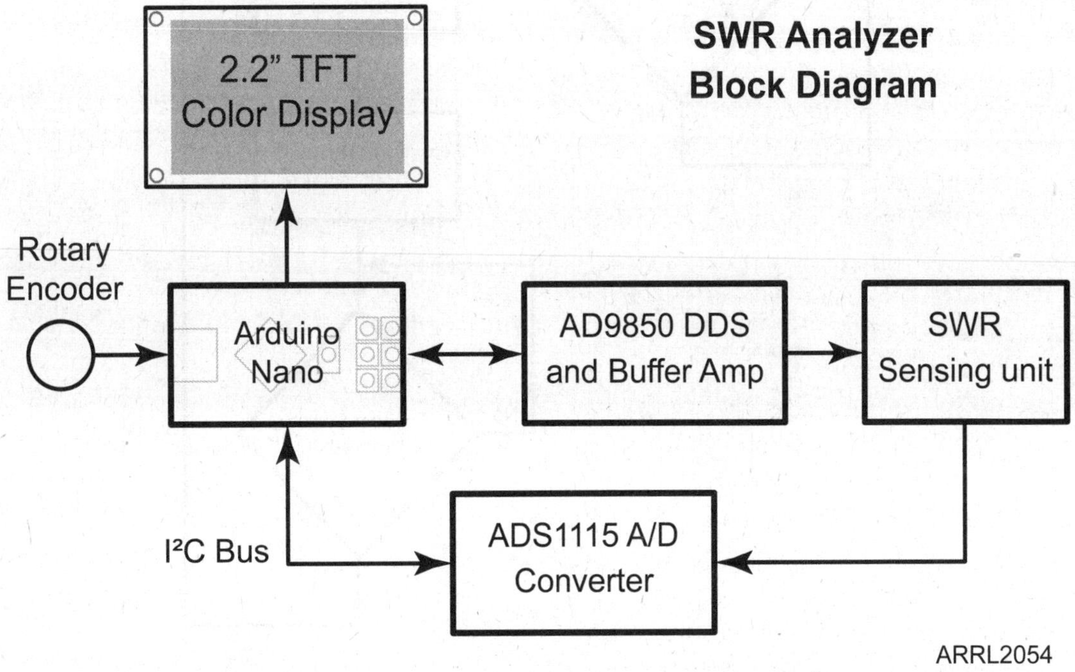

Figure 12.3 — The Antenna SWR Analyzer block diagram.

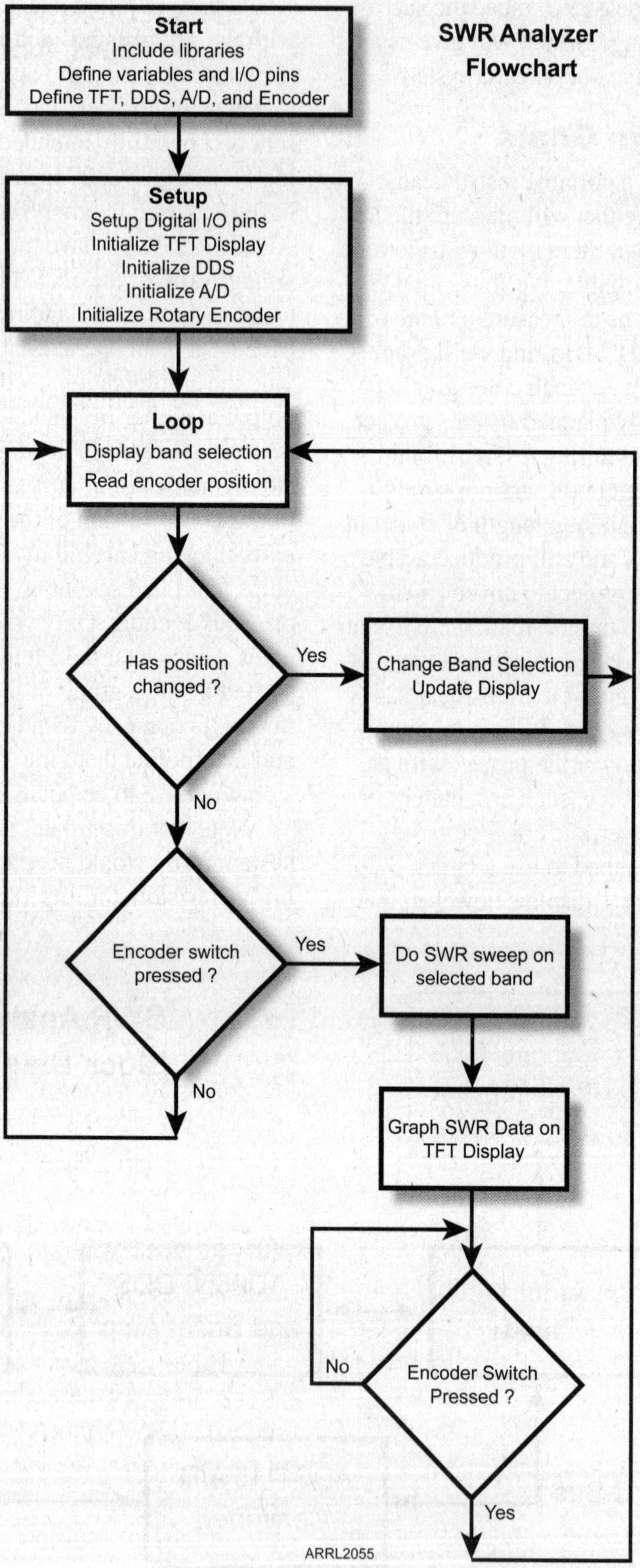

Figure 12.4 — Antenna SWR Analyzer flowchart.

The Antenna SWR Analyzer Block Diagram

Figure 12.3 shows the block diagram for the Antenna SWR Analyzer project. We'll use an Analog Devices AD9850 DDS module to generate the RF frequencies, and a TI ADS1115 four-channel programmable gain analog-to-digital converter. We'll use a rotary encoder with a built-in pushbutton switch to select the desired frequency range and to start the DDS module sweeping the desired frequency range. We'll display the results in an SWR graph on the color TFT display.

The Antenna SWR Analyzer Flowchart

The flowchart for the Antenna SWR Analyzer project is shown in **Figure 12.4**. We'll be using a lot of libraries in this sketch to support the Adafruit ILI9340 2.2-inch color TFT display, the TI ADS1115 A/D module, the AD9850 DDS module, and the rotary encoder. We'll start by including all of those libraries, and then we'll define the I/O pins used by the project. Next, we'll declare and initialize the variables we'll be using in the sketch. Finally, we'll create (instantiate) the color TFT display, DDS, A/D, and rotary encoder objects.

In the `setup()` loop, we'll set the pin modes for the I/O pins on the Arduino Nano that we'll need. Next, we'll initialize the TFT, DDS, A/D, and rotary encoders.

In the `main loop()`, we'll display the currently selected band. Because this project is designed for ham radio use, we'll select the frequency range by HF amateur band. As the rotary encoder is turned, we'll update the band selection settings and the TFT display.

When the pushbutton switch on the rotary encoder is pressed, we'll sweep 100 frequencies across the frequency range, read and calculate the SWR for each data point, and save the data into an array. Once the scan is complete, we'll plot the SWR graph on the TFT display. The data will be retained on the display until the rotary encoder is pressed a second time, and we're ready to select and scan another frequency range.

Constructing the Antenna SWR Analyzer

Figure 12.5, which shows the schematic design for the Antenna SWR Analyzer project, also includes the parts list in the caption. It is important to note that this project is designed without a level converter module for the TFT display, and I used the Adafruit 2.2-inch ILI9340-type TFT display because that module is designed to work on either a 3 or 5 V signal without a level converter. This was done to save space inside the enclosure. Unfortunately, the Adafruit 2.2-inch ILI9340 module is no longer available, so you'll need to upgrade to either a different ILI9340 or ILI9341-type display (available from eBay and other parts suppliers). However, be sure to incorporate a level converter module or voltage dropping resistors, with those displays. We'll be using the same AD826 instrumentation amplifier and transistor buffer amplifier that we used in the 1-30 MHz DDS VFO project. The project is built on a 2 × 4-inch piece of copper-clad perfboard that is mounted inside the aluminum enclosure, as shown in **Figure 12.6**. The Arduino Nano, the ADS1115 A/D, and ADS9850 DDS module are mounted in sockets made from two strips of DuPont-style 2.54-millimeter female header strips.

You'll notice that there is a heat sink on the LM7805 voltage regulator and on the AD9850 chip on the DDS module. I discovered that when the aluminum enclosure was closed, there was no ventilation and both components got warm, so I decided to err on the side of caution and use heat sinks.

Figure 12.7 shows the completed Antenna SWR Analyzer project. DuPont-style 2.54-millimeter header pins were used for the connections between the circuit board and the TFT display, and the rotary encoder was mounted in the top cover of the enclosure. The power switch, Powerpole dc power connector, and SO-239 connector for the antenna are mounted in the removable side plates of the enclosure.

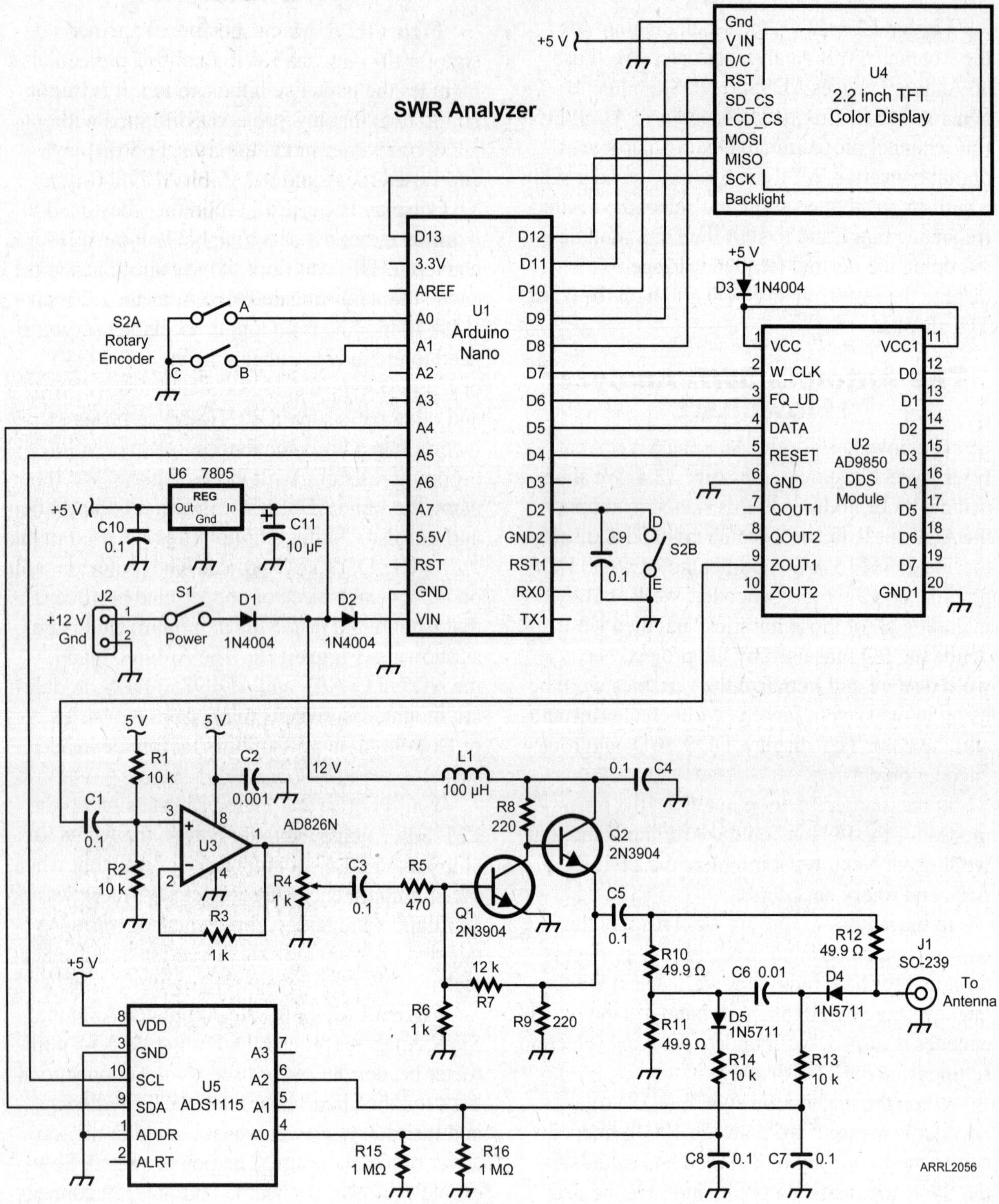

Figure 12.5 — Antenna SWR Analyzer schematic.

SWR Analyzer Parts List

Part	*Value*
C1, C3 – C5	0.1 µF ceramic capacitor, 50 V
C2	0.001 µF ceramic capacitor, 50 V
C6	.01 µF capacitor
C7 – C10	0.1µF ceramic capacitor, 50 V
C11	10 µF electrolytic capacitor, 35 V
D1 – D3	1N4004 diode
D4, D5	1N5711 Schottky diode
J1	SO-239 female coaxial connector
J2	Powerpole connectors
L1	100 µH inductor
Q1, Q2	2N3904 transistor
R1, R2, R13, R14	10 kΩ resistor, 1/8 W
R3, R6	1 kΩ resistor, 1/8 W
R4	1 kΩ potentiometer
R5	470 Ω resistor, 1/8 W
R7	12 kΩ resistor, 1/8 W
R8, R9	220 Ω resistor, 1/8 W
R10 – R12	49.9 Ω resistor, 1/8 W
R15, R16	1 MΩ resistor, 1/8 W
S1	SPST switch
S2	Rotary encoder with pushbutton switch
U1	Arduino Nano
U2	AD9850 DDS module
U3	AD826N high-speed dual-operational amplifier
U4	2.2-inch SPI color TFT LCD display
	(Adafruit Product ID: 1480 or similar. See text for more information.)
U5	ADS1115 four-channel 16-bit A/D converter module
U6	LM7805 5 V voltage regulator
Enclosure	Black aluminum project enclosure 124 × 88 × 38 millimeters

The Antenna SWR Analyzer Sketch

In the sketch for the Antenna SWR Analyzer, we'll first include all the libraries we'll need. Note that we've upgraded to the Adafruit ADS1X15 library for the ADS1115 from the previous projects. This library supports the 12-bit ADS1015 and the 16-bit ADS1115. The `Adafruit_ILI9340.h` library is the library that supports the 2.2-inch ILI9340-type TFT displays. The sketch and libraries for this project can be found at **kw5gp.com**.

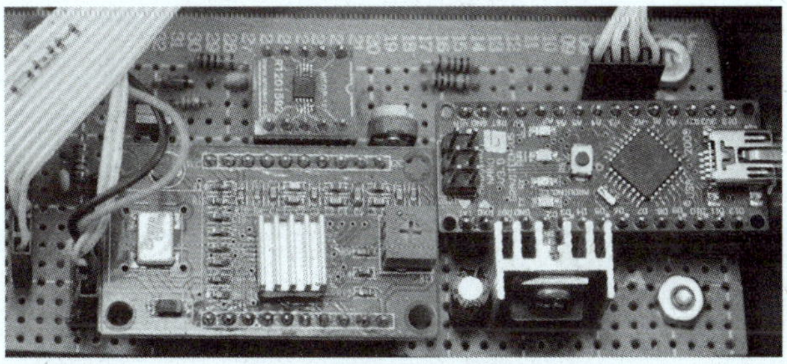

Figure 12.6 — The Antenna SWR Analyzer board.

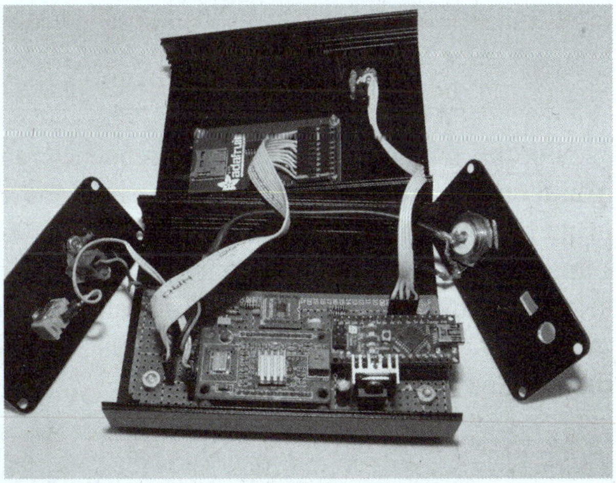

Figure 12.7 — The inside view of the Antenna SWR Analyzer.

```
#include <SPI.h>  // Include the Build-in SPI library
#include <Wire.h>  // Include the I2C Communication Library
#include <Adafruit_ADS1X15.h> // Include the ADS1015 16 bit A/D Library
#include <Adafruit_GFX.h>    // Include the Core graphics library
#include <Adafruit_ILI9340.h> // Include the Hardware-specific 2.2" TFT Library
#include <RotaryEncoder.h>  // Include the Rotary Encoder Library
#include <DDS.h>  // Include the M0XPD DDS Library
```

Next, we'll define the I/O pins that we'll need for this project. Because the color TFT display is an SPI bus device, we'll define it to use the SPI bus and its associated I/O pins.

```
// Define the AD9850 Module pin numbers
#define W_CLK  7
#define FQ_UD  6
#define DATA  5
#define RESET  4
// Define the TFT Hardware SPI pins for the UNO/Nano
#define _sclk 13
#define _miso 12
#define _mosi 11
#define _cs 10
#define _dc 9
#define _rst 8
```

For this project, we'll select the frequency range by amateur radio bands from 160 –15 meters. You'll note that we don't include any bands above 15 meters. The output of the DDS buffer amplifier is not high enough to provide reliable data over a long length of coax out to the antenna and back, so I limited the range to 160 – 15 meters. The band selection data is placed into arrays, allowing us to select the band and the associated high and low frequency ranges by rotating the rotary encoder, and indexing through the array variables. This also allows you to easily add your own ranges by changing the size and elements of the array.

```
#define select_array_size 8  // Define the size of the Band Selection array
#define swr_array_size 100  // Define the size of the SWR Data array
```

The number of data points scanned is also defined as an array. I chose to use 100 data points across the selected frequency range, but because the TFT display is a 320 × 240-pixel display, you can increase the number of data points as desired. This will result in a smoother output display, but the time needed to scan the frequency range will also increase. I chose 100 data points because it produced a usable display without sacrificing time to scan the band.

Next, we'll define the rotary encoder I/O pins, along with the pins for the rotary encoder's push-button switch. Notice that we use I/O pin numbers 14 and 15 for the rotary encoder's digital I/O pins, but there are no matching pin numbers on the Nano, and the schematic shows that the rotary encoder is connected to the Nano's analog input pins A0 and A1. When using an Arduino analog input as a digital I/O pin, the pins are actually numbered, starting at 14 for A0, 15 for A1, and so on.

```
//Define the Rotary Encoder pins
const int encoder_PinA = 14;  // Pin A of the Rotary Encoder (A0 on Nano)
const int encoder_PinB = 15;  // Pin B of the Rotary Encoder (A1 on Nano)
const int encoder_Switch_Pin = 2;  // Encoder Pushbutton Switch input pin (Pin 2)
```

In the next part of the initialization area, we'll declare and initialize all the variables used in the sketch, and in the last part of the initialization area, we'll create (instantiate) the TFT display, DDS, rotary encoder, and A/D objects.

```
int encoder_Pos = 0; // The current encoder position
int old_encoder_Pos = 0; // The previous encoder position
boolean pushbutton; // Flag to indicate encoder pushbutton was pressed
boolean encoder_change; // Flag to indicate the rotary encoder position has
changed
const String band[select_array_size] = { "160m", "80m", "60m", "40m", "30m",
"20m", "17m", "15m" }; // The Band Selection Menu array
const float start_freq[select_array_size] = { 1.8, 3.525, 5.3305, 7, 10.1, 14,
18.068, 21 }; // The Band Starting point data array
const float end_freq[select_array_size] = { 2, 4, 5.4035, 7.3, 10.15, 14.35,
18.168, 21.45 }; // The Band Ending point data array
byte SWR_array[swr_array_size];  // Initialize the SWR data array
float freq_step;  // Variable to hold the calculated frequency step size
int freq_select = 0;  // Variable to contain the band menu selection array index
pointer
int index;  // Variable to hold the current SWR array index pointer
int last_y; // Variable to contain the previous graph Y value for line starting
point
int current_y;  // variable to contain the graph Y value for line ending point
int last_x; // Variable to contain the previous graph X value for line starting
point
int current_x;  // variable to contain the graph X value for line ending point
double dds_freq = start_freq[freq_select];  // Variable to contain the DDS fre-
quency
double sweep_start, sweep_stop; // Variables to contain the frequency sweep
starting and ending point
float SWR_fwd, SWR_rev;  // Variables to contain the values read from the A/D
const int delay_time = 40;  // The A/D settling delay
float SWR; // Variable to hold SWR calculations
```

```
Adafruit_ILI9340 tft = Adafruit_ILI9340(_cs, _dc, _rst);  // Instantiate the TFT display object

DDS dds(W_CLK, FQ_UD, DATA, RESET); // Instantiate the DDS object

RotaryEncoder Enc(encoder_PinA, encoder_PinB);  // Instantiate the Rotary Encoder object

Adafruit_ADS1115 ads; // Instantiate the 16-bit A/D object
```

In the `setup()` loop, we'll start setting up the pin modes for the rotary encoder and enabling the pullup resistors on the rotary encoder I/O pins. Notice that this sketch uses the `INPUT_PULLUP` option instead of previously configuring the I/O pin as an input, and then uses a `digitalWrite()` to the input pin to enable the internal pullup resistor. This does it all in one step and is the newer, preferred method of enabling the internal pullup resistors on a digital input pin.

```
// Set up the Rotary Encoder and enable the Internal Pull-up resistor on the Encoder Inputs
  pinMode (encoder_PinA, INPUT_PULLUP);
  pinMode (encoder_PinB, INPUT_PULLUP);
  pinMode (encoder_Switch_Pin, INPUT_PULLUP);
```

Next, we'll attach an interrupt to the rotary encoder's pushbutton switch that will call the `read_pushbutton()` function when the pushbutton switch on the rotary encoder is pressed. Let's stop and take a look

```
//  attachInterrupt(digitalPinToInterrupt(2), read_pushbutton, LOW);
```

at the `read_pushbutton()` interrupt setup and the function that is executed when an interrupt occurs.

```
  attachInterrupt(digitalPinToInterrupt(2), read_pushbutton, LOW);  // Set up Interrupt for the Rotary Encoder pushbutton switch
```

The `read_pushbutton()` function will set the pushbutton flag that we'll use elsewhere in the sketch to let us know the pushbutton switch has been pressed.

```
// Interrupt Function to read the rotary encoder pushbutton switch
// Function to read the rotary encoder pushbutton switch
void read_pushbutton()
{
```

```
  if (digitalRead(encoder_Switch_Pin) == false)
  {
    pushbutton = true;   // Set the encoder pushbutton pressed flag

#ifdef debug
      Serial.println("Button Pushed");
#endif

  }
}
```

Returning to the `setup()` loop, we'll initialize the color TFT display, rotate the display 90° (because the display is actually mounted sideways), then we'll clear the TFT display by filling it with the color blue. Note that when using a TFT display you can't just erase the previous screen contents by overwriting it. You have to erase the area first by filling that section of the screen with the background color.

Next, we'll set the A/D gain to the 4× setting, which will read 0 to 1.024 V, with each A/D count equating to 0.03125 mV.

```
  ads.setGain(GAIN_FOUR); // Set the A/D gain to 4 (1.024V Range  1 bit =
0.03125mV)
```

Now, we'll start the AD9850 DDS module, along with the ADS1115 A/D module. We'll display a brief startup screen on the TFT display, and then continue on to the `main loop()`.

We start in the main loop by printing a screen display template and then filling in the data from the currently selected band by calling the `update_selection()` function. The `update_selection()` function will erase the lower portion of the TFT display screen, print the currently selected band, and then call the `update_common()` function to extract and display the starting and ending frequencies from the array variables. Returning to the `main loop()`, we'll turn off the pushbutton flag and set the index variable to the starting data point for the SWR data.

Next, we'll use a `Do` loop to repeatedly call the `read_encoder()` function to read the current position of the rotary encoder. The `read_encoder()` function will check to see if the rotary encoder position has changed, and we'll update the band array variable index to indicate we're selecting a different band. We let the rest of the sketch know that the rotary encoder position has changed by setting the `encoder_change` flag.

Back in the `main loop()`, we'll check to see if the rotary encoder position has changed, and if it has, we'll update the display with the newly selected band information. We'll repeat this `Do` loop continuously until the rotary encoder pushbutton switch is pressed and the `pushbutton` flag is set.

When the desired band has been selected and the rotary encoder pushbutton switch has been pressed, we'll start scanning the selected frequency range. First, we'll calculate the frequency step to cover the selected band based on the number of data points that we're collecting.

```
// The Band has been selected - start the frequency sweep
freq_step = ((end_freq[freq_select] - start_freq[freq_select]) / swr_array_
size) * 1000000;  // Calculate the frequency step value
```

Next, we'll clear the pushbutton flag, set the DDS to the starting frequency, and then set the scan starting and ending frequencies. Because the output of the transistor buffer amplifier on the DDS is not linear over frequency, we adjust the gain setting on the ADS1115 A/D based on the band we're scanning.

```
// Set the A/D gain based on selected band
if (dds_freq > 12000000)
{
  ads.setGain(GAIN_SIXTEEN);  // Set the A/D gain to 16 (0.256V Range  1 bit =
0.0078125mv) for fequencies above 12 MHz
}
else if (dds_freq > 7000000)
{
  ads.setGain(GAIN_EIGHT);  // Set the A/D gain to 8 (0.512V Range  1 bit =
0.015625mv) for fequencies between 7 MHz and 12 MHz
} else
{
  ads.setGain(GAIN_FOUR);  // Set the A/D gain to 4 (1.024V Range  1 bit =
0.03125mv) for fequencies below 7 MHz
}
```

For frequencies above 16 MHz, we set the A/D to the highest gain of 16×, which will read 0 to 256 mV with each A/D count equating to 0.0078125 mV. For frequencies between 7 and 12 MHz, we set the A/D gain to 8×, which will read 0 to 512 mV with each A/D count equating to 0.015625 mV. For frequencies below 7 MHz, we set the A/D gain to 4×, which will read 0 to 1.024V with each A/D count equating to 0.03125 mV.

Now we'll start scanning the selected band. Using a `Do` loop, we'll call the `read_adc()` function for each frequency step within the band. The `read_adc()` function will briefly delay to allow the A/D time to settle, then read the forward and reverse voltages coming from the SWR sensor three times each, then average each group of the three readings.

Next, we'll calculate the SWR based on the forward and reverse voltages. The formula for calculating SWR fails if the reverse SWR voltage is greater than the forward SWR voltage, so we'll force any such values to the highest SWR we use in this sketch, which is 5 to 1. For our calculations, we multiply the calculated SWR by 10 so we can save it as an integer variable to save space in the array but be able to display the SWR in tenths of a unit.

If we're not already at the maximum number of data points, we'll store the SWR value into the array variable and get ready for the next data point by incrementing the array index and changing the DDS to the next frequency step. We'll repeat the scanning loop until we reach the ending scan frequency.

Now that we have all the SWR scan values in the array, we can graph the SWR over the selected band. We'll start by clearing the TFT display and drawing the display template.

```
  // Prepare to graph the scan data
  last_x = 200 - (SWR_array[0] * 4);  // We need to store the first datapoint so
we can calculate the line starting point
  // Clear the TFT display and display the SWR graph template
  tft.fillScreen(ILI9340_BLUE);
  tft.drawFastVLine(150, 0, 200, ILI9340_BLACK);  // Draw a vertical line at the
center frequency
  tft.drawFastHLine(0, 200, 300, ILI9340_BLACK);  // Draw a horizontal line at
the SWR 1:1 point
  tft.drawFastHLine(0, 175, 300, ILI9340_BLACK);  // Draw a horizontal line at
the SWR 1.5:1 point
  tft.drawFastHLine(0, 150, 300, ILI9340_BLACK);  // Draw a horizontal line at
the SWR 2:1 point
  tft.drawFastHLine(0, 100, 300, ILI9340_BLACK);  // Draw a horizontal line at
the SWR 3:1 point
  tft.drawFastHLine(0, 50, 300, ILI9340_BLACK);  // Draw a horizontal line at
the SWR 4:1 point
  // Display the SWR value for the template lines
  tft.setTextSize(0);
  tft.setCursor(301, 0);
  tft.print(">5");
  tft.setCursor(301, 47);
  tft.print("4");
  tft.setCursor(301, 97);
  tft.print("3");
  tft.setCursor(301, 147);
  tft.print("2");
  tft.setCursor(301, 172);
  tft.print("1.5");
  tft.setCursor(301, 197);
  tft.print("1");
  tft.setCursor(0, 205);
  // Display the band starting and ending frequencies below the graph
  tft.print(start_freq[freq_select], 4);
  tft.setCursor(270, 205);
  tft.print(end_freq[freq_select], 4);
```

Next, we'll graph the SWR data from the array. The `Adafruit_GFX` library includes all the functions we'll need to draw lines, circles, and other graphic objects. For each of the data points, we'll draw a line between the two points until we have drawn the entire graph.

```
    // Figure out the line endpoints - TFT screen is x= 0 to 300, y= 0 to 200
    current_x = 200 - (((SWR_array[x] - 10) * 5)    ); // Calculate the X value
of the current line ending point
    last_x = 200 - (((SWR_array[x - 1] - 10 ) * 5)); // Calculate the X value
of the current line starting point
    if (swr_array_size == 100)
    {
      last_y = (x * 3) - 3; // Calculate the Y value of the current line start-
ing point (for 100 datapoints in the SWR array)
      current_y = x * 3;  // Calculate the X value of the current line ending
point (for 100 dataponts in the SWR array)
    } else {
      last_y = (x  - 1); // Calculate the Y value of the current line starting
point (for 100 datapoints in the SWR array)
      current_y = x;  // Calculate the X value of the current line ending point
(for 100 dataponts in the SWR array)
    }
    // Draw the graph line for the current data points
    tft.drawLine(last_y, last_x, current_y, current_x, ILI9340_GREEN); // Note
the X and Y values are transposed due to screen rotation
  }
```

After we have drawn the SWR graph, we'll wait until the rotary encoder pushbutton is pressed. When the pushbutton is pressed, we'll clear the display and get ready to scan the next band selection.

Figure 12.8 shows the output of the Antenna SWR Analyzer connected to my Mosley TA-33 beam on 20 meters. I compared the graph to the readings from my MFJ-269 SWR Antenna Analyzer, and both read virtually the same.

The sketch and libraries for this project can be found at **kw5gp.com**.

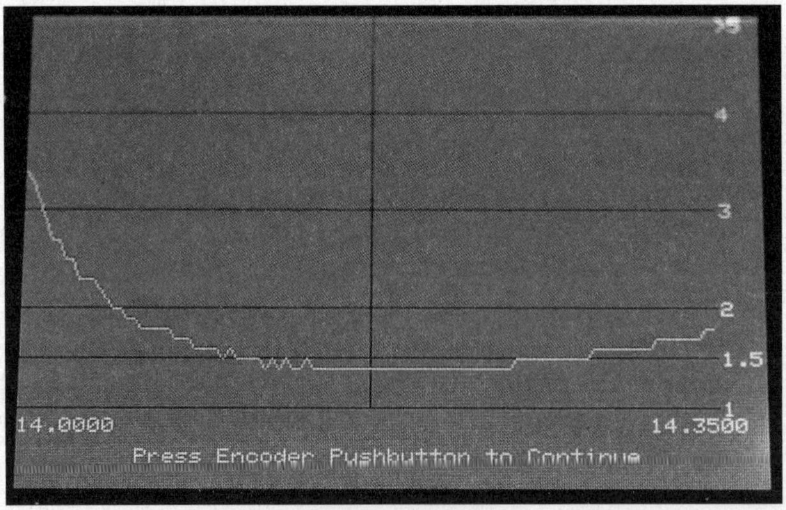

Figure 12.8 — Output display of the Antenna SWR Analyzer.

13 Yaesu G-450A/G800SA Rotator Controller Rebuild

Figure 13.1 — The Yaesu G450A rotator controller.

For the longest time, I used a GAP Challenger vertical for HF, and had my 6-meter, 2-meter, and 440-centimeter beams on a short mast on the roof. Naturally, I needed a light duty rotator for the beams, so I picked up a Yaesu G-450A rotator and controller (see **Figure 13.1**) at a hamfest. Since then, my antenna farm has taken root, and I added a Mosley TA-33 HF beam to the stack and moved it all to a tower. At a different hamfest, I was able to pick up a Yaesu G-800SA rotator and controller. My G-450A is now relegated to ARRL Field Day, or other portable operations where a trailer-mounted tower is used for the beams.

Unlike the CDE/Hy-Gain rotator controller, the Yaesu G-450A/G-800SA rotator controllers are a bit more complex on the inside. Instead of a meter to indicate the antenna position, the Yaesu rotator controllers use a dial indicator driven by a small motor inside the controller box. The control board on the Yaesu controllers do an electronic balancing act between the main antenna rotator position indicator and the dial indicator motor to keep the two in synchronization.

However, over time, both my G-450A, and one of my two G-800SA rotator controllers, developed what I call the "creeping crud." As time went on, I began to have calibration issues with the 450° calibration setting, as the dial indicator could just barely be adjusted to the 450° setting. This gradually became worse, and eventually I was no longer able to calibrate the rotator controller beyond even the 360° point, rendering it useless. I measured the 500 Ω position sensor on the rotator itself, checked the rotator control cable, and checked the rotator control box main board as best I could. I checked every resistor with an ohmmeter, and I even replaced the op amp chips on what appears to be a simple control board. I did everything I knew to try to fix the problem without any success whatsoever. By swapping the board from my known-good G-800SA controller box, I was able to positively verify that the problem was in the main control board. Since you can't order a replacement board from Yaesu, I decided to try to build an Arduino-powered replacement for the rotator controller main board. I ended up adding the ability to

Figure 13.2 — The Yaesu G450A/G800SA/G1000SA original main control board.

interface with *Ham Radio Deluxe* and to control the rotator from a computer.

The original control board inside the Yaesu G-450A is the exact same board used in the G-800SA, and the G-1000SA. There are some minor differences between the G-450A and the G-800/1000SA, but the same board (see **Figure 13.2**) is designed to work with all three rotator controller models. What's nice about this board is that everything inside the controller box connects to it using 2.54 mm spaced headers, so we can design our replacement board to be a plug-in replacement for the existing board. The rebuilt control board for the G-800SA should also be an exact replacement for the G-1000SA main control board, but since I don't have a GS-1000SA rotator controller box to test with, we'll only specifically cover the G-450A and the GS-800SA control board rebuild.

Design Goals

We'll replace the original G-450A/G-800SA rotator controller main board with a plug-in Arduino-powered one. We'll use a STMicroelectronics L298 Dual Full-Bridge Driver chip to control the positioning motor for the dial display on the rotator controller box. We'll use an ADS1115 programmable gain I^2C analog to digital converter (A/D) to read the position sensors for both the antenna rotator and the controller box's dial indicator motor. Since the Yaesu rotators can rotate from 0 to 450°, we'll also want to be able to control the **OVERLAP** LED on the rotator controller front panel. We'll also add computer-control capability, so that we can use software such as *Ham Radio Deluxe* to remotely control the antenna position. In order to do this, we'll use the USB port on the Arduino to emulate the Yaesu GS-232A rotator controller interface. To fit everything on a board that will physically replace the existing main control board, we'll use an Arduino Nano for this project. Since the G-450A and the G-800SA use the same main control board, with a minor design and sketch change, we can also use the new control board in my bad G-800SA. We'll use the Nano's onboard EEPROM to save our calibration settings so that they will be remembered after a reset or power cycle.

The STMicroelectronics L298N Dual Full-Bridge Driver

For this project, we'll be using the STMicroelectronics L298N Dual Full-Bridge Driver (**see Figure 13.3**) to drive the dial indicator position motor in the rotator controller box. Also known as an H-Bridge driver, the L298N can be used to drive the dial position indicator motor in both directions. While capable of driving two bi-directional motors from the same chip, we'll only be driving one motor for this project. The L298N can control DC motors up to 50 V and 2 A, so it's more than adequate to drive the small DC motor attached to the dial position indicator.

The Yaesu G-450A/G-800SA Rotator Controller Rebuild Block Diagram

Figure 13.4 shows the block diagram for the rebuilt main control board. Since there are only very minor electrical differences between the G-450A and G-800SA rotator controllers, the same block diagram applies to both. The connector designations are the ones as on the existing main control board, so we can remove the original control board and then drop the Arduino-powered replacement in the same place, using the existing cables and connectors. The only change we will need to make is replacing the larger 2-pin 18 V ac power connector with a 2-pin

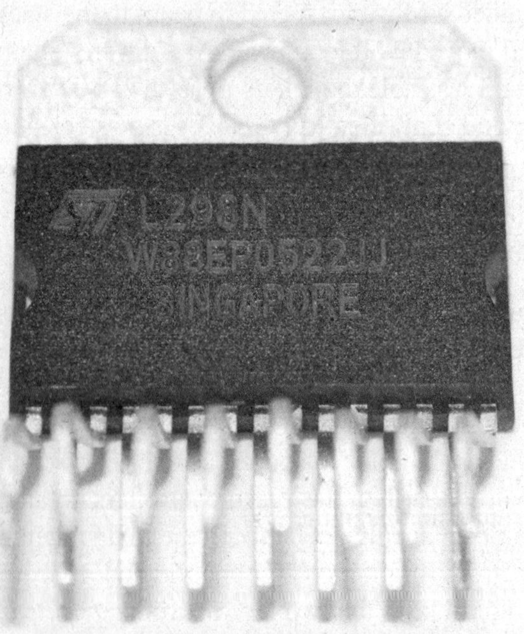

Figure 13.3 — The L298N dual full-bridge driver.

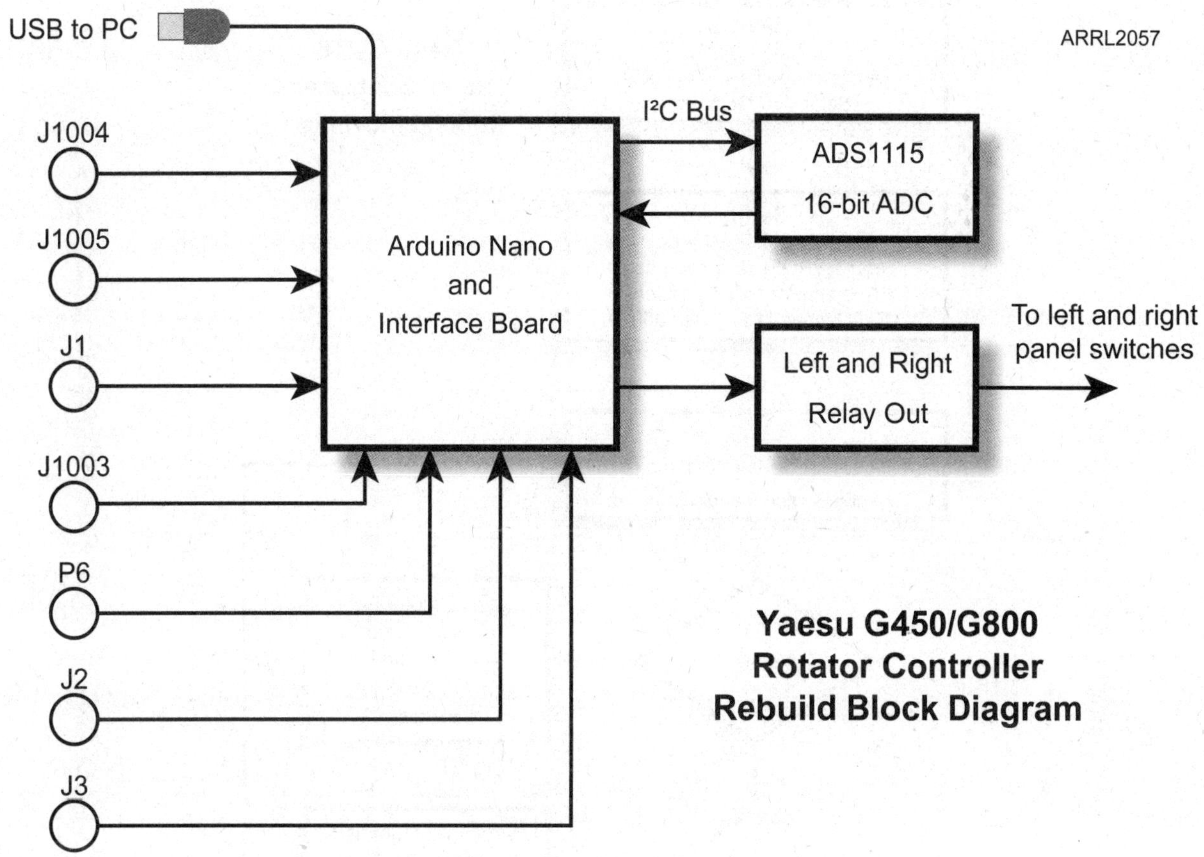

Figure 13.4 — Yaesu rotator controller rebuild block diagram.

Dupont-style header socket. I did this by simply splicing the new cable connector to the old 18 V ac lines and leaving the original connector intact in case I ever wanted to use the original main board again. The antenna rotator position sensor and the dial indicator motor position sensor will be read using two channels of the ADS1115 4-channel A/D converter. Relays will be connected in parallel with the controller box's front panel switches to control the **LEFT** (CCW) and **RIGHT** (CW) rotation.

We'll use the existing 18 V ac power supply in the controller box to generate the +5 V and +12 V dc that we need to power the replacement board and relays.

Figure 13.5 shows the flowchart for the rebuilt controller board. Once again, since the differences between the G-450A and the G-800SA controllers are very minor, the same flowchart applies to both.

We'll start by including the Adafruit_ADS_1X15 library for the ADS1115 A/D module, and the EEPROM library to allow us to save our calibration settings. We'll then define the I/O pins needed, and then declare and initialize the variables needed in the sketch. Then we'll create (instantiate) the A/D object.

In the `setup()` loop, we'll set the pin modes for the Arduino's I/O pins used in this project, and initialize the A/D module. Next, we'll read the calibration data from the EEPROM, then the current antenna rotator position, along with the current dial indicator position. In the last part of the `setup()` loop, we'll turn on the **OVERLAP** LED if the current position is over 360°.

In the `main loop()`, we'll first check to see if we have received a command on the Arduino's USB port, or if one of the rotator controller box's front panel rotate switches has been pressed. If we have received a rotate command, either from the USB serial port or the front panel switches, we'll energize the correct rotator motor relay, recalculate the current position while

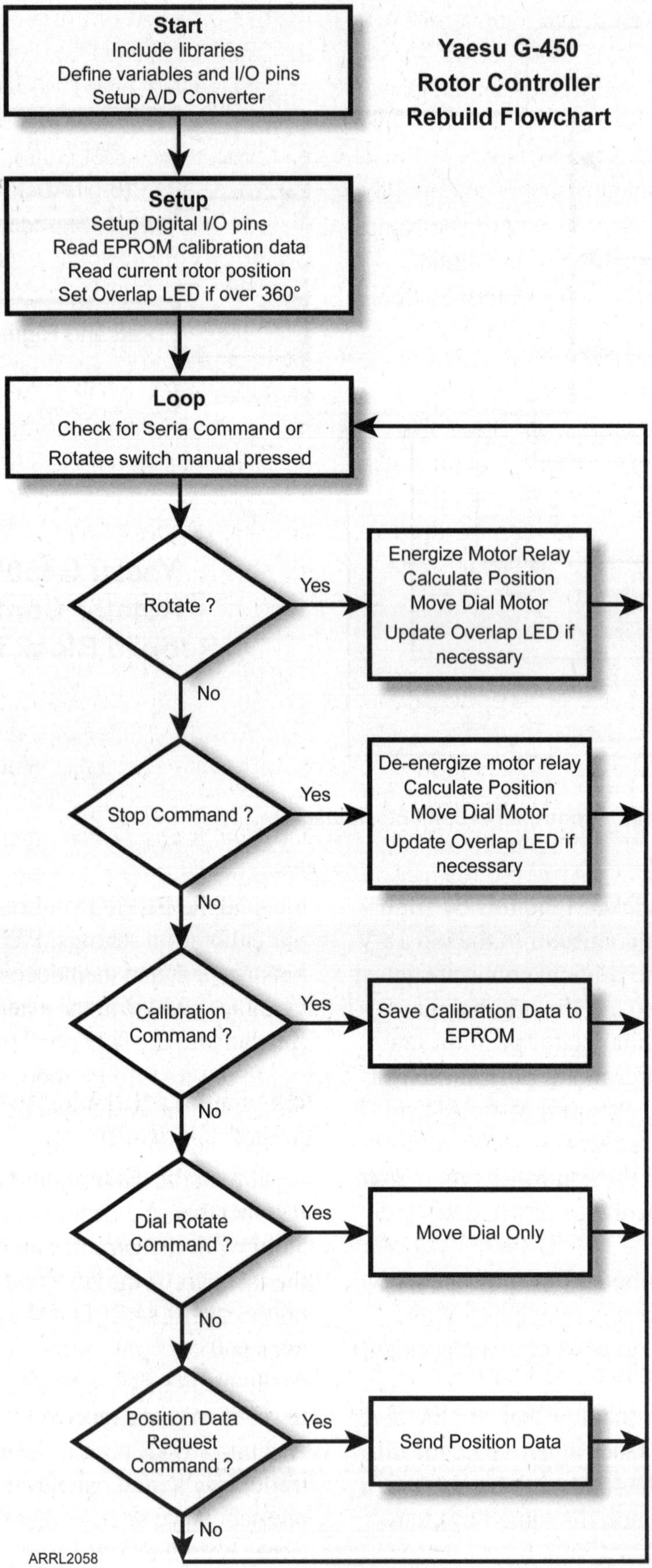

Figure 13.5 — Yaesu G-450 rotator controller rebuild flowchart.

we're rotating, and drive the dial indicator motor to match the current antenna position. We'll also keep track of the status of the **OVERLAP** LED and update it accordingly. If we receive a **STOP** command, or the front panel rotate switch is released, we'll de-energize the antenna rotator motor relay, recalculate the current position, and drive the dial indicator motor to match the current antenna position. Again, we'll keep track of the status of the **OVERLAP** LED and update it accordingly.

Our rebuilt rotator controller saves its calibration data in the Nano's EEPROM memory. In order to calibrate the rebuilt rotator control board, we'll use the same calibration commands used in the real Yaesu GS-232A Computer Control Interface. If we receive a calibration command, the calibration data is updated and saved to the Nano's onboard EEPROM.

Because we're actually dealing with two positioning motors, one on the actual antenna rotator motor and the other inside the rotator controller box, we need a way to calibrate the dial indicator position. We'll use the **UP** and **DOWN** GS-232A commands, which are normally used with the G-5400/5500 Elevation-Azimuth rotator controllers. Since the G-450A and G-800SA are azimuth only rotators, we can borrow these elevation command codes to give us control over the dial indicator motor so that we can set the calibration on it.

Finally, we'll check to see if this is an antenna position request received via the Nano's USB serial port. If it is, we'll respond with the current antenna position using the standard Yaesu GS-232A format. Since we have implemented the dial motor using the elevation command set, we can also use the **READ ELEVATION** commands to read the current position of the dial indicator motor.

Constructing the Rebuilt Main Control Board for the Yaesu G-450A

Figure 13.6, which shows the schematic for the G-450A version of the rebuilt main control board, also includes the parts list in the caption. We'll cover the G-800SA rebuilt main control board later in this chapter. First, we'll need to remove the original main control board and mount the LM7805 and LM7812 voltage regulators to the small metal tab that was used to support the original control board as shown in **Figure 13.7**. There is room to mount these voltage regulators on the rebuilt control board, but I did it this way to provide a built-in heat sink for the voltage regulators. You will also notice that a 7-pin circular twist lock connecter has been added to the rear of the controller box. This connector is in parallel with the original 6-pin Molex-style rotator cable connector. I have issues with loose connections on the original connector with just about every Yaesu rotator controller I have used; installing the twist-lock connector solves the loose connector issue once and for all. You may also notice that there are no additional holes cut or drilled in the back of the G-450 controller box, this is because the rebuilt board is constructed so that the Nano's USB port fits perfectly in the same hole used by the original main control board's calibration switch hole. The only additional hole we'll need to drill is the one for the **RESET DISABLE** switch, which we'll use to allow us to switch between the normal Arduino Programming mode and *Ham Radio Deluxe* Interface Mode (more on that in a bit).

We'll also need to solder wires across the two front panel rotation control switches to the contacts of the two rotator control relays, which are mounted to the top metal shelf on the inside of the controller box (see **Figure 13.8**).

Identifying the Existing Main Control Board Connectors

Unfortunately, the connectors that go to the main control board are not clearly marked. You could try remembering which connector was plugged into the original main control board which does have the connector designations, or you can use the pin count on the connector to determine which one is which. Here is a quick guide to determine the proper connector designation.

J1001 – larger 2-pin connector that is used to provide the 18 V ac power to the main control board. We replace this on our new board with a 2-pin Dupont-style 2.54 mm right-angle header.

J1003 – 4-pin connector

J1004 – 5-pin connector

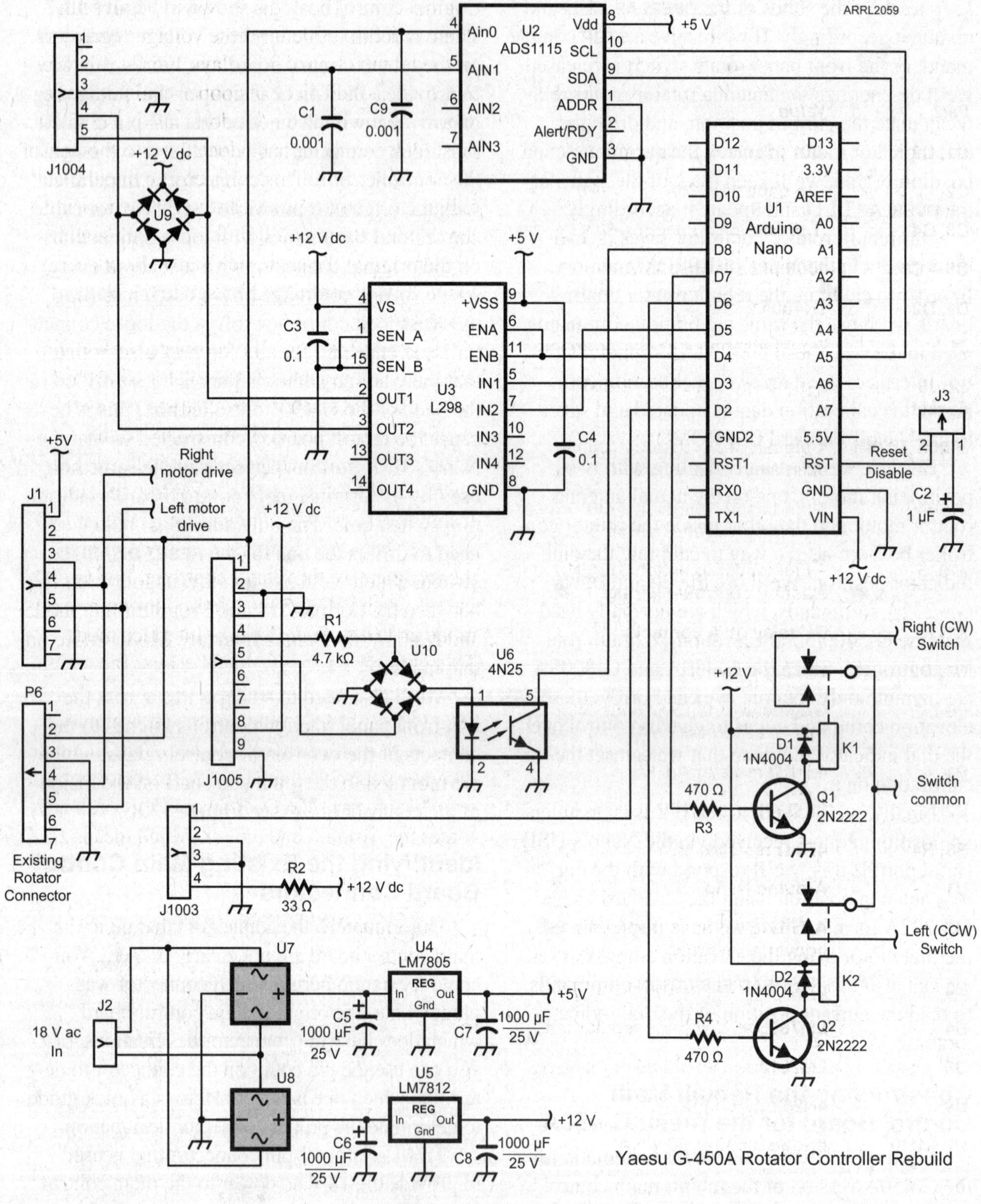

Figure 13.6 — Yaesu G-450 rebuild schematic.

Part	Value
C1, C9	.001 µf ceramic capacitor, 50 V
C2	1 µf capacitor, 35 V
C3, C4	.1 µf ceramic capacitor, 50 V
C5 – C8	1000 µf capacitor, 35 V
D1, D2	1N4004 diode
J1	7-pin twist-lock circular connector
J2, J3 socket	2-pin Dupont 2.54 mm header
J1003	4 pin header to match J1003 in existing controller wiring.
J1004	5 pin header to match J1004 in existing controller wiring.
J1005	9 pin header to match J1005 in existing controller wiring.
K1, K2	5 V relay, 10 A contacts
Q1, Q2	2N2222 transistor
P6	Original 6-pin Molex connector on rear of controller
R1	4.7 KΩ resistor, 1/8 W
R2	33 Ω resistor, 1/8 W
R3, R4	470 Ω resistor, 1/8 W
U1	Arduino Nano
U2	ADS1115 16-bit analog-to-digital converter module
U3	L298 dual full-bridge driver IC
U4	LM7805 +5 V voltage regulator IC
U5	LM7812 +12 V voltage regulator IC
U6	4N25 optocoupler IC
U7 – U10	bridge rectifier, 50 V 2 A

J1005 – 9-pin connector

Now it's finally time to build our replacement control board. The Arduino Nano and all of the control circuit, aside from the voltage regulators and antenna rotator drive relays, are mounted on a 2 3/4 × 5 ¼-inch piece of cooper-clad perfboard, which are mounted on standoffs inside the rotator controller box in the same location as the original Yaesu main control board. To ensure that the old cables to reach the new board without requiring any modifications, we'll install the right-angle header pins on the same side that will connect to the connectors that used to go to the original control board.

Figure 13.9 shows the completed new main control board mounted in the rotator controller box. I used 3/8 standoffs mounted to the metal shelf to provide the height necessary for the Nano's mini-USB connector to fit in the original board's calibration switch hole. While the schematic diagram shows a jumper for the Reset Disable pin, I chose to bring this out to a switch on the rear panel of the rotator controller box to allow me to switch between the Arduino Programming mode and *Ham Radio Deluxe* Interface mode. This jumper/switch is needed because the Arduino IDE uses the Serial/USB DTR signal to send a reset to the Arduino as part of the sketch upload process. When *Ham Radio Deluxe* (*HRD*) initiates communication with the GS-232A rotator controller, it also toggles the DTR signal. This ordinarily forces the Arduino into a reset, which means *HRD*

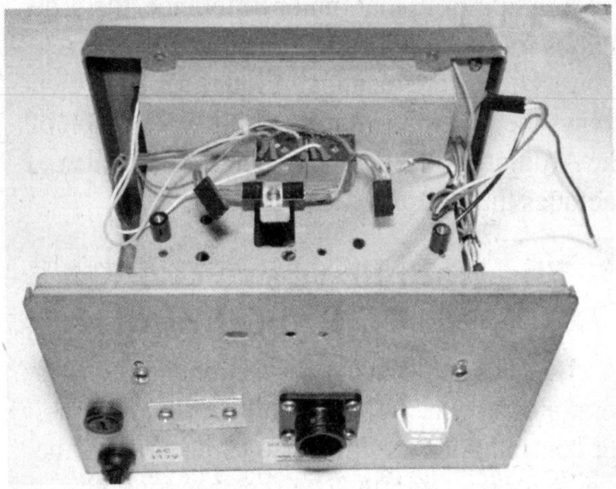

Figure 13.7 — Mounting the voltage regulators to the G-450A chassis.

Figure 13.8 — Yaesu G-450 rotation switch connections.

Figure 13.9 — The new Yaesu G-450 control board mounted on the chassis.

will never connect and time out. Adding the 1 μF capacitor to the Arduino's Reset line effectively disables the **RESET** command sent to the Arduino by *HRD*, allowing *HRD* to connect normally.

As discussed earlier, there is one minor electronic difference between the G450A and the G800/1000SA rotator controllers. On the G-450A, the "move sense" signal from the controller box comes in on J1005 pins 4 and 5. This signal is rectified using a bridge rectifier before being used to drive a 4N25 optocoupler attached to the Arduino's digital I/O pin 8. On the G-800/1000SA, the move sense signal comes in on J1005 pins 6 and 7, and instead of using a bridge rectifier, two diodes are used to provide the necessary rectification before being used to drive the 4N25 optocoupler that is attached to the Arduino's digital I/O pin 8. Aside from that minor difference, the schematics for the G-450A and the G-800/1000SA rebuilt boards are identical. **Figure 13.10**, which shows the schematic for the G-800SA version of the Arduino-powered main control board, also includes the parts list in the caption.

The Rebuilt Yaesu G-450A Rotator Controller Sketch

Because the sketches for the rebuilt G-450A rotator controller and the rebuilt G-800SA rotator controller are so similar, we'll just look at the sketch for the G-450A. The sketches for both rotator controllers are available on the **kw5gp.com** website.

We start out making the sketch by including the Adafruit_ADS1X15 library for the ADS1115 A/D module. We'll also include the EEPROM library so that we can save the calibration data to EEPROM.

Part	*Value*
C1, C9	.001 μf Ceramic Capacitor, 50 V
C2	1 μf capacitor, 35 V
C3, C4	.1 μf ceramic capacitor, 50 V
C5 – C8	1000 μf capacitor, 35 V
U7 – U9	bridge rectifier, 50 V 2 A
D1 – D4	1N4004 diode or equivalent
J1	7-pin twist-lock circular connector
J2, J3	2-pin Dupont 2.54 mm header socket
J1003	4 pin header to match J1003 in existing controller wiring
J1004	5 pin header to match J1004 in existing controller wiring
J1005	9 pin header to match J1005 in existing controller wiring
P6	Original 6 pin Molex connector on rear of controller
K1, K2	5 V relay, 10 A contacts
Q1, Q2	2N2222 transistor
R1, R3, R4	470 Ω resistor, 1/8 W
R2	33 Ω resistor, 1/8 W
U1	Arduino Nano
U2	ADS1115 16-bit analog-to-digital converter module
U3	L298 dual full-bridge driver
U4	LM7805 +5 V voltage regulator IC
U5	LM7812 +12 V voltage regulator IC
U6	4N25 optocoupler

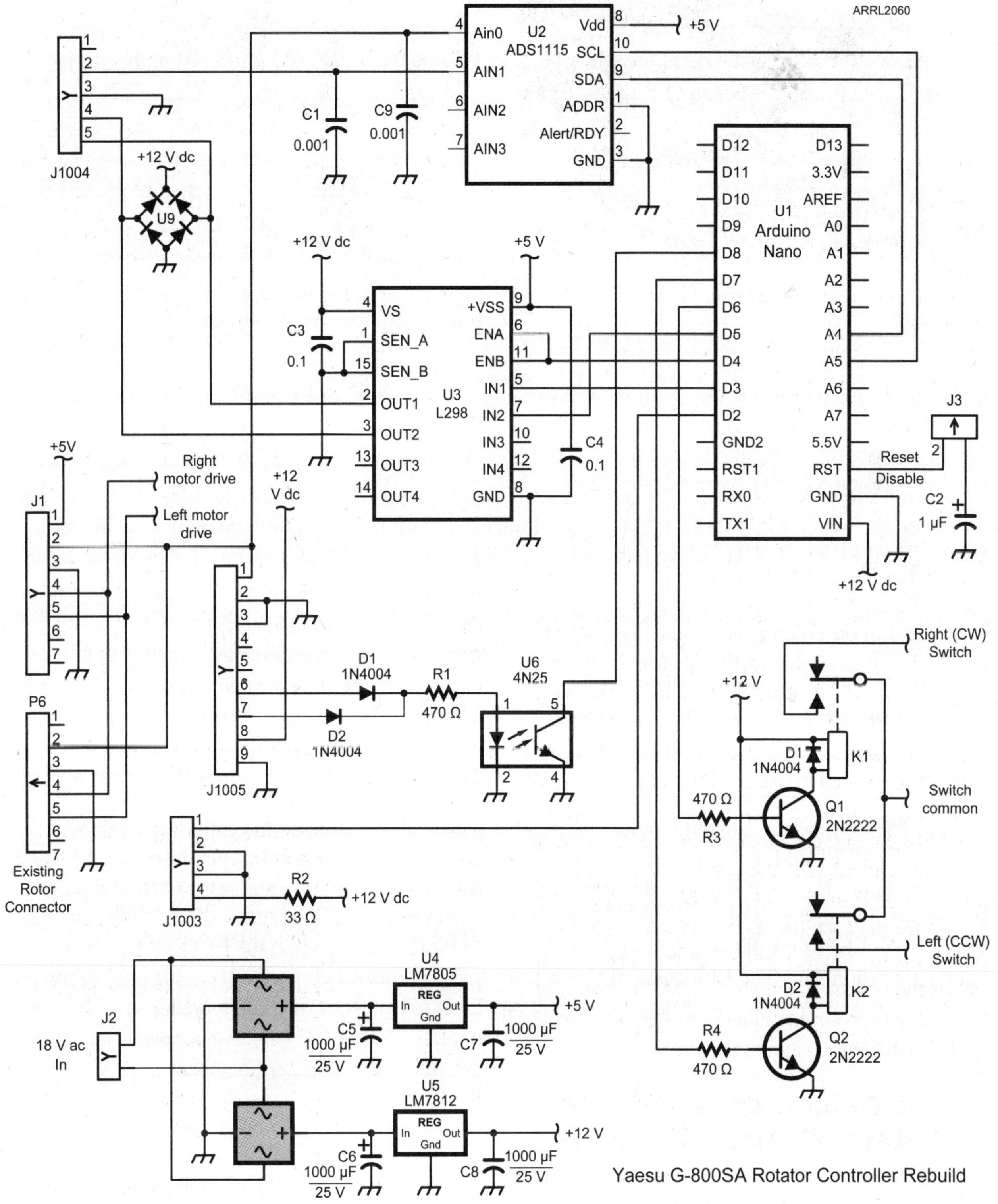

Figure 13.10 — Yaesu G-800SA rebuild schematic.

```
#include <EEPROM.h>  // EEPROM Library
#include <Adafruit_ADS1X15.h> // Include the Adafruit ADS1X15 A/D library
```

Next, we'll create (instantiate) the A/D object and define the digital I/O pins used in this project.

```
Adafruit_ADS1115 adc; // Define the ADS1115 as adc

#define dial_left 3  // Left (CCW) Dial Drive Pin
#define dial_right 5  // Right (CW) DIal Drive Pin
#define rotate_left 6  // Left (CCW) Rotor Drive Pin
#define rotate_right 7  // Right (CW) Rotor Drive Pin
#define overlap_led 2  // Overlap LED Pin
#define dial_enable 4  // Dial Drive Enable Pin
#define move_sense 8  // Rotor Movement Sense Pin
```

During sketch development, I found that the dial indicator motor moved faster than the rotator motor out on the tower. Not only that, but the dial indicator motor turned slightly faster in one direction as compared to the other. By using pulse width modulation on the digital I/O pins that drive the L298 H-Bridge and the dial indicator motor, we can get the dial motor to rotate at the same speed as the antenna rotator. This will prevent the dial indicator from rotating in a jerky, stop/start manner. You may need to adjust the Dial Speed PWM variables to match the speeds with your dial indicator motor.

```
int CWDialSpeed = 150; // PWM setting for Clockwise Dial Speed
int CCWDialSpeed = 148; // PWM setting for Counterclockwise Dial Speed
```

The EEPROM data cannot be randomly accessed. Instead, we store all the calibration values in consecutive EEPROM data locations and read the calibration values sequentially as a single EEPROM read command. Here, we'll define the initial calibration variables we store and retrieve from the EEPROM. The first byte of the EEPROM data is used to determine if the data is valid; if the first byte matches the EEPROM_ID value, in this case 56, we can safely assume that the EEPROM contains valid calibration data. The EEPROM_ID value can be any number, I chose the ID value used in this sketch randomly. We will also define the default baud rate for the Nano's USB port for use with *Ham Radio Deluxe*.

```
#define BAUD_RATE 9600
#define EEPROM_ID_BYTE 1   // EEPROM ID to validate EEPROM data location
#define EEPROM_ID  56  // EEPROM ID Value
#define EEPROM_AZ_CAL_0 2    // Azimuth Zero Calibration EEPROM location
#define EEPROM_AZ_CAL_MAX 4  // Azimuth Max Calibration Data EEPROM location
(450 degrees)
#define EEPROM_DIAL_CAL_0 6    // Dial Zero Calibration Data EEPROM location
```

```
#define EEPROM_DIAL_CAL_MAX 8  // Dial Max Calibration Data EEPROM location  (450
degrees)
```

Next, we'll define the default rotator calibration values. You will most likely need to adjust these for your particular rotator and control box. The tolerance values are used to set how accurate you want your controller to be. The settings used in the sketch are the ones I use in my controller.

```
#define AZ_CAL_0_DEFAULT 295  // Rotor Zero Calibration Point Default Value
#define AZ_CAL_MAX_DEFAULT 25350  // Rotor Max Calibration Point Default Value
#define DIAL_CAL_0_DEFAULT 300  // Dial Motor Zero Calibration Point Default
Value
#define DIAL_CAL_MAX_DEFAULT 25350  // Dial Motor Max Calibration Point Default
Value

#define AZ_Tolerance 1  // Rotor Positioning Tolerance in Degrees
#define DIAL_Tolerance 2  // Dial Motor Positioning Tolerance in Degrees
```

Finally, we'll declare and initialize all the variables we'll need for the sketch.

```
//variables
byte inByte = 0;  // incoming serial byte
byte serial_buffer[50];  // incoming serial byte buffer
int serial_buffer_index = 0;  // Pointer to the character in the serial buffer
int set_AZ;  // Azimuth set value
int set_DIAL;  // Dial set value
int current_AZ;  // Current Azimuth raw value
int current_DIAL;  // Current Dial raw value
String Serial_Send_Data; // Data to send to Serial Port
int AZ_0;  // Azimuth Zero Value from EEPROM
int AZ_MAX; // Azimuth Max Value from EEPROM
int DIAL_0;  // Dial 0 Value from EEPROM
int DIAL_MAX;  //Dial Max Value from EEPROM
int AZ_Degrees; // mapped AZ ADC value to Degrees
int DIAL_Degrees; // mapped DIAL ADC value to Degrees
String Requested_AZ; // RS232 Requested Azimuth - M and short W command
String Requested_DIAL; //RS232 Requested Azimuth and Dial - Full W command
int AZ_To; // Requested AZ Move
int DIAL_To; // Requested DIAL Move
int AZ_Distance; // Distance to move AZ
int DIAL_Distance; // Distance to move DIAL
```

```
int overlap_status;  // indicates we're past 360 Degrees
long manual_move;  // manual move flag

int map_az_0, map_az_max, map_dial_0, map_dial_max, map_az, map_dial;  // vari-
ables to map A/D Values to Degrees
```

In the `setup()` loop, we'll set the pin modes for the digital I/O pins. We'll also enable the internal pullup resistor for the move sense input pin and turn off all the motor control pins to make sure everything is stopped, and then we'll start the Nano's USB serial port and initialize the ADS1115 A/D module.

```
void setup()
{
  pinMode(dial_right, OUTPUT);  // Set the pin modes
  pinMode(dial_left, OUTPUT);
  pinMode(rotate_left, OUTPUT);
  pinMode(rotate_right, OUTPUT);
  pinMode(overlap_led, OUTPUT);
  pinMode(move_sense, INPUT_PULLUP);

  digitalWrite(dial_right, LOW);  // Turn off the motor drive
  digitalWrite(dial_left, LOW);
  digitalWrite(rotate_left, LOW);
  digitalWrite(rotate_right, LOW);

  Serial.begin(BAUD_RATE); // initialize serial communication

#ifdef debug
  Serial.println("Start Initializing");
#endif

  adc.begin(); // Start the ADS1115 A / D
```

We'll now set the ADS1115 mode to run continuously and set the gain to its highest range of `GAIN_TWOTHIRDS`, which is the 0 to 6.144 V range. Since we can't apply more than 5 V to the input of the A/D, this range is effectively 0 to 5 V with a value of .1875 mV per A/D count.

```
  adc.setGain(GAIN_TWOTHIRDS);  // 2/3x gain +/- 6.144V  1 bit = 3mV
0.1875mV (default)
```

Next, we'll turn off the rotator and dial indicator motor movement flags so that the sketch knows everything is stopped and ready for rotation commands.

Now we'll call the `read_eeprom_cal_data()` function to read the saved calibration data from the Nano's onboard EEPROM. Let's stop and take a look at that function now.

```
read_eeprom_cal_data();   // Readthe calibration data from EEPROM
```

The `read_eeprom_cal_data()` function will first check to see if the EEPROM_ID_BYTE read from the EEPROM matches the EEPROM_ID we defined in the initialization area. If it matches, we'll assign the calibration values to their matching sketch variables. Because the data is read from the EPPROM sequentially one byte at a time, we must reassemble the 2 bytes, comprising each of the integer values before storing them in their respective sketch variables.

```
void read_eeprom_cal_data()   // Read the Calibration Data from EEPROM
{
  if (EEPROM.read(EEPROM_ID_BYTE) == EEPROM_ID)
  {
    AZ_0 = (EEPROM.read(EEPROM_AZ_CAL_0)*256) + EEPROM.read(EEPROM_AZ_CAL_0 +
1);  // Rotor Zero Calibration Point Value
    AZ_MAX = (EEPROM.read(EEPROM_AZ_CAL_MAX)*256) + EEPROM.read(EEPROM_AZ_CAL_
MAX + 1);  // Rotor Max Calibration Point Value
    DIAL_0 = (EEPROM.read(EEPROM_DIAL_CAL_0)*256) + EEPROM.read(EEPROM_DIAL_
CAL_0 + 1);  // Dial Zero Calibration Point Value
    DIAL_MAX = (EEPROM.read(EEPROM_DIAL_CAL_MAX)*256) + EEPROM.read(EEPROM_DIAL_
CAL_MAX + 1);  // Dial Max Calibration Point Value
```

If the EEPROM ID byte read from the EEPROM does not match what we defined it to be, we'll assume that this is the first time we've run the sketch, initialize the EEPROM to the default values defined in the sketch initialization area, and call the `write_eeprom_cal_data()` function to write the calibration data to the EEPROM. We'll also use these default values for the calibration values used in the sketch.

Next, we'll turn off the rotator and dial indicator motor movement flags, so that the sketch knows everything is stopped and ready for rotation commands.

```
set_AZ = -1;   // Set rotor movement off
set_DIAL = -1;   // Set dial movement off
```

Now we'll call the read_eeprom_cal_data() function to read the saved calibration data from the Nano's onboard EEPROM. Let's stop and take a look at that function now.

```
void read_eeprom_cal_data()  // Read the Calibration Data from EEPROM
  {
    if (EEPROM.read(EEPROM_ID_BYTE) == EEPROM_ID)
    {
  #ifdef debug // Print the Calibration Data in Debug Mode
      Serial.println("Read EEPROM Calibration Data Valid ID");
      Serial.print("AZ_CAL_0 = ");
      Serial.println((EEPROM.read(EEPROM_AZ_CAL_0) * 256) + EEPROM.read(EEPROM_
AZ_CAL_0 + 1), DEC);
      Serial.print("AZ_CAL_MAX = ");
      Serial.println((EEPROM.read(EEPROM_AZ_CAL_MAX) * 256) + EEPROM.
read(EEPROM_AZ_CAL_MAX + 1), DEC);
      Serial.print("DIAL_CAL_0 = ");
      Serial.println((EEPROM.read(EEPROM_DIAL_CAL_0) * 256) + EEPROM.
read(EEPROM_DIAL_CAL_0 + 1 ), DEC);
      Serial.print("DIAL_CAL_MAX = ");
      Serial.println((EEPROM.read(EEPROM_DIAL_CAL_MAX) * 256) + EEPROM.
read(EEPROM_DIAL_CAL_MAX + 1), DEC);
  #endif

      AZ_0 = (EEPROM.read(EEPROM_AZ_CAL_0) * 256) + EEPROM.read(EEPROM_AZ_CAL_0
+ 1); // Rotor Zero Calibration Point Value
      AZ_MAX = (EEPROM.read(EEPROM_AZ_CAL_MAX) * 256) + EEPROM.read(EEPROM_AZ_
CAL_MAX + 1); // Rotor Max Calibration Point Value
      DIAL_0 = (EEPROM.read(EEPROM_DIAL_CAL_0) * 256) + EEPROM.read(EEPROM_
DIAL_CAL_0 + 1); // Dial Zero Calibration Point Value
      DIAL_MAX = (EEPROM.read(EEPROM_DIAL_CAL_MAX) * 256) + EEPROM.read(EEPROM_
DIAL_CAL_MAX + 1); // Dial Max Calibration Point Value

    } else
    { // initialize eeprom to default values

  #ifdef debug
      Serial.println("Read EEPROM Calibration Data Invalid ID - setting to de-
faults");
  #endif

      // If no valid EEPROM Data Set the Calibration Values to Default and save
to EEPROM
```

```
    AZ_0 = AZ_CAL_0_DEFAULT;
    AZ_MAX = AZ_CAL_MAX_DEFAULT;
    DIAL_0 = DIAL_CAL_0_DEFAULT;
    DIAL_MAX = DIAL_CAL_MAX_DEFAULT;
    write_eeprom_cal_data();
  }
}
```

The `write_eeprom_cal_data()` function will write the calibration data to the Nano's onboard EEPROM. Since the EEPROM data is written a byte at a time, and integer variables consist of 2 bytes, we must split the integer variables containing this data into their high and low order byte values before writing them to the EEPROM.

```
void write_eeprom_cal_data() // Write the Calibration Values to EEPROM
{

  EEPROM.write(EEPROM_ID_BYTE,EEPROM_ID);  // Write the EEPROM ID Byte used to
validate EEPROM data
  EEPROM.write(EEPROM_AZ_CAL_0,highByte(AZ_0));  // Write the Rotor Zero Cali-
bration Value High Order Byte
  EEPROM.write(EEPROM_AZ_CAL_0 + 1,lowByte(AZ_0));  // Write the Rotor Zero
Calibration Value Low Order Byte
  EEPROM.write(EEPROM_AZ_CAL_MAX,highByte(AZ_MAX));  // Write the Rotor Max
Calibration Value High Order Byte
  EEPROM.write(EEPROM_AZ_CAL_MAX + 1,lowByte(AZ_MAX));  // Write the Rotor Max
Calibration Value Low Order Byte
  EEPROM.write(EEPROM_DIAL_CAL_0,highByte(DIAL_0));  // Write the Dial Motor
Zero Calibration Value High Order Byte
  EEPROM.write(EEPROM_DIAL_CAL_0 + 1,lowByte(DIAL_0));  // Write the Dial Motor
Zero Calibration Value Low Order Byte
  EEPROM.write(EEPROM_DIAL_CAL_MAX,highByte(DIAL_MAX));  // Write the Dial Motor
Max Calibration Value High Order Byte
  EEPROM.write(EEPROM_DIAL_CAL_MAX + 1,lowByte(DIAL_MAX));  // Write the Dial
Motor Max Calibration Value Low Order Byte
}
```

Returning to the `setup()  loop`, we'll read both the antenna position and the dial indicator motor positions with the ADS1115 A/D by calling the `read_adc()` function. We'll stop and take a look at that function now.

```
read_adc();   // Read the A/D converter
```

The `read_adc()` function set the A/D gain to its maximum range of GAIN_TWOTHIRDS (0 to 5 V). After a brief delay to allow the A/D to settle, we'll read channel 0 in single-ended mode, and save this value as the current antenna position. We'll then set the A/D to read channel 1 in single-ended mode, and after a brief delay, we'll read channel 1 and save this as the current dial indicator motor position.

```
void read_adc() // Read the A/D Converter
{
#ifdef debug
    Serial.println("Read ADC Function  ");
#endif

  int RotorValue;

  adc.setGain(GAIN_TWOTHIRDS);   // 2/3x gain +/- 6.144V  1 bit = 3mV
0.1875mV (default)

  delay(10); // adc settling delay
  current_AZ = adc.readADC_SingleEnded(0);   // Read A/D Channel 0 (Rotator)

#ifdef debug
    Serial.println(current_AZ);
#endif
  delay(10); // adc settling delay
  current_DIAL = adc.readADC_SingleEnded(1); // Read A/D Channel 1 (Dial)

#ifdef debug
    Serial.println(current_DIAL);
#endif
}
```

Returning to the `setup()` loop, we'll map the current antenna rotator position to a value in degrees, set the `overlap_status` flag, and turn on the **OVERLAP** LED if the position is greater than 360°.

```
  // Map AZ to degrees

#ifdef debug
```

```
  Serial.println(current_AZ);
#endif

  AZ_Degrees = map(current_AZ, AZ_0, AZ_MAX, 0, 450);  // Map the A/D Value to
calibrated degrees

  if (AZ_Degrees > 360)
  {   // Turn on overlap LED if over 360 degrees
    overlap_led_on();
    overlap_status = 1;
  } else
  {
    overlap_led_off();  // Otherwise turn off overlap LED
    overlap_status = 0;
  }
```

In the last part of the `setup()` loop, we're ready to start synchronizing the dial indicator motor with the antenna rotator position.

```
  read_adc();  // Read the current rotor Azimuth
  set_DIAL = AZ_Degrees;  // Set the dial position to current Azimuth
  check_move();  // Move the dial to match Azimith
```

We'll read the A/D to get the current positions, then load the `set_dial` position variable to the current position of the antenna rotator. We'll then call the `check_move()` function to move the dial indicator motor to match. The `check_move()` function will first check to see if we're already rotating the antenna rotator and/or the dial indicator motor. If we are, it will read the A/D converter and map the current antenna rotator and dial indicator motor positions into degrees. If the antenna rotator position is greater than 360°, we'll also call the `overlap_led_on()` function. If the antenna rotator position is less than 360°, we'll call the `overlap_led_off()` function.

```
void check_move() // Manage Rotor and Dial movement
{
  if (set_AZ != -1 || set_DIAL != -1)
  {   // We're moving - check and stop as needed
    read_adc();  // Read the A/D Converter
    // Map AZ to degrees

#ifdef debug
      Serial.print("Current AZ: ");
      Serial.println(current_AZ);
```

```
#endif

    AZ_Degrees = map(current_AZ, AZ_0, AZ_MAX, 0, 450);  // Map the Rotor A/D
Value to Calibrated degrees
    if (AZ_Degrees > 360)    // Turn on overlap led if over 360
    { // Turn on overlap led
      overlap_led_on();
    } else
    {
      overlap_led_off();
    }
```

The `overlap_led_on()` function is used turn the front panel **OVERLAP** LED on and set the `overlap_status` flag if the antenna rotator position is greater than 360° and the **OVERLAP** LED and `overlap_status` is not already set.

```
void overlap_led_on() // Turn on the Overlap LED
{
  if (overlap_status == 0)
  {
    digitalWrite(overlap_led, HIGH);
    overlap_status = 1;
  }
}
```

The `overlap_led_off()` function does the reverse, turning off the **OVERLAP** LED and `overlap_status` flag if the antenna rotator position is less than 360° and the **OVERLAP** LED and `overlap_status` flag were not previously off.

Returning to the `check_move()` function, we check to see if we're turning the antenna rotator. We'll then calculate the distance in degrees we still need to rotate. If the distance we need to rotate is less than the defined tolerance value, we'll stop the rotation by calling the `az_rotate_stop()` function. This is done because the rotation could have been commanded by a USB serial command to rotate to a set position value, and we need the function to handle the entire rotation process automatically until we reach the desired position. If we're not at the desired position, we'll figure out which direction we need to rotate, and call either the `rotate_az_cw()` function to rotate clockwise, or the `rotate_az_ccw()` function to rotate counterclockwise.

```
    if (set_AZ != -1) // Calculate the rotor distance left to move if we're mov-
ing
    {
```

```
AZ_Distance = set_AZ - AZ_Degrees;
if (abs(AZ_Distance) <= AZ_Tolerance)  // We're there - No move needed
{
  az_rotate_stop();
  set_AZ = -1;
} else
{  // Move Azimuth - figure out which way
  if (AZ_Distance > 0)   //We need to move CW
  {
    rotate_az_cw();
  } else
  {
    rotate_az_ccw();
  }
}
```

The `az_rotate_stop()` function will de-energize the rotation control relays, and set the `set_AZ` variable to a -1, indicating that there is no active antenna rotation operation.

```
void az_rotate_stop() // Stop Rotation
{
  digitalWrite(rotate_right, LOW);
  digitalWrite(rotate_left, LOW);
  set_AZ = -1;
}
```

The `rotate_az_cw()` function will energize the clockwise rotation control relay, and make sure the counterclockwise rotation control relay is de-energized.

```
void rotate_az_cw() // Rotate Right (CW)
{
  digitalWrite(rotate_right, HIGH);
  digitalWrite(rotate_left, LOW);
}
```

The `rotate_az_ccw()` does the opposite of the `rotate_az_cw()` function by energizing the counterclockwise rotation control relay and making sure the clockwise rotation control relay is de-energized.

In the last part of the `check_move()` function, we'll handle the rotation for the dial indicator motor. First, we'll set the desired indicator position to the current position of the antenna rotator in order to keep the two tracking as close together as possible.

```
set_DIAL = AZ_Degrees;
    }
    if (set_DIAL != -1)  // Calculate the dial distance left to move if we're
moving
    {
      DIAL_Distance = set_DIAL - DIAL_Degrees;
      if (abs(DIAL_Distance) <= DIAL_Tolerance)   // We're there - No move need-
ed
      {
        dial_rotate_stop();
        set_DIAL = -1;
      } else
      { // Move Dial - figure out which way
        if (DIAL_Distance > 0)   //We need to move CW
        {
          rotate_dial_right();

#ifdef debug
            Serial.println("Need to move dial right");
#endif

        } else
        {
          rotate_dial_left();

#ifdef debug
            Serial.println("Need to move dial left");
#endif
```

As with the antenna rotator position, we'll stop the rotation when the dial position is within the dial tolerance value by calling the `dial_rotate_stop()` function.

```
void rotate_dial_right() // Rotate Dial Right (CW)
{
  analogWrite(dial_right, CWDialSpeed);  // Rotate using PWM
  digitalWrite(dial_left, LOW);
  digitalWrite(dial_enable, HIGH);  // Enable Dial Rotation
}
```

```
void rotate_dial_left() // Rotate Dial Left (CCW)
{
  analogWrite(dial_left, CCWDialSpeed);  // Rotate using PWM
  digitalWrite(dial_right, LOW);
  digitalWrite(dial_enable, HIGH);   // Enable Dial Rotation
}
```

If the dial indicator position does not match the current antenna position, we'll continue rotation in the proper direction by calling either the `rotate_dial_right()` or the `rotate_dial_left()` functions. If we are in a continuous rotation, the PWM settings for the dial indicator motor should allow the dial to rotate smoothly without any jerky starting and stopping as it tracks with the antenna rotator position.

```
void dial_rotate_stop() // Stop Dial Rotation
{
  digitalWrite(dial_right, LOW);
  digitalWrite(dial_left, LOW);
  digitalWrite(dial_enable, LOW);
}
```

The `dial_rotate_stop()` function will turn off both the drive to the L298N and the L298N Enable pin, which will then turn off the dial indicator motor. The `rotate_dial_right()` function will turn on the clockwise PWM drive to the L298N and turn on the Enable pin to the L298N, which will in turn rotate the dial indicator motor clockwise. The `rotate_dial_left` function, will turn on the counter-clockwise PWM drive to the L298N, and turn on the **ENABLE** pin to the L298N, which will in turn rotate the dial indicator motor counterclockwise.

Finally, the rotator controller is initialized, and the position indicator should now show the correct antenna position. Because we moved so many of the necessary operations into functions, the `main loop()` is actually very short. By having all the operations in functions, we can easily reuse this code for other projects. In fact, many of the functions in this sketch were originally created for the Satellite Tracker, Yaesu G5400/5500 Elevation-Azimuth rotator controller, and CDE Rotator Controller projects in the first book.

The first thing we do in the `main loop()` is check for a command on the Nano's USB serial port by calling the `check_serial()` function. This command can either be from the Arduino IDE's Serial Monitor, or from a rotator control software such as *Ham Radio Deluxe*. Since we designed this project to emulate the Yaesu GS-232A Computer Control Interface, we can use a subset of these commands sent via the Arduino IDE's Serial Monitor to test and calibrate the new rotator control board.

The `check_serial()` function will check the Nano's USB serial port for data. If there is data, it will read the incoming serial data one character at a time and add it to the `serial_buffer` variable until a Carriage Return (CR) character is received.

```
void check_serial() // Check the Serial port for incoming commands
{
  if (Serial.available() > 0)
  {
    inByte = Serial.read();  // Read the serial character

    if (inByte == 10)
    {
      return; // ignore Line Feeds
    }
    if (inByte != 13)
    { // Add to buffer if not CR
      serial_buffer[serial_buffer_index] = inByte;  // Add the character to the
serial command buffer

#ifdef debug // Display the character received in debug mode
      Serial.print("Received = ");
      Serial.println(serial_buffer[serial_buffer_index]);
#endif

      serial_buffer_index++;  // increment the serial data buffer pointer

    } else
    { // It's a CR, execute command

      if ((serial_buffer[0] > 96) && (serial_buffer[0] < 123))  // it's lower-
case, convert to uppercase
      {
        serial_buffer[0] = serial_buffer[0] - 32;
      }
```

Once a Carriage Return is received, we'll convert the incoming serial command to all uppercase characters to simplify decoding the command. Then we'll use a switch…case block of statements to decode the first character of the command string.

```
switch (serial_buffer[0])   // Decode first character of command
      {

        case 65:  // "A" Command - All Stop
```

```
#ifdef debug
        Serial.println("A Command Received");
#endif

        az_rotate_stop();
        break;

      case 66:  // "B" Command - Send current dial position

#ifdef debug
        Serial.println("B Command Received");
#endif

        send_current_dial();
        break;

      case 67:      // "C" command - return current azimuth

#ifdef debug
        Serial.println("C Command Received");
        Serial.println(serial_buffer_index);
#endif

        if ((serial_buffer_index == 2) & (serial_buffer[1] == 50))
        {
#ifdef debug
          Serial.println("C2 Command Received");
#endif

          send_current_az_dial();  // Send the dial position if C2 command
        } else
        {
          send_current_az();
        }
        break;

      case 68:  // "D" command - rotate dial left

#ifdef debug
```

```
          Serial.println("D Command Received");
#endif

          rotate_dial_left();
          break;

        case 69:  // "E" command - Stop Dial Rotation

#ifdef debug
          Serial.println("E Command Received");
#endif

          dial_rotate_stop();
          break;

        case 70:  // "F" command - Set full scale calibration value

#ifdef debug
          Serial.println("F Command Received");
#endif

          if ((serial_buffer_index == 2) & (serial_buffer[1] == 50))
          {
#ifdef debug
            Serial.println("F2 Command Received");
#endif

            set_dial_max_cal();  // Set the dial full scale calibration value if
F2 command
          } else
          {
            set_az_max_cal();
          }
          break;

        case 76:  // "L" Command - Rotate Left (CCW)

#ifdef debug
          Serial.println("L command Received");
```

```
#endif

          rotate_az_ccw();
          break;

        case 77:  // "M" command - Rotate to Azimuth xxx degrees

#ifdef debug
          Serial.println("M Command Received");
#endif

          rotate_az_to();
          break;

        case 79:  // "O" command - Set zero calibration point value

#ifdef debug
          Serial.println("O Command Received");
#endif

          if ((serial_buffer_index == 2) & (serial_buffer[1] == 50))
          {
#ifdef debug
              Serial.println("02 Command Received");
#endif

            set_dial_0_cal();  // Set Dial zero calibration point value if 02
          } else
          {
            set_az_0_cal();
          }
          break;

        case 82:  // "R" command - Rotate Right (CW)

#ifdef debug
            Serial.println("R Command Received");
#endif
```

```
            rotate_az_cw();
            break;

          case 83:  // "S" command - Stop all rotation

#ifdef debug
              Serial.println("S Command Received");
#endif

            az_rotate_stop();
            dial_rotate_stop();
            break;

          case 85:  // "U" command - Rotate Dial Right (CW)

#ifdef debug
              Serial.println("U Command Received");
#endif

            rotate_dial_right();
            break;

          case 87:  // "W" command - Rotate Azimuth to xxx degrees

#ifdef debug
              Serial.println("W Command Received");
#endif

            rotate_az_to();
            break;
        }
        serial_buffer_index = 0;  // Clear the serial buffer
        serial_buffer[0] = 0;
      }
    }
}
```

For this project, only a subset of the Yaesu GS-232A Computer Interface commands are decoded and used. Note that the elevation commands for the actual GS-232A will control the dial indicator motor instead.

Valid Yaesu GS-232A Computer Interface Commands

A – All Stop. Stops all CW/CCW Rotation

B – Send the current dial indicator motor position value to the workstation

C – Return Current Azimuth – Send the current antenna rotator position value to the workstation

C2 – Returns the Current Dial Position

D – Rotate Dial Left – Commands the dial indicator motor to rotate left, or counterclockwise. This is used to calibrate the dial indicator motor position.

E – Stop Dial Rotation – Stops all dial indicator motor rotation

F – Sets the antenna rotator position full scale (450°) calibration value

F2 – Sets the dial indicator position full scale (450°) calibration value

L – Rotate Left (CCW) – Commands the antenna rotator to rotate left (counterclockwise)

M – Rotate to XXX degrees. Commands the antenna rotator to rotate to the specified heading

O – Sets the zero-calibration value for the 0° antenna rotator position

O2 – Sets the zero-calibration value for the 0° dial indicator position

S – Rotate Right (CW) – Commands the antenna rotator to rotate right (clockwise)

U – Rotate Dial Right (CW) – Commands the dial indicator motor to rotate right (clockwise)

W – Same as the M command – Commands the antenna rotator to rotate to the specified heading

If we receive an **A** command, we'll stop all antenna rotation by calling the same `az_rotate_stop()` function we used in the `setup()` loop. If we receive the **B** command, we'll call the `send_current_dial()` function, which will return the current dial position in degrees to the workstation in the Yaesu GS-232A format. The `send_current_dial()` function will read the A/D converter, and map the dial current position to degrees. It will then pad the degree value into a three-character format with leading zeroes as called for in the GS-232A format, and then add a "+0" ahead of the position value for the proper GS-232A response to the B command. The function will then send the formatted response to the workstation via the Nano's serial port.

If we receive a **C** command, we must handle things a little differently, since we can have a **C** command or a **C2** command. The **C** command is used to return the current antenna rotator position value in degrees by calling the `send_current_az()` function. The `send_current_az()` function operates very similarly to the `send_dial_az()` function, except that it returns the current antenna rotator position using the same +0XXX format, where the XXX represents the current antenna rotator position in degrees.

If we receive a **C2** command, we'll send both the current antenna rotator position and the dial indicator motor position values back in a single response to the workstation by calling the `send_current_az_dial()` function. The function combines the current antenna position and current dial indicator into a single response back to the workstation in the GS-232A format of "+0XXX+0YYY", where XXX is the current antenna rotator position in degrees, and YYY is the current dial position in degrees.

If we received a D command, we'll rotate the dial left by calling the `rotate_dial_left()` function we looked at earlier during the `setup()` loop. This command is most often used to move the dial indicator to the zero-calibration point. Do not let the dial position motor rotate too far beyond the zero point on the dial when calibrating the dial position, as you could literally wrap the wires coming from the potentiometer around each other if you allow the counterclockwise rotation to continue unchecked.

If we receive an **E** command, we'll stop all dial rotation by calling the `dial_rotate_stop()` function we looked at earlier.

The **F** command is another two-function command. If we receive an **F**, we'll set the full scale calibration value for the antenna rotator position sensor, and write this value to the Nano's EEPROM memory by calling the `set_az_max_cal()` function. The `set_az_max_cal` function will read the A/D converter and write the current antenna rotator position value into the Nano's EEPROM as the maximum (450°) calibration value for the antenna rotator position.

If we receive an **F2** command, we'll set the full scale calibration value for the dial indicator position motor by calling the `set_dial_max_cal()` function. This function will read the A/D converter and then write the current dial indicator motor position value into the Nano's EEPROM as the maximum (450°) calibration value for the dial indicator motor position.

If we receive an **L** command, we'll rotate the antenna rotator left (counterclockwise) by calling the `rotate_az_ccw()` function.

If we receive an M command, we'll rotate the antenna rotator to the heading specified in the M command by calling the `rotate_az_to()` function. This function will decode the heading information from the M or W command and rotate the antenna rotator to the specified heading. After decoding the heading, the function will read the A/D converter to get the current antenna rotator position in degrees and set the **OVERLAP** LED if we're beyond 360°.

Next, we'll calculate how far we need to move the antenna rotator and the dial indicator motors. We'll set the specified heading as the new antenna rotator and dial indicator positions, and start the rotation in the correct direction by calling either the `rotate_az_cw()` or `rotate_az_ccw()` functions. The rotation will continue automatically until we reach the specified heading on both the antenna rotator and the dial indicator.

The O command is another two-part command. If we receive an **O**, we'll set the antenna rotator zero calibration point by calling the `set_az_0_cal()` function. This function will read the A/D to get the current position of the antenna rotator, then write the current position to the EEPROM as the antenna rotator zero calibration point by calling the `write_eeprom_cal_data()` function.

If we receive an **O2** command, we'll do the same thing, except for the dial indicator zero calibration point, by calling the `set_dial_0_cal()` function. The `set_dial_0_cal()` function will read the A/D converter and set the current dial indicator motor position as the zero calibration point, then write the calibration information to the EEPROM by calling the `write_eeprom_cal_data()` function as before.

If we receive an **R** command, we'll rotate the antenna rotator right (clockwise) by calling the `rotate_az_cw()` function.

If we receive an **S** command, we will stop rotation on both the antenna rotator and the dial indicator motor by calling the `az_rotate_stop()` and the `dial_rotate_stop()` functions.

If we receive a U command, we'll rotate the dial indicator motor right (clockwise) by calling the `rotate_dial_right()` function. This command is commonly used to rotate the dial indicator to the 450° position for calibration.

And finally, if we receive a **W** command, we will rotate the antenna rotator to the specified heading by calling the `rotate_az_to()` function as we did with the **M** command. In the real Yaesu GS-232A, the W command actually performs a rotation based on time intervals, but rather than implement that function, we'll have it do the same thing as an **M** command since some rotator control software uses the Yaesu GS-232A **W** command instead of the **M** command to rotate the antenna.

As the final step of the `check_serial()` function, we'll clear the serial command buffer in preparation for the next command.

Returning to the `main loop()`, we'll call the `check_move function()` to process any ongoing rotation commands. Then, we'll handle any manual rotation operations. These operations occur when you press the rotator controller's **ROTATE LEFT** (CCW) or **ROTATE RIGHT** (CW) front panel switches.

First, we'll read the move sense digital I/O pin from the controller box. Whenever a switch is pressed, it energizes the antenna rotator relays on the power supply/relay board located on the bottom of the rotator controller box. This will also cause the move sense I/O line to go to a logic low, indicating one of the front panel **ROTATE** buttons is pressed. If we're not executing a rotation command, and the move sense I/O line is low, then we know that a front panel rotation button has been pressed, and the antenna is being rotated manually. We'll track the antenna rotator position by reading the A/D converter and moving the dial indicator motor to track the current antenna rotator position. We'll also set a two second delay that will start when the manual rotation button is released. After the two second delay to allow the antenna to stop rotating, we'll read the A/D converter to get the current antenna position and update the dial position indicator for the final position update after all rotation has stopped.

And that's the end of the `main loop()`. As you can see, a lot of the action takes place in the function calls, but you also must remember that there is a lot involved in handling and tracking the antenna and dial position indicators, as well as allowing for computer-controlled operation. The good news is that because everything is handled by functions, the `main loop()` for the sketch is kept short and simple. This also means that the `main loop()` executes quickly, and the rotator responds quickly and smoothly to both computer-directed and manual antenna rotation commands.

By moving the majority of the rotator control operations into functions, we can easily reuse these same functions in other sketches, as you will see in the next chapter, where we will modify a working Yaesu G-800SA rotator controller. By having all the functions in the sketch for this project available, we can quickly adapt the sketch to work with a G-800SA rotator controller that has a functional Yaesu main control board.

Calibrating the New Rotator Main Control Board

Before you can begin using your new Arduino-powered rotator controller, you'll most likely need to calibrate the antenna position and dial indicator settings. First, we'll calibrate the antenna position settings. Using the Arduino IDE's Serial Monitor, send the **L** (left), **R** (right), and **A** or **S** (Stop) commands, or manually rotate the antenna rotator until the rotator motor reaches the counterclockwise end stop. Using the Arduino IDE's Serial Monitor, send an **O** command to set the antenna rotator position zero calibration point.

Next, using the Arduino IDE's Serial Monitor, send the **L** (left), **R** (right), and **A** or **S** (Stop) commands, or manually rotate the antenna rotator to the right (clockwise) until the antenna rotator reaches the 450° end stop. Using the Arduino IDE's Serial Monitor, send an **F** command to set the antenna rotator position 450° calibration point.

Now we'll do the same for the dial indicator motor position sensor. First, rotate the antenna rotator to the left (counterclockwise) end stop. Next, using the Arduino IDE's Serial Monitor, send the **D** (left), **U** (right), and **A** or **S** (Stop) commands to move the dial indicator motor to indicate 0°. Don't let the dial indicator move too far beyond the 0° point or you may hit a dead spot on the dial position potentiometer, and you also don't want to allow the dial indicator motor to rotate continually beyond the 0° or 450° point, as you will twist the wires going to the dial position sensor. You may need to alternate between the **U**, **D**, and **A** or **S** commands to get the dial indicator to the 0° point. You'll probably have to play with this a few times until you get used to the dial indicator motor's coasting distance when it receives the stop command. Once you have the dial indicator at 0°, send the **O2** command to save the dial indicator position sensor zero calibration setting.

Now, rotate the antenna to the far right (clockwise) end stop, and adjust the dial indicator motor to indicate 450° using the same method as we used to set the dial zero calibration point. Remember that for the indicator to properly indicate 450° it must rotate past the 0° point on the front panel dial once on its way to 450°. The dial on the controller box should now indicate 90°, and the **OVERLAP** LED should be lit. At this point, send the **F2** command to the Arduino, and the dial maximum calibration data will be saved to the Nano's EEPROM.

Congratulations, your rotator controller should now be calibrated, and you're good to go! You will also need to compile and upload the sketch with the debug mode turned off, as rotator controller software such as *Ham Radio Deluxe* will get confused when it sees anything other than valid GS-232A command responses. Remember, if you plan to use *Ham Radio Deluxe*, or other rotator control software, you need to select the Yaesu GS-232A as the controller in their software, and have the Reset Disable jumper in place, or the **RESET DISABLE** switch in the on position for the rotator control software to connect to the Arduino.

The rebuild process is the same for the Yaesu G-800SA, with only the minor electronic change on the board for the move sense signal coming in from the controller box. For the G-800SA, the Nano is positioned in the area where there was a small plate on the back of the G-800SA, as shown in **Figures 13.11** and **13.12**. Note also that the voltage regulators are mounted in a slightly different position inside the controller box, as shown in Figure 13.11. Aside from that, the only other difference is in the sketch itself. Since the sketch for the G-800SA is nearly identical to the sketch for the G-450A, we won't go into the sketch detail.

The sketches and libraries for both the Yaesu G-450A and the G-800SA main control board rebuild projects can be downloaded from **kw5gp.com**.

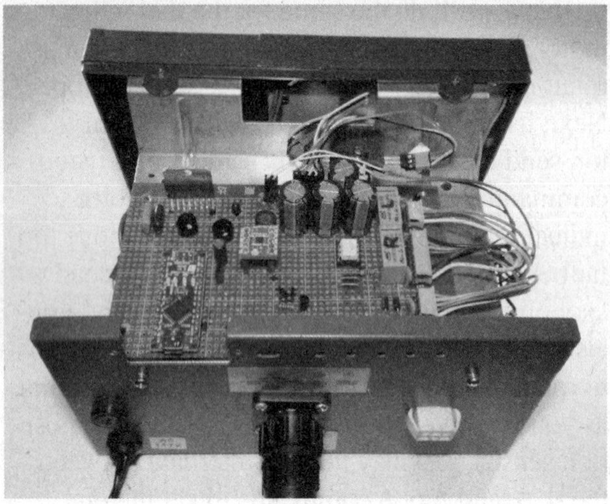

Figure 13.11 — The new controller board mounted in the Yaesu G-800SA.

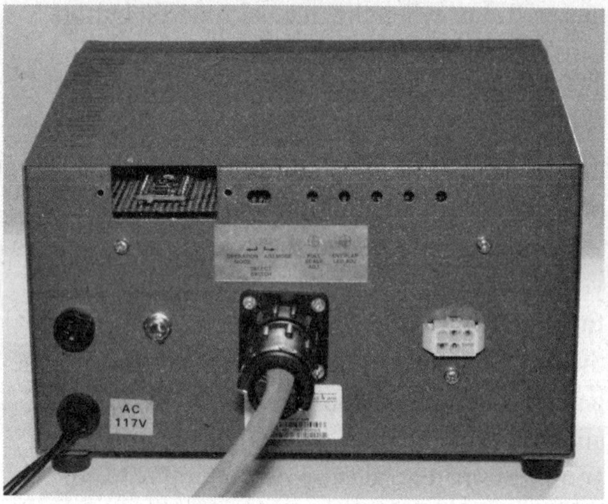

Figure 13.12 — Rear view of the new controller board in a Yaesu G-800SA.

14 Yaesu Rotator Controller Modification

Figure 14.1 — The Yaesu G-800 controller.

For those of you who are lucky enough to not have experienced the "creeping crud" with your Yaesu G-450A, G-800SA, or G-1000SA rotator controller, you can modify it for use with your workstation and rotator control software such as *Ham Radio Deluxe (HRD)*. This is a much easier project than rebuilding the entire main control board, as we did in the previous project, although it does use many of the same components and sketch functions that we previously used. This is yet another example of the versatility of the Arduino; we are able to re-use many pieces from one project to create an entirely different one. With a simple add-on Arduino board, we can quickly add computer control to the Yaesu G-450A, G-800SA, and G-1000SA rotator controllers. Since my version of this project was built with a Yaesu G-800SA (see **Figure 14.1**), we'll use that rotator controller in the project discussion, although this modification and sketch applies equally to the G-450A and G-1000SA series of rotator controllers as well.

Design Goals

For this project, we'll want to take a working Yaesu G-800SA rotator controller and modify it for use with rotator control software such as *HRD*. We want to build a small add-on board that will fit inside the rotator controller box and connect to a workstation using the Arduino's USB serial port, this will emulate the Yaesu GS-232A Computer Control Interface for Antenna Rotators. Two relays will be used to control the antenna rotator motor, and the relay contacts will be connected in parallel with the appropriate controller box front panel rotate switches. As far as the workstation and the rotator software are concerned, they think that they are communicating with a Yaesu GS-232A interface. While we're at it, we'll also replace the Molex-type connector for the rotator cable with a more secure and reliable 7-pin twist-lock connector.

The Modified Yaesu G-800SA Rotator Controller Block Diagram

Figure 14.2 shows the block diagram for the modified G-800SA rotator controller project. Since we start out with a functioning rotator controller box, this is a much simpler modification than the total rebuild that we did in the last chapter. We'll build a small interface board with an Arduino Nano that will interface with the existing antenna rotator position sensor and the controller box front panel switches. A Texas Instruments ADS1115 16-bit I^{2}C analog-to-digital converter A/D will be used to read the antenna rotator position sensor. Since this project is designed to be interfaced with a workstation for computer control, we'll power the add-on board and rotator control relays using power from the Nano's USB serial port.

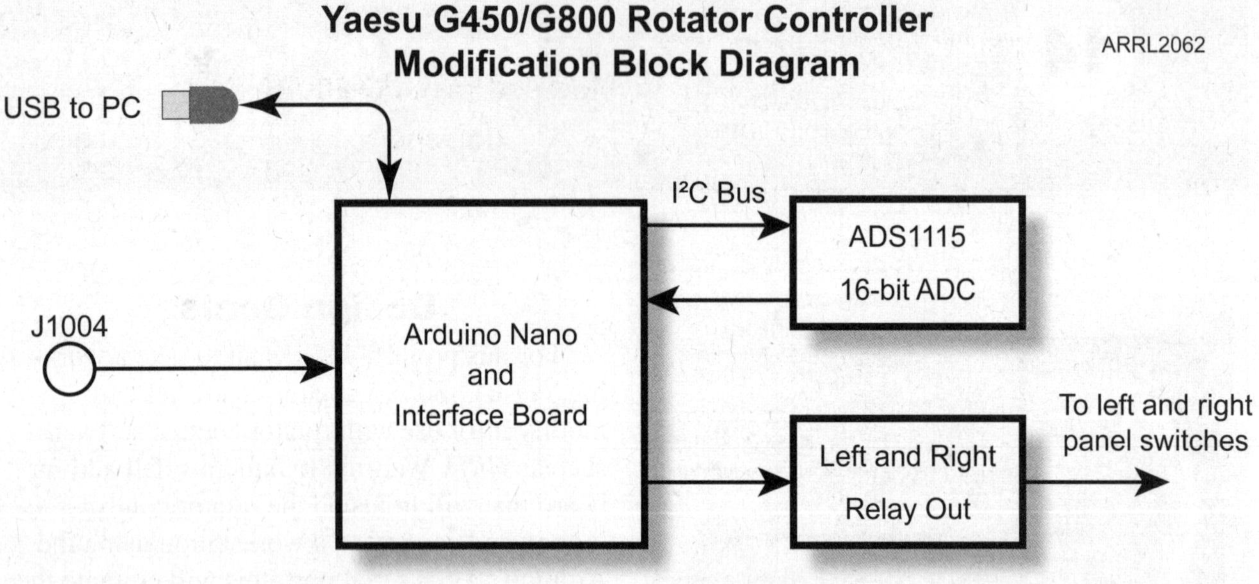

Figure 14.2 — Yaesu rotor controller modification block diagram.

The Modified Yaesu G-800SA Rotator Controller Flowchart

Figure 14.3 shows the flowchart for the modified Yaesu rotator controller project. This is very similar to the flowchart for the rotator controller rebuild project, although it is actually much simpler since we don't have to control the dial indicator motor because we have a working Yaesu main control board. Since we used functions for most of the operations in the rebuilt controller sketch, we'll be re-using many of them in the sketch for this project. We'll start out by including the libraries we'll need, then define the I/O pins, the EEPROM data format, and the default calibration values.

Next, we'll declare and initialize the variables needed in the sketch, and then create (instantiate) the A/D converter object.

In the `setup()` loop, we'll set the pin modes for the digital I/O pins we use in this project. Then we'll turn off the rotation control relays to make sure all computer-controlled rotation is stopped until we're ready. Next, we'll initialize the A/D converter module and read the calibration data from the Nano's onboard EEPROM.

In the `main loop()`, we'll check to see if a computer control command has been received on the Nano's USB serial port. Note that we don't have to keep track of any manual rotation operations with this project since that part of the controller box is already functioning correctly. All we have to do with this sketch is respond based on the incoming commands received on the Nano's USB serial port.

If we receive a rotate command, we'll energize the correct motor relay, which is in parallel with the appropriate front panel switch, and rotate in the commanded direction, or to the heading specified in the rotate command.

If we receive a stop command, we'll de-energize the motor control relays and stop computer-controlled antenna rotation.

If we receive a calibration command, we'll read the current antenna rotator position and update the calibration data before saving it to the Nano's onboard EEPROM.

If we receive an antenna position request, we'll calculate the current antenna rotator position and send the position data back to the workstation in the Yaesu GS-232A format.

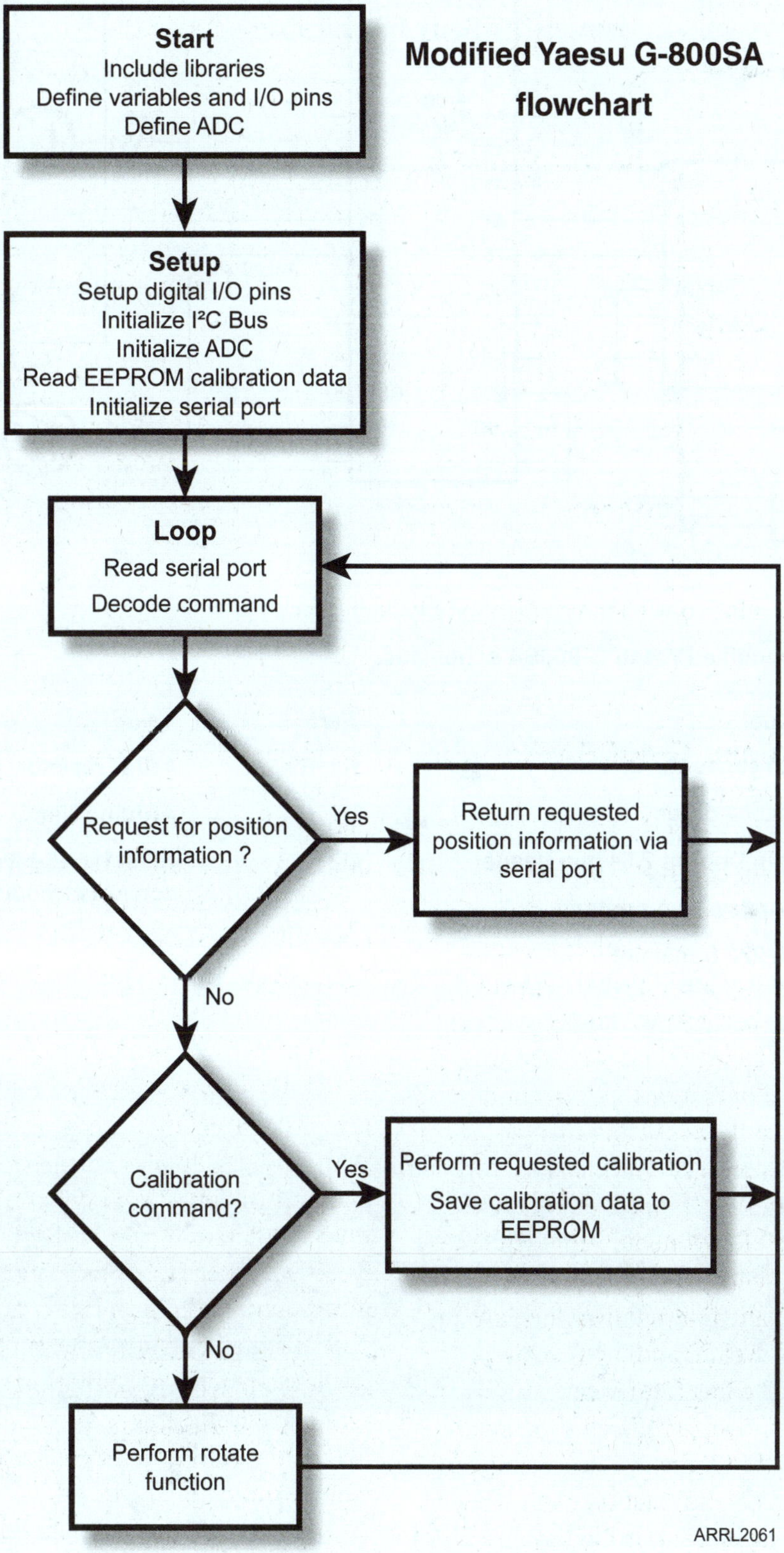

Figure 14.3 — Modified CDE rotor controller flowchart.

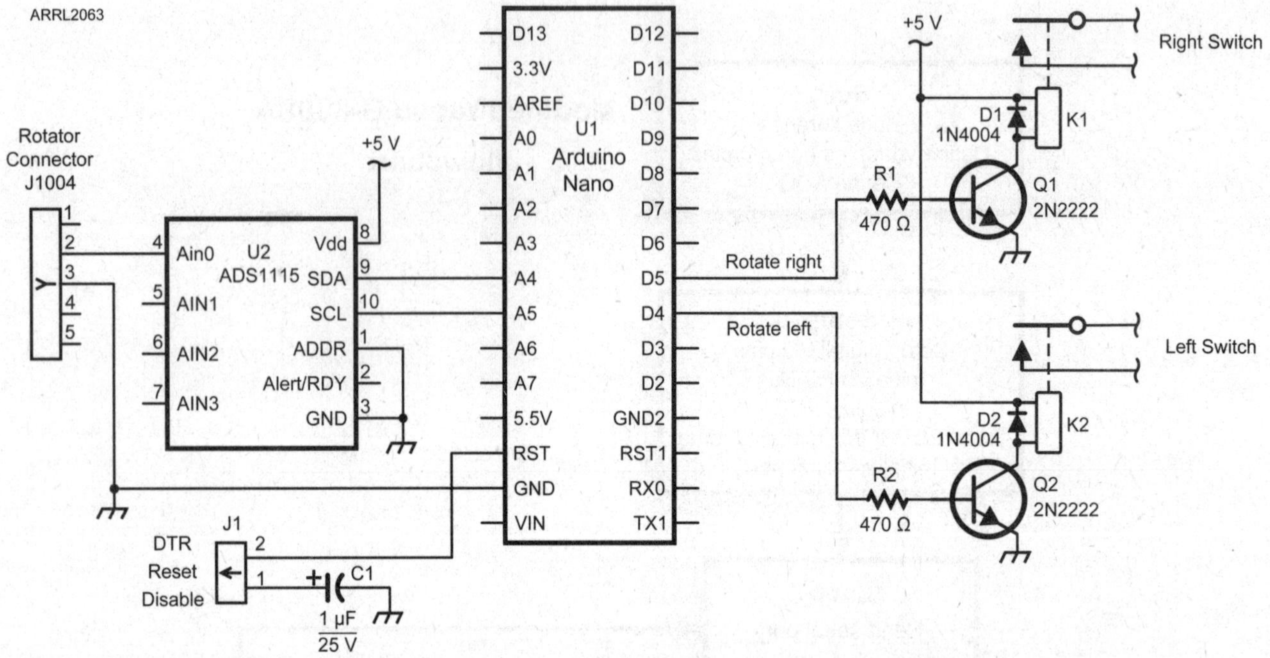

Figure 14.4 — The modified Yaesu G-800SA schematic.

Part	*Value*	*Part*	*Value*
C1	**1 µf capacitor, 25 V**	**R1, R2**	**470 Ω resistor, 1/8 W**
D1, D2	**1N4004 diode**	**U1**	**Arduino Nano**
J1	**2-pin Dupont 2.54 mm header**	**U2**	**ADS1115 16-bit analog-to-digital converter module**
K1, K2	**5 V relay, 10 A contacts**		
Q1, Q2	**2N2222 transistor**		

Figure 14.5 — The Arduino rotator control board.

Figure 14.6 — The Arduino rotator control board mounted in the Yaesu G-800 controller.

Figure 14.7 — The Yaesu G-800 switch modification wiring.

Constructing the Modified Yaesu G-800SA Rotator Controller Project

Figure 14.4, which shows the schematic for the modified Yaesu rotator controller project, also contains the parts list. This is a much simpler schematic than for the rotator controller rebuild we did in the previous chapter. The Arduino Nano and associated electronics are assembled on a 1-3/4 × 4 inch piece of copper-clad perfboard and are mounted on the top metal shelf inside the rotator controller box, as shown in **Figures 14.5** and **14.6**. The Arduino Nano is mounted in a location on the perfboard that places it behind the small removable metal piece on the G-800SA controller rear panel. The antenna rotator position sensor is connected to the ADS1115 A/D by splicing it into the position sensor input on J1004 pin 2 that connects to the Yaesu main control board, and we get the ground for A/D input and the add-on board from J1004 pin 3. For this project, we use 5 V relays mounted on the copper-clad perfboard instead of 12 V relays mounted to the chassis as we did with the controller box rebuild project. The relay contacts are connected across each of the front panel rotate switches as shown in **Figure 14.7**. As with the rebuilt controller project, we'll use a reset disable jumper or switch attached to the Nano's reset line to allow us to switch back and forth between programming mode and *HRD* interface mode. As you can see, it is far easier to modify a working Yaesu G-800SA controller than it is to rebuild the main control board from scratch, and the sketch is also much simpler.

The Modified Yaesu Rotator Controller Sketch

While it may not appear so at first, the sketch for the modified Yaesu rotator controller project is much simpler than the rebuilt main control board version sketch. Since we have a fully functional Yaesu main control board, we don't have to keep track of the **OVERLAP** LED and the dial indicator motor to keep them synchronized with the actual antenna rotator position. That saves a lot of code and headaches. However, there is an issue with calculating the correct antenna rotator position because we're spliced into the existing position sensing circuit, rather than having that sensor and signal all to ourselves. Because we splice into the existing position control circuit, the analog position voltage we read on the A/D is not linear across the entire 450° range, so we must break it down into 90° zones for our position calibration settings. It's not a big thing, but we do need to be aware that we have multiple calibration settings to deal with. Also, when using rotator controller software such as *Ham Radio Deluxe*, you'll need to be sure that the sketch is compiled with the debug mode disabled, otherwise *HRD* will see the debug information coming from the Nano and get confused when trying to communicate with the add-on board.

Starting out with the sketch for the modified Yaesu rotator controller, we'll include the wire library for I^2C communication, the Adafruit_ADS1X15.h library to support the ADS1115 A/D converter, and the EEPROM library that will allow us to save our calibration values in the Arduino Nano's onboard EEPROM.

```
#include <Wire.h>  // Include the I2C Communication Library
#include <EEPROM.h>  // Include the EEPROM Library
#include <Adafruit_ADS1X15.h> // Include the Adafruit ADS1X15 A/D library
```

Next, we'll create (instantiate) the A/D object and define both the I/O pins and the Nano's USB serial port baud rate. It is important to note that the baud rate defined here is also the speed you will select in your rotator control software on your workstation.

```
Adafruit_ADS1115 adc; // Define the ADS1115 as adc

#define rotate_left 4  // Define Rotate Left as Pin 4
#define rotate_right 5  // Define Rotate Right as Pin 5

#define BAUD_RATE 9600  // Set the Serial Port Baud rate to 9600
```

In the next part of the initialization area, we define the EEPROM data format, and the default calibration values for the sketch. Because the EEPROM data is written and read sequentially, we must prepare the calibration data in a specific format so that we can retrieve it in a sequential manner. Because of the non-linearity in the antenna rotator positioning signal, you'll see that we have a separate calibration value for each of the 90° zones for the antenna position calculations.

```
#define EEPROM_ID_BYTE 1   // EEPROM ID to validate EEPROM data location
#define EEPROM_ID  56  // EEPROM ID Value
#define EEPROM_AZ_CAL_0 2     // Azimuth Zero Calibration EEPROM location
#define EEPROM_AZ_CAL_90 4    // Azimuth Zero Calibration EEPROM location
#define EEPROM_AZ_CAL_180 6   // Azimuth Zero Calibration EEPROM location
#define EEPROM_AZ_CAL_270 8   // Azimuth Zero Calibration EEPROM location
#define EEPROM_AZ_CAL_360 10  // Azimuth Zero Calibration EEPROM location
#define EEPROM_AZ_CAL_MAX 12  // Azimuth Max Calibration Data EEPROM location

#define AZ_CAL_0_DEFAULT 740        // Preset the Azimuth Zero Calibration Point
to 0
#define AZ_CAL_90_DEFAULT 5940      // Preset 90 Degree Calibration point
#define AZ_CAL_180_DEFAULT 10550    // Preset 180 Degree Calibration point
```

```
#define AZ_CAL_270_DEFAULT  14700    // Preset 270 Degree Calibration point
#define AZ_CAL_360_DEFAULT  18200    // Preset 360 Degree Calibration point
#define AZ_CAL_MAX_DEFAULT 19000    // Preset the Azimuth Max (450 degree)
Calibration Point
```

In the final portion of the initialization area, we'll define the accuracy tolerance value used in the antenna rotation command and positioning part of the sketch. Typically, this is set to 1, but if you have excessive noise or jitter on your antenna positioning sensor, you may want to increase this value. Then, we'll declare and initialize the variables used in the sketch.

```
#define AZ_Tolerance 1  // Set the Azimuth Accuracy Tolerance

//variables
byte inByte = 0;  // incoming serial byte
byte serial_buffer[50];  // incoming serial byte buffer
int serial_buffer_index = 0;  // The index pointer variable for the Serial buf-
fer
int set_AZ;  // Azimuth set value
int current_AZ;  // Current Azimuth raw value
String Serial_Send_Data; // Data to send to Serial Port
int AZ_0;  // Azimuth Zero Value from EEPROM
int AZ_90;  // Azimuth 90 degree value from EEPROM
int AZ_180;  // Azimuth 180 degree value from EEPROM
int AZ_270;  // Azimuth 270 degree value from EEPROM
int AZ_360;  // Azimuth 360 degree value from EEPROM
int AZ_MAX; // Azimuth Max Value from EEPROM
int AZ_Degrees; // mapped AZ ADC value to Degrees
String Requested_AZ; // RS232 Requested Azimuth - M and short W command
int AZ_To; // Requested AZ Move
int AZ_Distance; // Distance to move AZ
```

In the `setup()` loop, we'll set the pin modes for the rotation control relay pins, and turn the rotation control relays off to ensure that we are not accidentally commanding an antenna rotation. Then, we'll start the Nano's USB serial port, start the I²C interface bus, and initialize the ADS1115 A/D converter.

```
void setup()
{
  pinMode(rotate_left, OUTPUT); // Define the Control Pins as Outputs
  pinMode(rotate_right, OUTPUT);
```

```
  digitalWrite(rotate_left, LOW);  // Turn off all the relays just to be sure
  digitalWrite(rotate_right, LOW);

  Serial.begin(BAUD_RATE); // initialize serial communication

  Wire.begin();  // join I2C bus

  adc.begin();  // Start the ADS1115 A/D

  adc.setGain(GAIN_TWO);  // 2.048V  1 bit = .0625mv


  set_AZ = -1;  // Preset the Azimuth and Elevation Move Variables
```

Next, we'll set the A/D mode to run at a gain setting of GAIN_TWO, which is a range of 0 to 2.048 V, with each A/D count representing 62.5 mV. Then, we'll set the `set_AZ` variable to a -1 in order to indicate that there is no computer-controlled rotation occurring yet.

In the last part of the `setup()` loop, we'll read the saved calibration values from the Nano's EEPROM and store them in the calibration variables used in the sketch by calling the `read_eeprom_cal_data()` function.

```
void read_eeprom_cal_data()  // Function to Read the Azimuth and Elevation Cali-
bration Data
{
  if (EEPROM.read(EEPROM_ID_BYTE) == EEPROM_ID) // Verify the EEPROM has valid
data
  {
#ifdef debug // If in Debug Mode Print the Calibration Values
    Serial.println("Read EEPROM Calibration Data Valid ID");
    Serial.println((EEPROM.read(EEPROM_AZ_CAL_0) * 256) + EEPROM.read(EEPROM_AZ_
CAL_0 + 1), DEC);
    Serial.println((EEPROM.read(EEPROM_AZ_CAL_90) * 256) + EEPROM.read(EEPROM_
AZ_CAL_90 + 1), DEC);
    Serial.println((EEPROM.read(EEPROM_AZ_CAL_180) * 256) + EEPROM.read(EEPROM_
AZ_CAL_180 + 1), DEC);
    Serial.println((EEPROM.read(EEPROM_AZ_CAL_270) * 256) + EEPROM.read(EEPROM_
AZ_CAL_270 + 1), DEC);
    Serial.println((EEPROM.read(EEPROM_AZ_CAL_360) * 256) + EEPROM.read(EEPROM_
AZ_CAL_360 + 1), DEC);
    Serial.println((EEPROM.read(EEPROM_AZ_CAL_MAX) * 256) + EEPROM.read(EEPROM_
```

```
AZ_CAL_MAX + 1), DEC);
#endif

    AZ_0 = (EEPROM.read(EEPROM_AZ_CAL_0) * 256) + EEPROM.read(EEPROM_AZ_CAL_0 +
1); // Read the Azimuth Zero Calibration Value from EEPROM
    AZ_90 = (EEPROM.read(EEPROM_AZ_CAL_90) * 256) + EEPROM.read(EEPROM_AZ_CAL_90
+ 1); // Read the Azimuth Zero Calibration Value from EEPROM
    AZ_180 = (EEPROM.read(EEPROM_AZ_CAL_180) * 256) + EEPROM.read(EEPROM_AZ_
CAL_180 + 1); // Read the Azimuth Zero Calibration Value from EEPROM
    AZ_270 = (EEPROM.read(EEPROM_AZ_CAL_270) * 256) + EEPROM.read(EEPROM_AZ_
CAL_270 + 1); // Read the Azimuth Zero Calibration Value from EEPROM
    AZ_360 = (EEPROM.read(EEPROM_AZ_CAL_360) * 256) + EEPROM.read(EEPROM_AZ_
CAL_360 + 1); // Read the Azimuth Zero Calibration Value from EEPROM
    AZ_MAX = (EEPROM.read(EEPROM_AZ_CAL_MAX) * 256) + EEPROM.read(EEPROM_AZ_CAL_
MAX + 1); // Read the Azimuth Maximum Calibration Value from EEPROM

  } else
  { // initialize eeprom to default values

#ifdef debug
    Serial.println("Read EEPROM Calibration Data Invalid ID - setting to de-
faults");
#endif

    AZ_0 = AZ_CAL_0_DEFAULT;  // Set the Calibration To Default Values
    AZ_90 = AZ_CAL_90_DEFAULT;  // Set the Calibration To Default Values
    AZ_180 = AZ_CAL_180_DEFAULT;  // Set the Calibration To Default Values
    AZ_270 = AZ_CAL_270_DEFAULT;  // Set the Calibration To Default Values
    AZ_360 = AZ_CAL_360_DEFAULT;  // Set the Calibration To Default Values
    AZ_MAX = AZ_CAL_MAX_DEFAULT;
    write_eeprom_cal_data();  // Write the Default Values to EEPROM
  }
}
```

First, we'll verify that the calibration data in the Nano's onboard EEPROM is valid by checking the `EEPROM_ID` variable contained in the first byte of the EEPROM data. If the EEPROM ID is valid, we'll store the saved calibration information into the variables used in the sketch to hold the calibration values. If the EEPROM ID is not valid, this usually indicates that we have not previously saved any calibration data. We'll set the calibration values to the default values and write them to the Nano's onboard EEPROM by calling the `write_eeprom_cal_data()` function.

The `write_eeprom_cal_data()` function will format the calibration values in the sketch calibration variables in preparation to be written to the EEPROM. Since the EEPROM library writes the data to the EEPROM sequentially one byte at a time, we must break the integer calibration variables into a high and low order byte structure before they can be written to the EEPROM.

```
void write_eeprom_cal_data() // Function to Write the Calibration Values to EE-
PROM
{
  Serial.println("Writing EEPROM Calibration Data");

#ifdef debug // If in Debug Mode Print the Calibration Values
    Serial.println("Write EEPROM Calibration Data");
    Serial.println(AZ_0, DEC);
    Serial.println(AZ_90, DEC);
    Serial.println(AZ_180, DEC);
    Serial.println(AZ_270, DEC);
    Serial.println(AZ_360, DEC);
    Serial.println(AZ_MAX, DEC);
#endif

  EEPROM.write(EEPROM_ID_BYTE, EEPROM_ID); //   Write the EEPROM ID
  EEPROM.write(EEPROM_AZ_CAL_0, highByte(AZ_0)); // Write the Azimuth Zero Cali-
bration High Order Byte
  EEPROM.write(EEPROM_AZ_CAL_0 + 1, lowByte(AZ_0));   // Write the Azimuth Zero
Calibration Low Order Byte
  EEPROM.write(EEPROM_AZ_CAL_90, highByte(AZ_90)); // Write the Azimuth Zero
Calibration High Order Byte
  EEPROM.write(EEPROM_AZ_CAL_90 + 1, lowByte(AZ_90));   // Write the Azimuth
Zero Calibration Low Order Byte
  EEPROM.write(EEPROM_AZ_CAL_180, highByte(AZ_180)); // Write the Azimuth Zero
Calibration High Order Byte
  EEPROM.write(EEPROM_AZ_CAL_180 + 1, lowByte(AZ_180));   // Write the Azimuth
Zero Calibration Low Order Byte
  EEPROM.write(EEPROM_AZ_CAL_270, highByte(AZ_270)); // Write the Azimuth Zero
Calibration High Order Byte
  EEPROM.write(EEPROM_AZ_CAL_270 + 1, lowByte(AZ_270));   // Write the Azimuth
Zero Calibration Low Order Byte
  EEPROM.write(EEPROM_AZ_CAL_360, highByte(AZ_360)); // Write the Azimuth Zero
Calibration High Order Byte
  EEPROM.write(EEPROM_AZ_CAL_360 + 1, lowByte(AZ_360));   // Write the Azimuth
```

```
Zero Calibration Low Order Byte
  EEPROM.write(EEPROM_AZ_CAL_MAX, highByte(AZ_MAX)); // Write the Azimuth Max
Calibration High Order Byte
  EEPROM.write(EEPROM_AZ_CAL_MAX + 1, lowByte(AZ_MAX)); // Write the Azimuth Max
Calibration Low Order Byte
}
```

Now let's take a look at the main loop—all two statements of it. Because most of the operations of this project are contained in the functions we adapted from the rebuilt Yaesu rotator controller project, the `main loop()` consists of only two statements that call the `check_serial()` and `check_move()` functions.

```
void loop()
{
  check_serial();  // Check the Serial Port for Data
  check_move();  // Check to see if executing move command
} // End Main Loop
```

The `check_serial()` function is used to read and decode any commands received on the Nano's USB Serial port. First, we'll read the incoming serial data one character at a time and add each character to the `serial_buffer` variable until we receive a carriage return (CR) character that indicates the end of the command data.

```
void check_serial() // Check the Serial port for incoming commands
{
  if (Serial.available() > 0)
  {
    inByte = Serial.read();  // Read the serial character

    if (inByte == 10)
    {
      return; // ignore Line Feeds
    }
    if (inByte != 13)
    { // Add to buffer if not CR
      serial_buffer[serial_buffer_index] = inByte;  // Add the character to the
serial command buffer

#ifdef debug // Display the character received in debug mode
      Serial.print("Received = ");
```

```
        Serial.println(serial_buffer[serial_buffer_index]);
#endif

        serial_buffer_index++;  // increment the serial data buffer pointer

      } else
      { // It's a CR, execute command

        if ((serial_buffer[0] > 96) && (serial_buffer[0] < 123))  // it's lower-
case, convert to uppercase
        {
          serial_buffer[0] = serial_buffer[0] - 32;
        }
```

You will also notice that provision has been made in the sketch to echo each received character back to the workstation as it is received. Some rotator control applications, such as SatPC32, require that each character sent be echoed back to the rotator control program. Since this block of code originally came from the Yaesu GS-232A emulation code in the Yaesu G-5400/5500 Elevation-Azimuth rotator controller project, I left it in, but commented it out, just in case it might be needed later. See what I mean about reusing functions? So far, this particular function has been used in virtually every rotator controller project I've built, with only a few minor changes needed to make it work for the new project, regardless of the rotator manufacturer.

Once we have the complete command in the `serial_buffer` variable, the first thing we'll do is convert the data to all uppercase in order to make our command decoding easier. Now, we'll decode the command based on the first character contained in the `serial_buffer` using the Arduino's switch…case code block.

```
        switch (serial_buffer[0]) {  // Decode first character of command

          case 65:  // A Command - Stop the Azimuth Rotation

 #ifdef debug
                Serial.println("A Command Received");
#endif
            az_rotate_stop();
            break;

          case 67:       // C - return current azimuth

#ifdef debug  // Return the Buffer Index Pointer in Debug Mode
```

```
            Serial.println("C Command Received");
            Serial.println(serial_buffer_index);
#endif

          send_current_az();  // Return Azimuth if C Command
          break;

        case 70:  // F - Set the Max Calibration

#ifdef debug
            Serial.println("F Command Received");
            Serial.println(serial_buffer_index);
#endif

          if (serial_buffer_index == 2)
          {
#ifdef debug
              Serial.println("Fx command Received");
#endif

            // It's a secondary calibration F1=90 F2=180 F3=270 F4=360

            if (serial_buffer[1] == 49)
            {
              // It's an F1 - Calibrate the 90 degree value

#ifdef debug
                Serial.println("F1 command Received");
#endif

              set_90_cal();
            }
            if (serial_buffer[1] == 50)
            {
              // It's an F2 - Calibrate the 180 degree value

#ifdef debug
                Serial.println("F2 command Received");
#endif
```

```
            set_180_cal();
          }
          if (serial_buffer[1] == 51)
          {
            // It's an F3 - Calibrate the 270 degree value

#ifdef debug
              Serial.println("F3 command Received");
#endif
            set_270_cal();
          }
          if (serial_buffer[1] == 52)
          {
            // It's an F4 - Calibrate the 180 degree value

#ifdef debug
              Serial.println("F4 command Received");
#endif

            set_360_cal();
          }
        } else {
          set_max_az_cal();  // F - Set the Max Azimuth Calibration
        }
        break;

      case 76:  // L - Rotate Azimuth CCW

#ifdef debug
          Serial.println("L Command Received");
#endif

        rotate_az_ccw();  // Call the Rotate Azimuth CCW Function
        break;

      case 77:  // M - Rotate to Set Point

#ifdef debug
          Serial.println("M Command Received");
```

```
#endif

        rotate_to();  // Call the Rotate to Set Point Command
        break;

      case 79:  // O - Set Zero Calibration

#ifdef debug
          Serial.println("O Command Received");
          Serial.println(serial_buffer_index);
#endif

        set_0_az_cal();  // O - Set the Azimuth Zero Calibration
        break;

      case 82:  // R - Rotate Azimuth CW

#ifdef debug
          Serial.println("R Command Received");
#endif

        rotate_az_cw();  // Call the Rotate Azimuth CW Function
        break;

      case 83:  // S - Stop All Rotation

#ifdef debug
          Serial.println("S Command Received");
#endif

        az_rotate_stop();  // Call the Stop Azimith Rotation Function
        break;

    }

    serial_buffer_index = 0;  // Clear the Serial Buffer and Reset the Buffer
Index Pointer
    serial_buffer[0] = 0;
  }
```

We'll be using a subset of the Yaesu GS-232A computer control interface for antenna rotators command set, so when you use rotator control software with this project, be sure to select the Yaesu GS-232A as the rotator to use in the software.

If the first character of the command is an **A**, we'll stop the antenna rotation by calling the `az_rotate_stop()` function.

```
void az_rotate_stop() // Stop Rotation
{
  digitalWrite(rotate_right, LOW);
  digitalWrite(rotate_left, LOW);
  set_AZ = -1;
}
```

This function will de-energize both rotation control relays and stop any computer-controlled rotation in progress by setting the `set_AZ` variable to –1 to indicate there is no rotation command currently in progress.

If a **C** command is received, we'll return the current antenna rotator position to the workstation by calling the `send_current_az()` function. This function will read the A/D and map the A/D value into degrees. The function will then send the current position in the Yaesu GS-232A format of "+0XXX," where XXX is the current antenna rotator position in degrees.

```
void send_current_az() // Send the Current Azimuth Function
{
  read_adc();   // Read the ADC

  // Map Azimuth to degrees

#ifdef debug
    Serial.println(current_AZ);
#endif

  AZ_Degrees = Degrees(current_AZ);

#ifdef debug
    Serial.println(AZ_Degrees);
#endif

  // Send it back via serial
  Serial_Send_Data = "";
  if (AZ_Degrees < 100)  // pad with 0's if needed
  {
    Serial_Send_Data = "0";
```

```
  }
  if (AZ_Degrees < 10)
  {
    Serial_Send_Data = "00";
  }
  Serial_Send_Data = "+0" + Serial_Send_Data + String(AZ_Degrees);  // Send the 
Azimuth in Degrees
  Serial.println(Serial_Send_Data);  // Return value via RS-232 port
}

void set_max_az_cal() // Set the Max Azimuth (450 degree) Calibration Function
{
  Serial.println("Calibrate Max AZ (450 Degree) Function");

  read_adc();  // Read the ADC

  // save current az and el values to EEPROM - Zero Calibration
  Serial.println(current_AZ);

  AZ_MAX = current_AZ;  // Set the Azimuth Maximum (450 degree) Calibration to 
Current Azimuth Reading
  write_eeprom_cal_data();  // Write the Calibration Data to EEPROM
  Serial.println("Max Azimuth (450 degree) Calibration Complete");
}
```

The `read_adc()` function will set the A/D converter gain to `GAIN_TWO`, or 2.048 V, for full scale, which equates to 62.5 mV per A/D count. We'll then read the A/D converter in single-ended mode to get the current antenna rotator position.

```
void read_adc() // Function to read the ADC
{
#ifdef debug
    Serial.println("Read ADC Function  ");
#endif

  int RotorValue;  // Variable to store the rotor value
  adc.setGain(GAIN_TWO); // Set the ADC gain to 0.0625 Volts per step
  delay(10); // adc settling delay

  current_AZ = adc.readADC_SingleEnded(0);
}
```

After we have read the A/D converter, we'll need to apply the correct calibration values for the 90° zone we are currently in by calling the `Degrees()` function and passing the A/D value, in degrees, to the `Degrees()` function.

```
int Degrees(int az_value)
{
  int corrected_degrees;

  // Corrects Azimuth degrees because of non-linearity in rotor position sensing
  if (az_value <= AZ_0)
  {
    corrected_degrees = 0;
  }
  if ((az_value >= AZ_0) && (az_value <= AZ_90))
  {
    corrected_degrees = map(az_value, AZ_0, AZ_90, 0, 90);
  }
  if ((az_value > AZ_90) && (az_value <= AZ_180))
  {
    corrected_degrees = map(az_value, AZ_90, AZ_180, 91, 180);
  }
  if ((az_value > AZ_180) && (az_value <= AZ_270))
  {
    corrected_degrees = map(az_value, AZ_180, AZ_270, 181, 270);
  }
  if ((az_value > AZ_270) && (az_value <= AZ_360))
  {
    corrected_degrees = map(az_value, AZ_270, AZ_360, 271, 360);
  }
  if ((az_value > AZ_360) && (az_value <= AZ_MAX))
  {
    corrected_degrees = map(az_value, AZ_360, AZ_MAX, 361, 450);
  }
  if (az_value > AZ_MAX)
  {
    corrected_degrees = 450;
  }

  return (corrected_degrees);
}
```

Based on the 90° zone that we are currently in, this function will adjust the calculated antenna rotator position using a map statement with the correct calibration data for the current 90° zone.

The **F** command is used to set the full-scale calibration point on the rotator controller. Because we have to break the calibration data into 90° zones due to the nonlinearity of the antenna position sensor, the **F** command was expanded by adding a number to the end of the command string to tell us which zone we are calibrating. We'll handle the standard **F** command at the end of this section.

If the command is a two character command, and the command is decoded as an **F1**, we are calibrating the 0 – 90° zone, and we'll call the `set_90_cal()` function to save the first zone's calibration data to the Nano's onboard EEPROM.

```
void set_90_cal() // Set the 90 degree Azimuth Calibration Function
{
  Serial.println("Calibrate 90 Degree AZ Function");

  read_adc();  // Read the ADC

  // save current Azimuth value to EEPROM - 90 degree Calibration
  Serial.println(current_AZ);

  AZ_90 = current_AZ;  // Set the Azimuth 90 degree Calibration to Current Azi-
muth Reading
  write_eeprom_cal_data();  // Write the Calibration Data to EEPROM
  Serial.println("90 degree Azimuth Calibration Complete");
}
```

The `set_90_cal()` function will read the A/D converter to get the current antenna position, save the current antenna position in the 90° calibration variable, and write the calibration to the EEPROM by calling the `write_eeprom_cal_data()` function.

If an **F2** command is received, we'll repeat the same process that we did with the **F1** command, only this time we'll set the calibration for the 91 – 180° zone, and call the `set_180_cal()` function. This function will do the same thing as the `set_90_cal()` function, except for the 91 – 180° zone. We'll repeat this process for each of the remaining 90°, 181 – 270°, and 271 – 360° zones using the **F3** and **F4** commands to calibrate each and calling their respective calibration functions. For the 361 – 450° zone, we'll use the standard full-scale calibration command, **F**, which will call the `set_max_az_cal()` function.

While this calibration method may seem a bit cumbersome, we need to do it this way to maintain accuracy and synchronization between the actual antenna rotator position and the position that we report back to the workstation. The good news is that we should only need to perform the calibration process once.

When we receive an **L** command, we'll start the antenna rotator turning left, or counterclockwise, by calling the `rotate_az_ccw()` function.

```
void rotate_az_ccw() // Function to Rotate Azimuth CCW
{
  digitalWrite(rotate_left, HIGH);  // Set the Rotate Left Pin High
  digitalWrite(rotate_right, LOW);  // Make sure the Rotate Right Pin is Low
}
```

This function will energize the rotate left (counterclockwise) control relay. The relay will remain energized until a stop or rotate right (clockwise) command is received.

If an **M** command is received, we'll rotate the antenna rotator to the heading specified in the command by calling the `rotate_to()` function. This function will first verify that the remainder of the command contains a valid heading within the range of 0 – 450°.

```
void rotate_to() // Function to Rotate to Set Point
{
#ifdef debug
    Serial.println("M Command -  Rotate Azimuth To Function");
#endif

  // Decode Command - Format Mxxx - xxx = Degrees to Move to

#ifdef debug
    Serial.println(serial_buffer_index);
#endif

  if (serial_buffer_index == 4)  // Verify the Command is the proper length
  {
#ifdef debug
      Serial.println("Value in [1] to [3]?");
#endif

    Requested_AZ = (String(char(serial_buffer[1])) + String(char(serial_buf-
fer[2])) + String(char(serial_buffer[3]))) ;  // Decode the Azimuth Value
    AZ_To = (Requested_AZ.toInt()); // AZ Degrees to Move to as integer
    if (AZ_To < 0) // Make sure we don't go below 0 degrees
    {
      AZ_To = 0;
    }
    if (AZ_To > 450) // Make sure we don't go over 450 degrees
    {
      AZ_To = 450;
    }
```

Next, we'll read the A/D converter to get the current antenna rotator position and calculate the distance in degrees that we need to rotate the antenna rotator. We'll rotate the antenna rotator in the proper direction until we reach the specified heading by calling the `rotate_az_cw()`, `rotate_az_ccw()`, or `az_rotate_stop()` functions.

```
    read_adc();   // Read the ADC

    // Map it to degrees

#ifdef debug
      Serial.println(current_AZ);
#endif

    AZ_Degrees = Degrees(current_AZ);

#ifdef debug
      Serial.println(AZ_Degrees);
#endif

    AZ_Distance = AZ_To - AZ_Degrees;   // Figure out far we have to move
    set_AZ = AZ_To;
    if (abs(AZ_Distance) <= AZ_Tolerance)  // No move needed if we're within the
Tolerance Range
    {
      az_rotate_stop();   // Stop the Azimuth Rotation
      set_AZ = -1;   // Turn off the Move Command
    } else {   // Move Azimuth - figure out which way
      if (AZ_Distance > 0)    //We need to move CW
      {
        rotate_az_cw();   // If the distance is positive, move CW
      } else {
        rotate_az_ccw();   // Otherwise, move counterclockwise
      }
```

If we receive an **O** command, we'll set the antenna rotator position calibration for the zero calibration point by calling the `set_0_az_cal()` function.

```
void set_0_az_cal() // Set Azimuth Zero Calibration
{
  Serial.println("Calibrate Zero Function");
```

```
  read_adc();  // Read the ADC
  // save current Azimuth value to EEPROM - Zero Calibration
  Serial.println(current_AZ);

  AZ_0 = current_AZ;  // Set the Azimuth Zero Calibration to current position
  write_eeprom_cal_data();  // Write the Calibration Data to EEPROM
  Serial.println("Zero Azimuth Calibration Complete");
}
```

This function will read the current antenna rotator position and copy the value into the sketch variables used to hold the calibration data. The calibration data is then written to the Nano's onboard EEPROM by calling the `write_eeprom_cal_data()` function.

If the received command is an **R** command, we'll command the antenna rotator to rotate right (clockwise) by calling the `rotate_az_cw()` function, and energizing the right rotation control relay. The relay will remain energized until a stop or rotate left (counterclockwise) command is received.

```
void rotate_az_cw() // Function to Rotate Azimuth CW
{
  digitalWrite(rotate_right, HIGH);    // Set the Rotate Right Pin High
  digitalWrite(rotate_left, LOW);    // Make sure the Rotate Left Pin Low
}
```

If we receive an **S** command, we'll stop all computer-controlled rotation by calling the `az_rotate_stop()` function, which will de-energize both of the rotation control relays.

After processing all the received commands, we'll clear the serial buffer in preparation for the next serial command.

Returning to the `main loop()`, the only other command we execute is the `check_move()` function. This function is used to perform the computer-controlled rotation commands. If the `set_AZ` variable contains a heading instead of a –1, we need to rotate the antenna rotator until we reach the heading specified in the `set_AZ` variable. If the `set_AZ` variable is a –1, it means that there is no active computer-controlled rotation in progress.

If we are processing a rotation command, the first thing we do in the `check_move()` function is read the A/D converter to get the current antenna rotator position.

```
void check_move() // Check to see if we've been commanded to move
{
  if (set_AZ != -1) {   // We're moving - check and stop as needed
    read_adc();  // Read the ADC
    // Map AZ to degrees
```

```
#ifdef debug
      Serial.println(current_AZ);
#endif

    AZ_Degrees = Degrees(current_AZ);

#ifdef debug
      Serial.println(AZ_Degrees);
#endif

    if (set_AZ != -1) // If Azimuth is moving
    {
      AZ_Distance = set_AZ - AZ_Degrees;  // Check how far we have to move
      if (abs(AZ_Distance) <= AZ_Tolerance)  // No move needed if we're within
the tolerance range
      {
        az_rotate_stop();  // Stop the Azimuth Rotation
        set_AZ = -1;  // Turn off the Azimuth Move Command
      } else {  // Move Azimuth - figure out which way
        if (AZ_Distance > 0)   //We need to move CW
        {
          rotate_az_cw();  // Rotate CW if positive
        } else {
          rotate_az_ccw();  // Rotate CCW if negative
        }
```

If we are already rotating, we'll calculate how far we are from the specified heading, and if we have reached the specified heading, we'll call the `az_rotate_stop()` function and set the `set_AZ` variable to a –1 to indicate that we have stopped all computer-controlled rotation. If we have not yet reached the specified heading, we'll energize the correct rotation control relay for the specified heading and return to the main loop to repeat the cycle.

Calibrating the Modified Yaesu Rotator Controller

To calibrate your modified Yaesu rotator controller, first rotate the antenna rotator to 0°, then using the Arduino IDE's serial monitor, send an **O** command to set the zero-calibration point. Then rotate the antenna rotator to 90° and send an **F1** command. Next, rotate to 180° and send an **F2** command, then to 270° and send the **F3** command, then to 360° and send the **F4** command. Finally, rotate to 450° and send the **F** command. Calibration is now complete for your modified Yaesu rotator controller.

Operating the Modified Yaesu Rotator Controller

You're now ready to use your modified Yaesu rotator controller. Be sure that you have compiled and uploaded the sketch with the debug mode disabled so that the debug information sent to the serial port won't confuse your computer's antenna rotator control software. Also make sure that the reset disable jumper is in place, or, if you used a switch for the reset disable line, make sure the switch is in the on position. You should now be able to connect to the controller using your workstation's rotator control software as shown in **Figure 14.8** and control your antenna rotator using the software. If you are using software other than *HRD's* rotator control software, and your software will not connect to the modified rotator controller's Arduino, remember that you may need to uncomment out the line in the sketch that will echo the received command back to the workstation.

The sketch and libraries for this project can be downloaded from **kw5gp.com.**

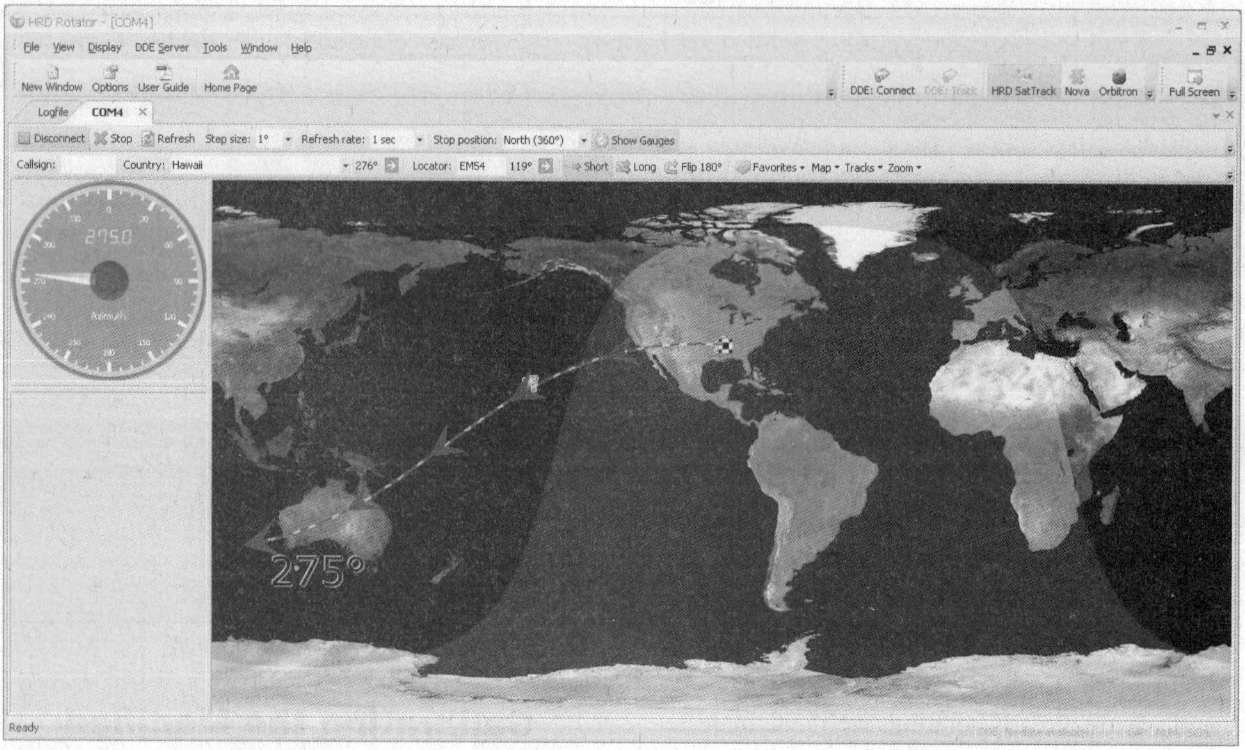

Figure 14.8 — The Yaesu G-800 rotator controller connected to and communicating with *Ham Radio Deluxe*.

15 Modified CDE/Hy-Gain Rotator Controller

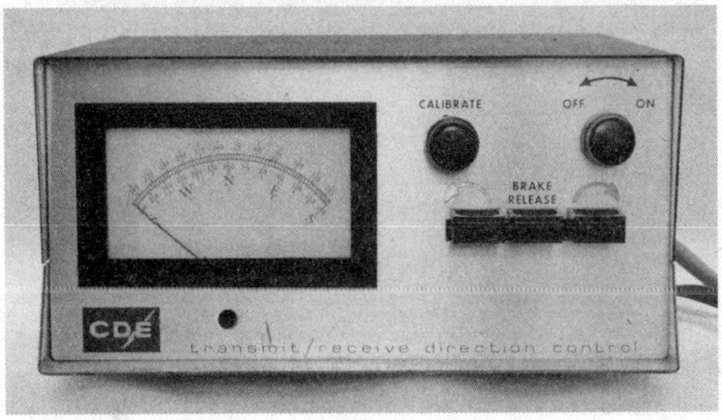

Figure 15.1 — The CDE/Hy-Gain rotator controller.

The CDE/Hy-Gain series of antenna rotators is probably the most popular antenna rotator ever made. However, computer control for the CDE/Hy-Gain rotators didn't exist until the Ham-V series in 1994, which doesn't help those of us that have the older versions. By now you may have guessed that I have a thing for the Arduino and rotator controllers, so it would be a natural thing for me to modify an existing CDE/Hy-Gain rotator controller (see **Figure 15.1**) and give it Arduino-powered computer control. The plan is to modify the CDE/Hy-Gain rotator controller so that it can be controlled remotely by *Ham Radio Deluxe* (*HRD*) and rotator control programs.

In researching this project, I found only one similar Arduino project for the CDE/Hy-Gain controllers, but that project required modifications inside the rotator bell housing assembly. This is due to the basic design of the CDE/Hy-Gain rotator assembly. Because the positioning sensor inside the rotator bell housing shares its dc ground with the rotator brake solenoid and drive motor ac ground, there is about a half of a volt of ac hum on the positioning signal that I was unable to filter out.

Design Goals

After playing with my CDE/Hy-Gain controller, measuring voltages, and looking at signals all through the rotator positioning circuit, I found that by measuring the position sensor voltage between the rear terminal pin 3 and the chassis ground on rear terminal pin 1, I got a usable positioning signal that ranged from about 0 – 13.5 V dc. While this voltage is too high to be used directly with the Arduino analog inputs, or the ADS1115 A/D converter, we can use a voltage divider to bring the voltage range down to a safer 0 – 3.5 V dc. With the addition of a filtering capacitor across the input to the ADS1115 A/D, the analog to digital converter is now able to read the rotator position signal reliably and safely, even with the rotator turning.

You can also tap off this positioning signal inside the controller on the rear of the calibration

Figure 15.2 — The CDE/Hy-Gain rotator metering point.

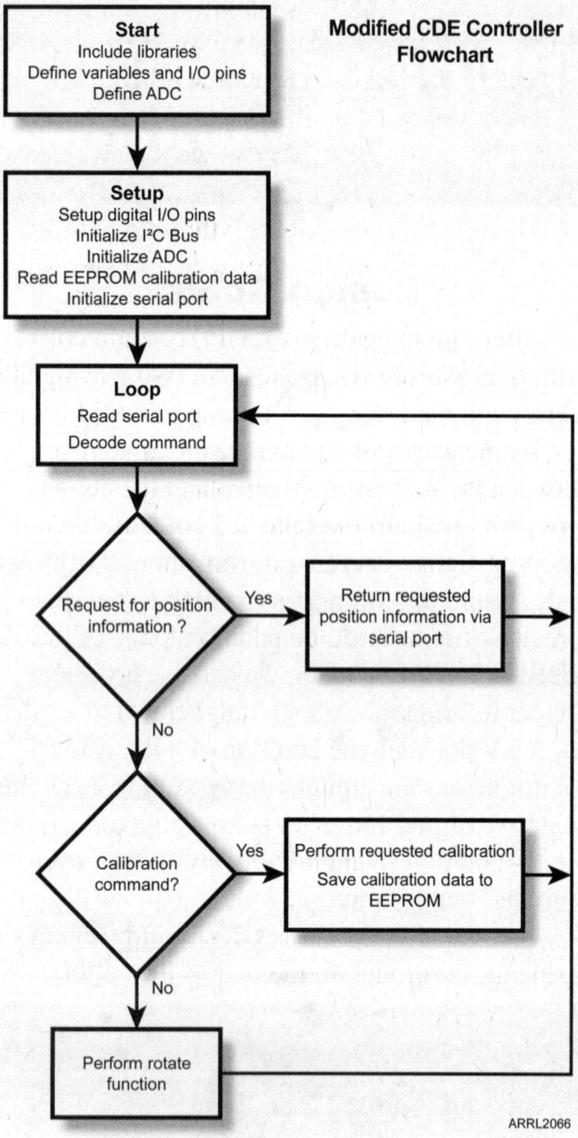

Figure 15.3 — The modified CDE/Hy-Gain rotator controller block diagram.

Figure 15.4 — The underside of the controller box showing where we can install the Arduino Uno.

potentiometer for chassis ground, and the fuse lug nearest to the calibration potentiometer, as shown in **Figure 15.2**. Be sure to verify with an ohmmeter that you have these connections and the polarity correct before connecting to the voltage divider on the Arduino prototyping board.

At its lowest gain setting of GAIN_TWOTHIRDS, the A/D can read from 0 – 6.144 V, for a resolution of 0.1875 mV per bit. This gain is set at 2/3 since you cannot apply more than 5 V to the ADS1115 input, but the values are scaled to the 6.144 range. The 0 – 3.5 V input from the positioning sensor will yield A/D counts from 0 – ~18,667 for full scale. This will give us an overall resolution of about 52 A/D counts per degree of rotation, more than adequate to accurately position the rotator.

Since I wanted this project to be able to interface with the rotator controller in *HRD*, we'll have the Arduino emulate the Yaesu GS-232A in azimuth-only mode. Since we already created a sketch that emulates the Yaesu GS-232A when we built the Yaesu Azimuth-Elevation controller project, we'll be able to reuse most of that sketch here, reducing the sketch development process drastically. In fact, so much of that sketch's code was able to be reused that the initial test sketch for this project was up and running in just a few minutes. The sketch for the Yaesu Azimuth-Elevation controller project was created using groups of function calls, and with just a few minor changes, those same functions are adaptable for an azimuth-only controller like the CDE/Hy-Gain. This is yet another reason the Arduino has become so popular.

Figure 15.3 shows the block diagram for the modified CDE/Hy-Gain rotator controller project. The actual modification to the rotator control box is minimal, consisting of a small board to hold the three control relays, and an Arduino Uno with a prototyping shield to interface to the relays and position sensor. We'll use the Arduino's USB port to interface with the PC running *HRD*.

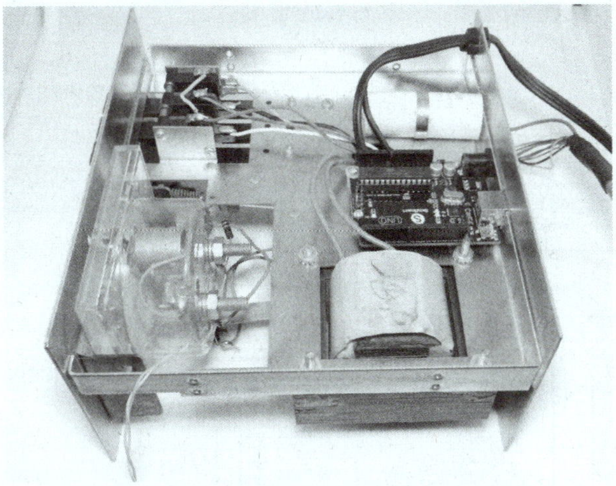

Figure 15.5 — Mounting the Arduino Uno in the controller.

Constructing the Modified CDE/ Hy-Gain Rotator Controller

Starting out, we must find a place inside the CDE/HY-Gain rotator controller box to mount the Arduino Uno. The only place with enough room is on the underside of the controller box, between the 26 V ac power transformer and the motor capacitor (**Figure 15.4**). We'll cut out a small hole in the back cover to allow access to the Arduino's USB port for programming and interfacing to the PC. The downside to mounting the Arduino here, despite it being the only available space, is that the cutout for the USB port is also where the control box chassis has a double layer of metal, which makes cutting the hole slightly more difficult (**Figure 15.5**). Fortunately, the hole we need is small, so it can be done quickly. **Figure 15.6** shows the Arduino Uno and the prototyping shield mounted to the underside of the control box chassis.

Next, we'll glue the three miniature relays to a strip of perfboard with epoxy and mount it to the ends of two existing screws in the control box chassis (**Figure 15.7**). You can mount the relay board anywhere you want on the underside of the chassis, but the two existing screws were convenient in this case.

Now it's time to wire everything together. Because we need to wire the switches and position sensing circuit into the rotator control box, we won't build a test circuit on the breadboard. Instead, we'll wire everything into the rotator control box and hope for the best. Be sure to unplug your control box, as there is 120 V ac all over the inside of the controller box. **Figure 15.8**, which shows the schematic for building this project, also contains the parts list. We'll first mount the resistors and transistors used to drive the relays, along with the ADS1115 A/D converter and the reset disable jumper, on an Arduino ProtoShield. Then we'll use the Dupont-style 2.54 mm header and socket connectors to connect the relays and rotator position sensing.

By adding a 470 µf filtering capacitor across the positioning input, any ac hum on the input is filtered out, and we get a nice, stable position sense voltage. I also added a ferrite choke to the A/D around the position sense input wires for added noise protection.

Figure 15.6 — The Arduino UNO and prototyping shield mounted in the controller.

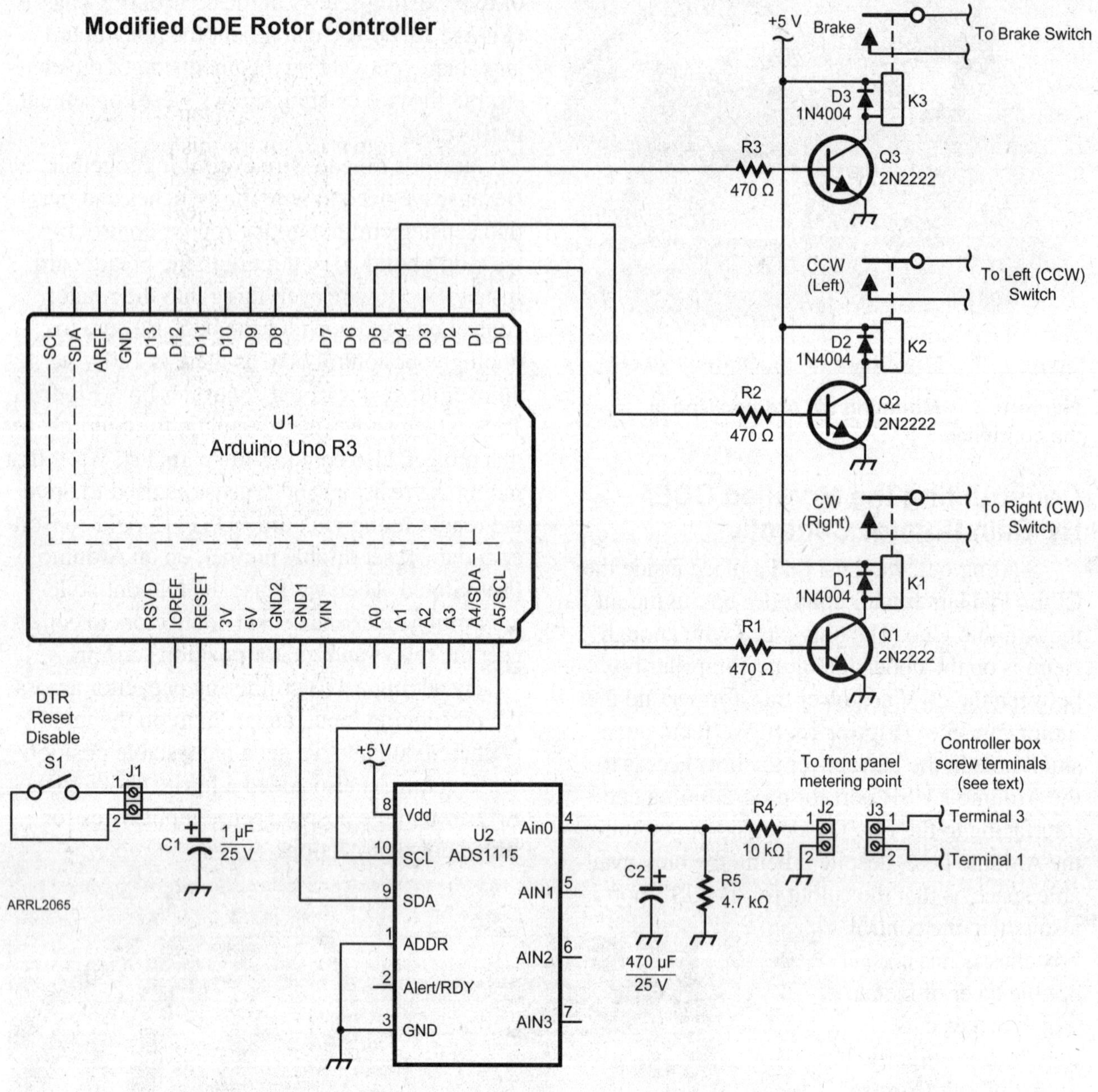

Figure 15.8 — The modified CDE/Hy-Gain rotor controller schematic.

Part	*Value*	*Part*	*Value*
C1	1 µf capacitor, 25 V	R1-R3	470 Ω resistor
C2	470 µF capacitor, 25 V	R4	10 kΩ
D1-D3	1N4004 diode or equivalent	R5	4.7 kΩ resistor
J1-J3	2-pin 2.54 mm Dupont-style header	U1	Arduino Uno
K1-K3	5 V SPST relay	U2	ADS1115 16-bit A/D converter module
Q1-Q3	2N2222A transistor		

Figure 15.7 — The relay board mounted in the controller.

The relays are wired across the contacts on the motor and brake switches on the front panel. Be careful with the brake switch wiring; the brake switch actually controls the 120 V ac power to the 26 V ac power transformer, and the brake is released by applying power to the motor drive circuit. **Figure 15.9** shows the completed modifications to the CDE rotator control box.

The Modified CDE/Hy-Gain Rotator Controller Sketch

Now that the modifications to the CDE/Hy-Gain Rotator Controller box are complete, we can start creating our sketch. You will see that the flowchart (**Figure 15.10**) for this project is very similar the one we used for the Yaesu azimuth-elevation rotator controller project. We'll use the EEPROM on the Arduino to save our azimuth calibration data. To allow interfacing with *HRD* for rotator control, we'll have the Arduino emulate the Yaesu GS-232A in azimuth-only mode. This will allow us to use most of the same sketch we used for the Yaesu azimuth-elevation rotator controller project. All we must do is make a few minor changes and remove the portions of the sketch that deal with elevation control. The Yaesu GS-232A command codes needed to communicate with *HRD* are listed in **Table 15.1**. Since the Yaesu GS-232A command set includes the commands needed to calibrate the zero and maximum azimuth values, we'll use those commands to perform our position sensor calibration.

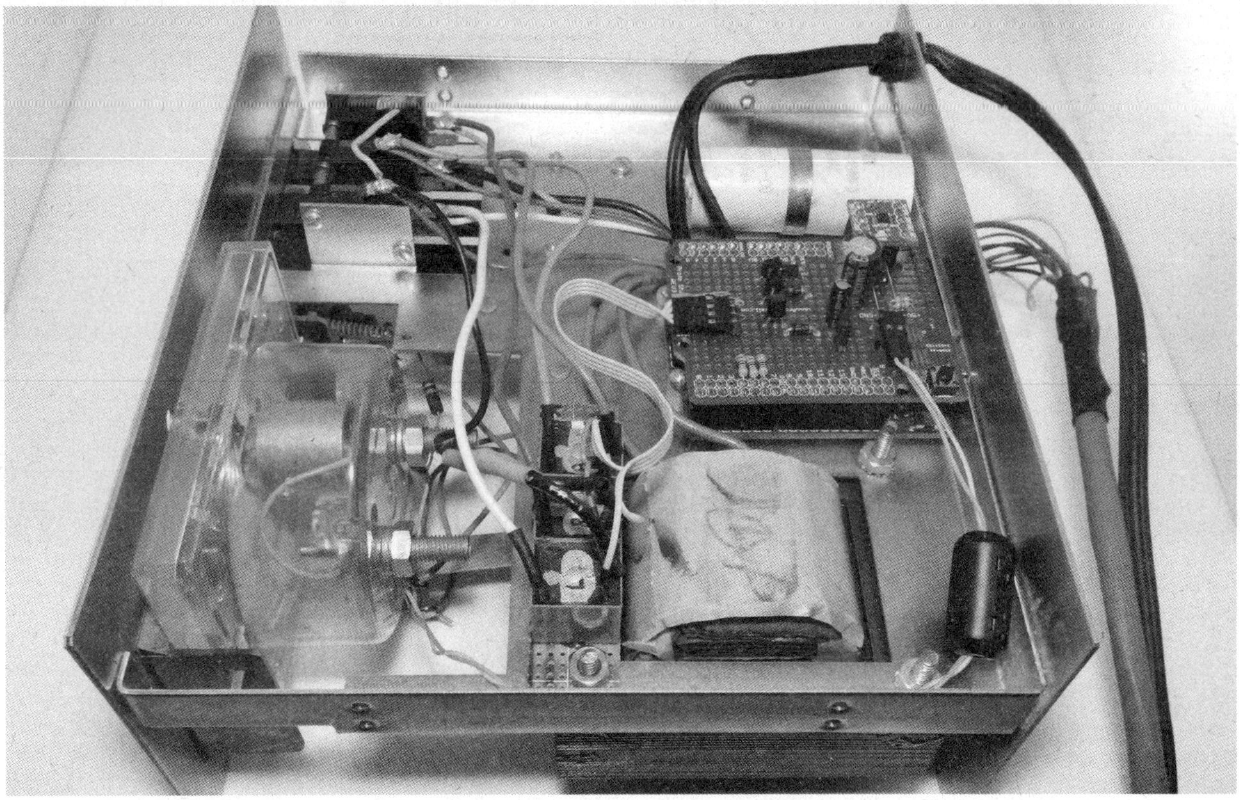

Figure 15.9 — The completed CDE/Hy-Gain rotator controller modification.

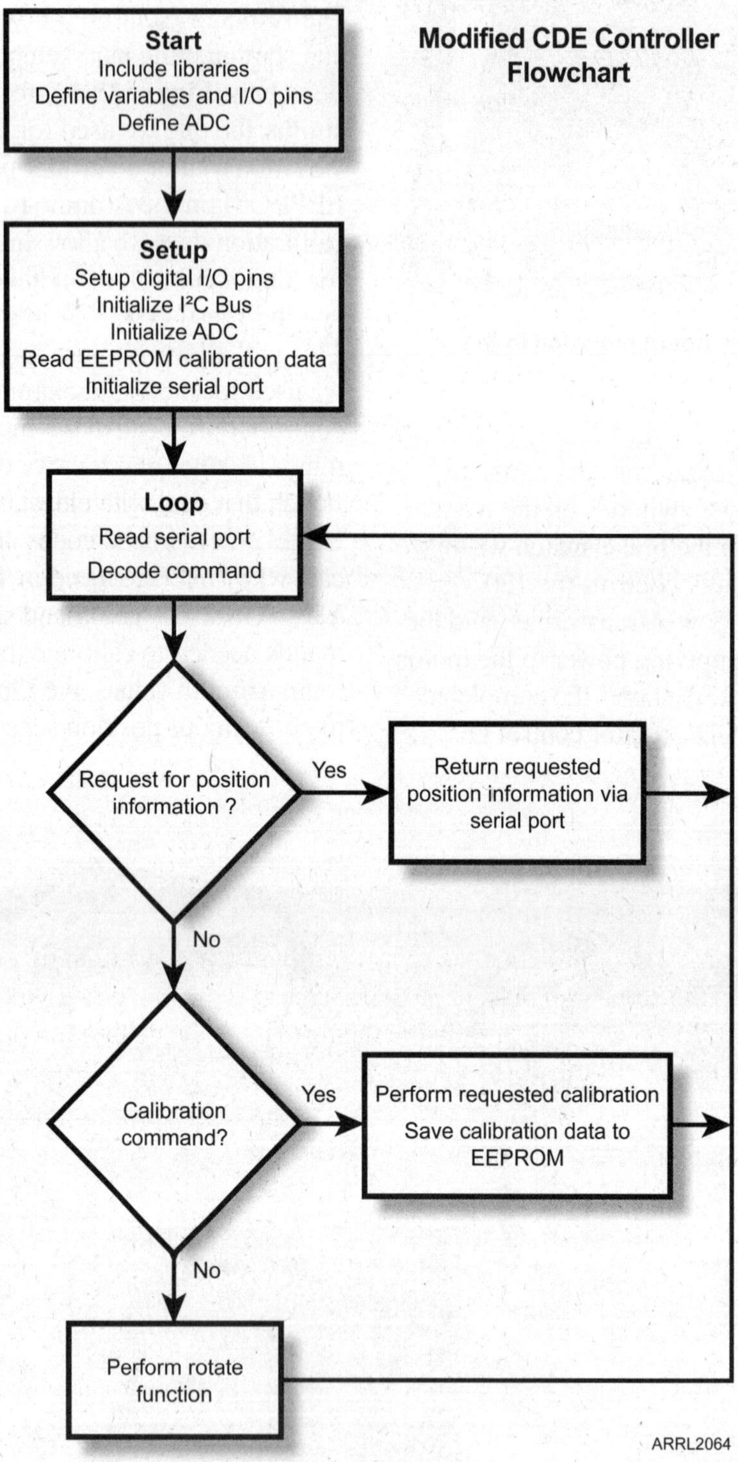

Figure 15.10 — The modified CDE/Hy-Gain rotator controller flowchart.

Table 15.1

CDE Rotator Controller Commands (Subset of Yaesu GS-232A Commands)

Command	*Action*
A	Stop azimuth rotation
C	Return current azimuth value in degrees (format +0aaa)
F	Set azimuth full scale calibration
L	Rotate azimuth counter-clockwise
Maaa	Rotate azimuth to aaa degrees
O	Set azimuth zero calibration
R	Rotate azimuth clockwise
S	Stop all rotation

The complete sketch and libraries for the modified CDE/Hy-Gain rotator controller can be found online at **kw5gp.com**. Starting out with the sketch, we'll include the libraries we need, and initialize the ADS1115 A/D converter object. When your sketch is tested and debugged, don't forget to add the reset disable jumper on the ProtoShield and disable the debug mode, otherwise the *HRD* rotator controller software will not connect to the Arduino.

```
#include <EEPROM.h>  // Include the EEPROM Library
#include <Adafruit_ADS1X15.h> // Include the Adafruit ADS1X15 A/D library

Adafruit_ADS1115 adc; // Define the ADS1115 as adc
```

The next thing we need to do is define the relay pins, our default positioning calibration data, and define the baud rate on the serial port. You need to be sure whatever baud rate you choose matches the settings you plan to use in *HRD*. We'll also define the format for the calibration data that will be kept in the Arduino's onboard EEPROM.

```
#define rotate_left 4  // Define Rotate Left as Pin 4
#define rotate_right 5  // Define Rotate Right as Pin 5
#define brake 6  // Define the Brake pin

#define BAUD_RATE 9600  // Set the Serial Port Baud rate to 9600

#define EEPROM_ID_BYTE 1   // EEPROM ID to validate EEPROM data location
#define EEPROM_ID  55  // EEPROM ID Value
#define EEPROM_AZ_CAL_0 2    // Azimuth Zero Calibration EEPROM location
#define EEPROM_AZ_CAL_MAX 4  // Azimuth Max Calibration Data EEPROM location
#define AZ_CAL_0_DEFAULT 0        // Preset the Azimuth Zero Calibration Point
to 0
```

```
#define AZ_CAL_MAX_DEFAULT 23000        // Preset the Azimuth Max Calibration
Point to 27000

#define AZ_Tolerance 2  // Set the Azimuth Accuracy Tolerance
```

As the last step in the initialization portion of our sketch, we'll define all the variables used in this sketch:

```
//variables
byte inByte = 0;  // incoming serial byte
byte serial_buffer[50];  // incoming serial byte buffer
int serial_buffer_index = 0;  // The index pointer variable for the Serial buffer
int set_AZ;  // Azimuth set value
int current_AZ;  // Current Azimuth raw value
String Serial_Send_Data; // Data to send to Serial Port
int AZ_0;  // Azimuth Zero Value from EEPROM
int AZ_MAX; // Azimuth Max Value from EEPROM
int AZ_Degrees; // mapped AZ ADC value to Degrees
String Requested_AZ; // RS232 Requested Azimuth - M and short W command
int AZ_To; // Requested AZ Move
int AZ_Distance; // Distance to move AZ
```

In the `setup()` loop, we'll start out by setting the digital I/O pin modes for the brake solenoid and motor drive relays. Then we'll turn off the relays to make sure everything is turned off before we begin. Next, we'll start the serial port and the ADS1115 A/D converter. We'll then configure the A/D gain to `GAIN_TWOTHIRDS`, or the 6.144 V range, where one bit equals .01875 mV.

In the last portion of the `setup()` loop, we'll turn off the flag we use to indicate rotator movement and read the rotator position calibration data from the Arduino's EEPROM. If there are no calibration values saved, the sketch will use the defined default calibration values.

```
  pinMode(rotate_left, OUTPUT); // Define the Control Pins as Outputs
  pinMode(rotate_right, OUTPUT);
  pinMode(brake, OUTPUT);

  digitalWrite(rotate_left, LOW);  // Turn off all the relays just to be sure
  digitalWrite(rotate_right, LOW);
  digitalWrite(brake, LOW);

  adc.begin();  // Start the ADS1115 A/D
  adc.setGain(GAIN_TWOTHIRDS);  // 2/3x gain +/- 6.144V  1 bit = 3mV
0.1875mV (default)
```

```
  set_AZ = -1;  // Preset the Azimuth and Elevation Move Variables

  read_eeprom_cal_data();  // Read the Azimuth and Elevation Calibration Values
from EEPROM
```

As with the Yaesu Azimuth-Elevation rotator controller project, there are only two statements in the `main loop()`. The major portion of the work is done by function calls. The `check_serial()` function checks the serial port for incoming commands from *HRD* and executes any commands it receives. The `check_move()` function takes care of the actual rotator movement.

```
  check_serial();  // Check the Serial Port for Data
  check_move();  // Check to see if executing move command
```

The functions in this sketch handle most of the work. The first function, `read_eeprom_cal_data()`, reads the calibration data stored in the Arduino's onboard EEPROM and places the variables we use to map the raw rotator position data into the actual azimuth in degrees.

```
void read_eeprom_cal_data()  // Function to Read the Azimuth and Elevation Cali-
bration Data
{
  if (EEPROM.read(EEPROM_ID_BYTE) == EEPROM_ID) // Verify the EEPROM has valid
data
  {
#ifdef debug // If in Debug Mode Print the Calibration Values
    Serial.println("Read EEPROM Calibration Data Valid ID");
    Serial.println((EEPROM.read(EEPROM_AZ_CAL_0) * 256) + EEPROM.read(EEPROM_AZ_
CAL_0 + 1), DEC);
    Serial.println((EEPROM.read(EEPROM_AZ_CAL_MAX) * 256) + EEPROM.read(EEPROM_
AZ_CAL_MAX + 1), DEC);
#endif

    AZ_0 = (EEPROM.read(EEPROM_AZ_CAL_0) * 256) + EEPROM.read(EEPROM_AZ_CAL_0 +
1); // Read the Azimuth Zero Calibration Value from EEPROM
    AZ_MAX = (EEPROM.read(EEPROM_AZ_CAL_MAX) * 256) + EEPROM.read(EEPROM_AZ_CAL_
MAX + 1); // Read the Azimuth Maximum Calibration Value from EEPROM

  } else
  { // initialize eeprom to default values

#ifdef debug
    Serial.println("Read EEPROM Calibration Data Invalid ID - setting to
defaults");
```

```
#endif

    AZ_0 = AZ_CAL_0_DEFAULT;  // Set the Calibration To Default Values
    AZ_MAX = AZ_CAL_MAX_DEFAULT;
    write_eeprom_cal_data();  // Write the Default Values to EEPROM
  }
}
```

If there is no valid calibration data, the default values are used and saved in the Arduino's EEPROM. The `write_eeprom_cal_data()` function is used to write the calibration data to the Arduino's onboard EEPROM:

```
void write_eeprom_cal_data() // Function to Write the Calibration Values to EE-
PROM
{
  Serial.println("Writing EEPROM Calibration Data");

  EEPROM.write(EEPROM_ID_BYTE, EEPROM_ID); //   Write the EEPROM ID
  EEPROM.write(EEPROM_AZ_CAL_0, highByte(AZ_0)); // Write the Azimuth Zero Cali-
bration High Order Byte
  EEPROM.write(EEPROM_AZ_CAL_0 + 1, lowByte(AZ_0));   // Write the Azimuth Zero
Calibration Low Order Byte
  EEPROM.write(EEPROM_AZ_CAL_MAX, highByte(AZ_MAX)); // Write the Azimuth Max
Calibration High Order Byte
  EEPROM.write(EEPROM_AZ_CAL_MAX + 1, lowByte(AZ_MAX)); // Write the Azimuth Max
Calibration Low Order Byte
}
```

The `check_serial()` function will read the incoming characters from the PC, decode, and execute the Yaesu GS-232A command.

```
void check_serial() // Function to check for data on the Serial port
  {
  if (Serial.available() > 0) // Get the Serial Data if available
  {
    inByte = Serial.read();  // Get the Serial Data

    // You may need to uncomment the following line if your PC software
    // will not communicate properly with the controller
```

```
  // Serial.print(char(inByte));  // Echo back to the PC

  if (inByte == 10)  // ignore Line Feeds
  {
    return;
  }
  if (inByte != 13) // Add to buffer if not CR
  {
    serial_buffer[serial_buffer_index] = inByte;

#ifdef debug // Print the Character received if in Debug mode
    Serial.print("Received = ");
    Serial.println(serial_buffer[serial_buffer_index]);
#endif

    serial_buffer_index++;  // Increment the Serial Buffer pointer

  } else
  { // It's a Carriage Return, execute command

    if ((serial_buffer[0] > 96) && (serial_buffer[0] < 123))  //If first char-
acter of command is lowercase, convert to uppercase
    {
      serial_buffer[0] = serial_buffer[0] - 32;
    }
```

If you are not using *HRD* to control your rotator, you may need to uncomment the statement that echoes the received data back to the PC. *HRD* will not connect to the rotator controller if debug mode is on, or the received characters are echoed back to the PC. You will also need to have the DTR reset disable jumper or switch enabled to prevent *HRD* from resetting the Arduino when it tries to connect.

Once a serial command has been received, we'll use a `switch...case()` statement to decode and execute the command.

```
switch (serial_buffer[0]) {  // Decode first character of command

      case 65:  // A Command - Stop the Azimuth Rotation


#ifdef debug
        Serial.println("A Command Received");
#endif
```

```
            az_rotate_stop();
            break;

        case 67:        // C - return current azimuth

#ifdef debug
            Serial.println("C Command Received");
            Serial.println(serial_buffer_index);
#endif

            send_current_az();  // Return Azimuth if C Command
            break;

        case 70:  // F - Set the Max Calibration

#ifdef debug
            Serial.println("F Command Received");
            Serial.println(serial_buffer_index);
#endif

            set_max_az_cal();  // F - Set the Max Azimuth Calibration
            break;

        case 76:  // L - Rotate Azimuth CCW

#ifdef debug
            Serial.println("L Command Received");
#endif

            rotate_az_ccw();  // Call the Rotate Azimuth CCW Function
            break;

        case 77:  // M - Rotate to Set Point

#ifdef debug
            Serial.println("M Command Received");
#endif

            rotate_to();  // Call the Rotate to Set Point Command
            break;
```

```
        case 79:  // O - Set Zero Calibration

#ifdef debug
          Serial.println("O Command Received");
          Serial.println(serial_buffer_index);
#endif

          set_0_az_cal();  // O - Set the Azimuth Zero Calibration
          break;

        case 82:  // R - Rotate Azimuth CW

#ifdef debug
          Serial.println("R Command Received");
#endif

          rotate_az_cw();  // Call the Rotate Azimuth CW Function
          break;

        case 83:  // S - Stop All Rotation

#ifdef debug
          Serial.println("S Command Received");
#endif

          az_rotate_stop();  // Call the Stop Azimith Rotation Function
          break;

      }

      serial_buffer_index = 0;  // Clear the Serial Buffer and Reset the Buffer
Index Pointer
      serial_buffer[0] = 0;
    }
```

After the decoded command has been executed, the serial data buffer is cleared, and the sketch is ready to receive the next command.

The `send_current_az()` function will read the A/D converter and convert the raw A/D value to a calibrated azimuth value.

```
void send_current_az() // Send the Current Azimuth Function
{
  read_adc();  // Read the ADC

  // Map Azimuth to degrees

#ifdef debug
  Serial.println(current_AZ);
#endif

  AZ_Degrees = map(current_AZ, AZ_0, AZ_MAX, 0, 360);  // Map the Current Azi-
muth to Degrees
  if (AZ_Degrees > 180) // Correction Since Azimuth Reading starts at Meter Cen-
ter Point
  {
    AZ_Degrees = AZ_Degrees - 180;
  } else {
    AZ_Degrees = AZ_Degrees + 180;
  }

#ifdef debug
  Serial.println(AZ_Degrees);
#endif

  // Send it back via serial
  Serial_Send_Data = "";
  if (AZ_Degrees < 100)  // pad with 0's if needed
  {
    Serial_Send_Data = "0";
  }
  if (AZ_Degrees < 10)
  {
    Serial_Send_Data = "00";
  }
  Serial_Send_Data = "+0" + Serial_Send_Data + String(AZ_Degrees);  // Send the
Azimuth in Degrees
  Serial.println(Serial_Send_Data);  // Return value via RS-232 port
}
```

The azimuth data is adjusted to allow for the 0/360° position being at midscale instead of at the minimum and maximum values on a north-centered rotator configuration. You'll need to modify this function if you are using a south-centered rotator. The azimuth value in degrees is then sent to the PC via the USB port.

```
void set_max_az_cal() // Set the Max Azimuth Calibration Function
{
  Serial.println("Calibrate Max AZ Function");

  read_adc();  // Read the ADC

  // save current az value to EEPROM - Zero Calibration
  Serial.println(current_AZ);

  AZ_MAX = current_AZ;  // Set the Azimuth Maximum Calibration to Current Azi-
muth Reading
  write_eeprom_cal_data();  // Write the Calibration Data to EEPROM
  Serial.println("Max Azimuth Calibration Complete");
}
```

The `set_max_az_cal()` function sets the calibration value for the maximum azimuth rotation. This value is then saved to the `AZ_MAX` calibration variable, and to the Arduino's onboard EEPROM.

```
void rotate_az_ccw() // Function to Rotate Azimuth CCW
{
  digitalWrite(brake, HIGH);  // Release the brake
  digitalWrite(rotate_left, HIGH);  // Set the Rotate Left Pin High
  digitalWrite(rotate_right, LOW);  // Make sure the Rotate Right Pin is Low
}

void rotate_az_cw() // Function to Rotate Azimuth CW
{
  digitalWrite(brake, HIGH);  // Release the brake
  digitalWrite(rotate_right, HIGH);    // Set the Rotate Right Pin High
  digitalWrite(rotate_left, LOW);    // Make sure the Rotate Left Pin Low
}
```

The `rotate_az_cw()` and `rotate_az_ccw()` functions control the brake solenoid and the rotator motor drive relays.

```
void az_rotate_stop() // Function to Stop Azimuth Rotation
{
  digitalWrite(rotate_right, LOW);  // Turn off the Rotate Right Pin
  digitalWrite(rotate_left, LOW);   // Turn off the Rotate Left Pin
  set_AZ = -1;
  delay(1000);
  digitalWrite(brake, LOW);  // Engage the brake
}
```

The `az_rotate_stop()` function stops all rotation. The motor drive relays are de-energized, and a one second delay is added to allow the rotator to coast to a stop. After the delay expires, the brake solenoid is de-energized, and the brake is engaged.

```
void rotate_to() // Function to Rotate to Set Point
{
#ifdef debug
  Serial.println("M Command -  Rotate Azimuth To Function");
  // Decode Command - Format Mxxx - xxx = Degrees to Move to
  Serial.println(serial_buffer_index);
#endif

  if (serial_buffer_index == 4)  // Verify the Command is the proper length
  {
#ifdef debug
    Serial.println("Value in [1] to [3]?");
#endif

    Requested_AZ = (String(char(serial_buffer[1])) + String(char(serial_buf-
fer[2])) + String(char(serial_buffer[3]))) ;  // Decode the Azimuth Value
    AZ_To = (Requested_AZ.toInt()); // AZ Degrees to Move to as integer
    if (AZ_To < 0) // Make sure we don't go below 0 degrees
    {
      AZ_To = 0;
    }
    if (AZ_To > 360) // Make sure we don't go over 360 degrees
    {
      AZ_To = 360;
    }
    if (AZ_To > 180) // Adjust for Meter starting at Center
    {
```

```
        AZ_To = AZ_To - 180;
      } else {
        AZ_To = AZ_To + 180;
      }

#ifdef debug
    Serial.println(Requested_AZ);
    Serial.println(AZ_To);
#endif

    // set the move flag and start
    read_adc();  // Read the ADC

    // Map it to degrees

    AZ_Degrees = map(current_AZ, AZ_0, AZ_MAX, 0, 360);  // Map the Azimuth
Value to Degrees

#ifdef debug
    Serial.println(current_AZ);
    Serial.println(AZ_Degrees);
#endif

    AZ_Distance = AZ_To - AZ_Degrees;  // Figure out far we have to move
    set_AZ = AZ_To;
    if (abs(AZ_Distance) <= AZ_Tolerance)  // No move needed if we're within the
Tolerance Range
    {
      az_rotate_stop();  // Stop the Azimuth Rotation
      set_AZ = -1;  // Turn off the Move Command
    } else {  // Move Azimuth - figure out which way
      if (AZ_Distance > 0)   //We need to move CW
      {
        rotate_az_cw();  // If the distance is positive, move CW
      } else {
        rotate_az_ccw();  // Otherwise, move counterclockwise
      }
```

The `rotate_to()` function moves the rotator to the commanded position. The rotator will turn automatically until it reaches the desired position or is given a stop command.

```
void set_0_az_cal() // Set Azimuth Zero Calibration
{
  Serial.println("Calibrate Zero Function");

  read_adc();  // Read the ADC
  // save current Azimuth value to EEPROM - Zero Calibration
  Serial.println(current_AZ);

  AZ_0 = current_AZ;  // Set the Azimuth Zero Calibration to current position
  write_eeprom_cal_data();  // Write the Calibration Data to EEPROM
  Serial.println("Zero Azimuth Calibration Complete");
}
```

The `set_0_az_cal()` function sets the azimuth zero calibration value. This value is then stored to the `AZ_0` calibration variable and saved to the Arduino's onboard EEPROM.

```
void read_adc() // Function to read the ADC
{
#ifdef debug
  Serial.println("Read ADC Function  ");
#endif

  int RotorValue;  // Variable to store the rotor value
  adc.setGain(GAIN_TWOTHIRDS);
  delay(10); // adc settling delay

  current_AZ = adc.readADC_SingleEnded(0);
}
```

The `read_adc()` function reads the ADS1115 module, and returns the raw uncalibrated rotator position digital value.

```
void check_move() // Check to see if we've been commanded to move
{
  if (set_AZ != -1) {   // We're moving - check and stop as needed
    read_adc();  // Read the ADC
    // Map AZ to degrees

    AZ_Degrees = map(current_AZ, AZ_0, AZ_MAX, 0, 360);  // Map the Current Azi-
muth reading to Degrees
```

```
#ifdef debug
    Serial.println(current_AZ);
    Serial.println(AZ_Degrees);
#endif

    if (set_AZ != -1) // If Azimuth is moving
    {
      AZ_Distance = set_AZ - AZ_Degrees;  // Check how far we have to move
      if (abs(AZ_Distance) <= AZ_Tolerance)  // No move needed if we're within
the tolerance range
      {
        az_rotate_stop();  // Stop the Azimuth Rotation
        set_AZ = -1;  // Turn off the Azimuth Move Command
      } else {  // Move Azimuth - figure out which way
        if (AZ_Distance > 0)   //We need to move CW
        {
          rotate_az_cw();  // Rotate CW if positive
        } else {
          rotate_az_ccw();  // Rotate CCW if negative
        }
```

Finally, the `check_move()` function handles the actual rotation. If the sketch has received a move command, the `check_move()` function will manage the rotation until the commanded rotator position has been reached or a stop command is received.

You can now see how much of the Yaesu azimuth-elevation controller sketch we were able to reuse. This is one reason why working with the Arduino is so much fun. Once you have built your own Arduino sketch and library collection, you can reuse them endlessly in your future projects, greatly reducing your sketch development time. Being able to reuse huge blocks of code made this sketch very easy to create, since all we had to do was modify a little here and there. Once you have the sketch ready for testing and debugging, you can use the serial monitor to send the Yaesu GS-232A commands to the rotator controller.

After you have everything debugged and working, don't forget to recompile and upload the sketch with the debug mode turned off, and the reset disable jumper installed on the Arduino ProtoShield, before trying to control the rotator with *HRD*.

Calibrating the Modified CDE/ Hy-Gain Rotator Controller

Now we're ready to have some real fun. With the sketch complete, we're ready to test the interface and set the azimuth calibration values. To test the interface and set the azimuth calibration values, you can use the Arduino IDE's serial monitor to send the Yaesu GS-232A commands directly to the Arduino. Manually rotate the azimuth to zero (the front panel analog meter at minimum deflection) and use the **O** command to set the zero-calibration point. Then, manually rotate the azimuth to full-scale on the front panel analog meter and use the **F** command to set the azimuth full-scale value. Once you have done that, your controller is calibrated, and the calibration values are stored in the Arduino's onboard

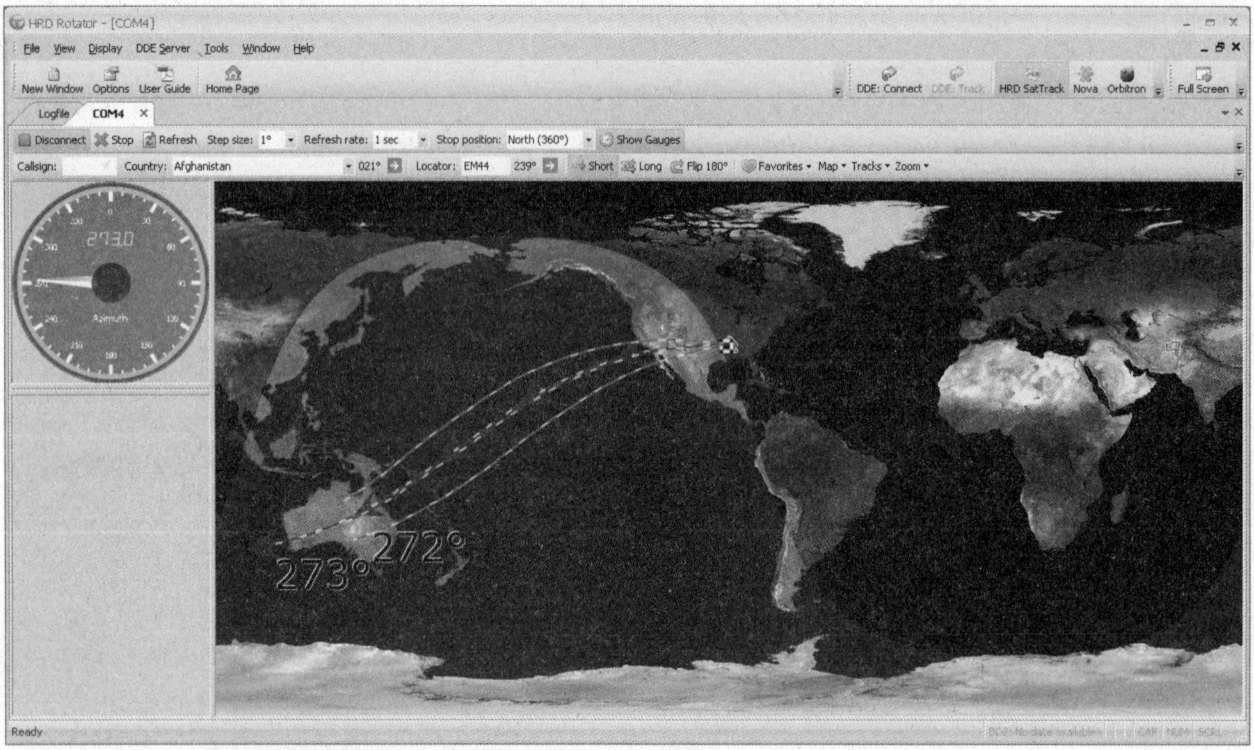

Figure 15.11 — *Ham Radio Deluxe* communicating with the modified CDE/Hy-Gain controller.

EEPROM. You can recalibrate your controller at any time by repeating the calibration process. You can now send the various Yaesu GS-232A controller commands listed in **Table 15.1** to test all the functions on your controller.

Operating the Modified CDE/ Hy-Gain Rotator Controller

Now that your controller is calibrated, you're ready to use it with the *HRD* rotator controller software on your PC. Don't forget to install the reset disable jumper and have the sketch loaded with the debug mode turned off. Configure the *HRD* rotator controller to use the Arduino's COM port, set the baud rate to match the serial port speed setting in the sketch, and select the Yaesu GS-232A/Az rotator controller in *HRD*. Now your rotator can be controlled by *HRD*, and the current rotator position will be displayed on the *HRD* rotator controller screen as shown in **Figure 15.11**. If the rotator turns correctly, but *HRD* doesn't show anything on the rotator dial, you need to verify that the Arduino does not have the command echo statement uncommented in the `check_serial()` function. *HRD* doesn't like it when you echo the command back to it. If everything went well, you should now have a fully operational CDE/Hy-Gain rotator controller that can be controlled by *HRD* or other rotator software on your PC.

16 Yaesu Azimuth-Elevation Rotator Controller

The Yaesu azimuth-elevation rotator controller (see **Figure 16.1**) is one of my favorite Arduino projects. With a simple design, I was able to control my satellite antenna system to automatically track the amateur radio satellites and the International Space Station as they moved across the sky. It has always been a project that I wanted to revisit and incorporate more features into. For this version, we'll upgrade to an Arduino Nano and add a 2.2-inch color TFT display for the elevation and azimuth angles.

Based on the success of my ARRL Field Day satellite tracker prototype, the next step was to build a full-scale Arduino controller so that my Yaesu G-5400B series azimuth-elevation dual rotator controller could be controlled from either SatPC32 or *Ham Radio Deluxe (HRD)* software. The Yaesu G-5400B rotator controller is designed to be controlled by the Yaesu GS-232 rotator computer interface. There are several different models of the Yaesu GS-232 rotator computer interface with a few slight differences between them. For this project, we will be emulating the Yaesu GS-232A interface.

The Yaesu G-5400B series dual rotator controllers have a built in 8-pin DIN connector for external control (see **Figure 16.2**). This connector provides an analog voltage output for the azimuth and elevation values, and relay control inputs for azimuth and elevation control. Ordinarily, you would connect the Yaesu GS-232 rotator computer interface unit to this connector, but we're going to replace the GS-232 with an Arduino, and have the Arduino emulate the GS-232 interface. We'll use the Arduino's USB port to interface with SatPC32 or *HRD* on the PC side of things. We'll also be using the Arduino's onboard EEPROM to save our calibration settings. This way, the azimuth-elevation rotator controller will remember the calibration data even after a reset or power-off.

Figure 16.1 — The Yaesu G-5400B azimuth-elevation rotator controller.

Figure 16.3 shows the block diagram for this project. The external hardware needed is minimal, comprised of six resistors, four transistors, three capacitors, an Arduino Nano, an ILI9341-type color TFT display, a TXB0108-type level converter, and a Texas Instruments ADS1115 four-channel 16-bit I²C analog-to-digital converter. You could use the Arduino's built in 10-bit analog-to-digital converter instead of the ADS1115, but you will not have the resolution that the extra 6-bits of the ADS1115 provides. The Arduino's A/D converter would only read 3 A/D counts per degree of rotation versus the 148 A/D counts per degree of the ADS1115. Any noise on the positioning signals could cause positioning errors. Since we're tracking satellites, it's worth using the 16-bit A/D to get the extra accuracy.

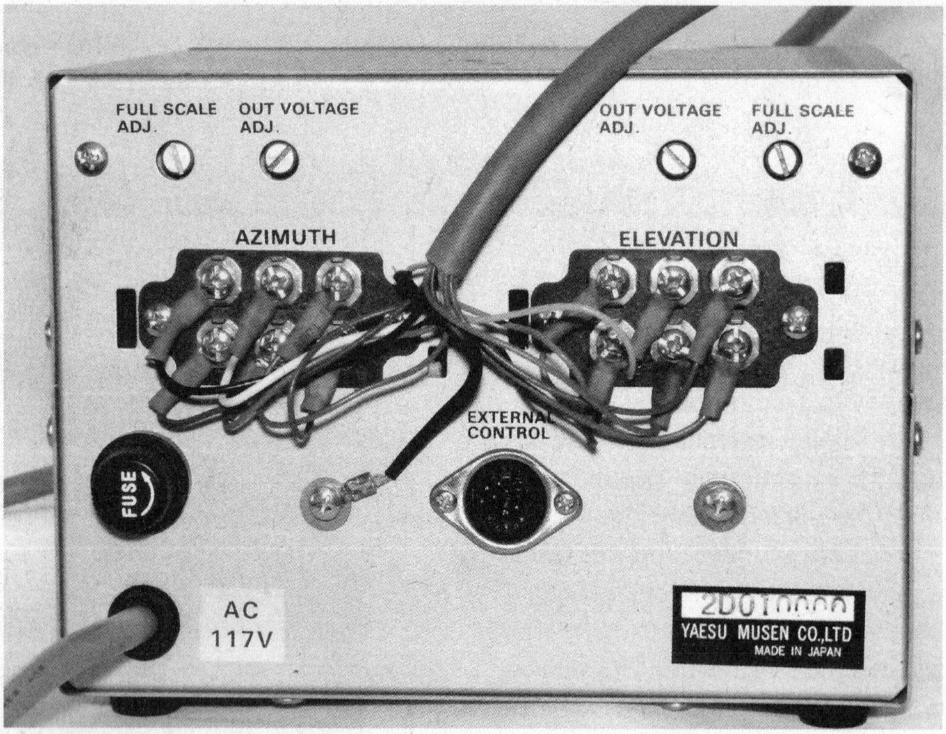

Figure 16.2 — Azimuth elevation rotator controller rear view.

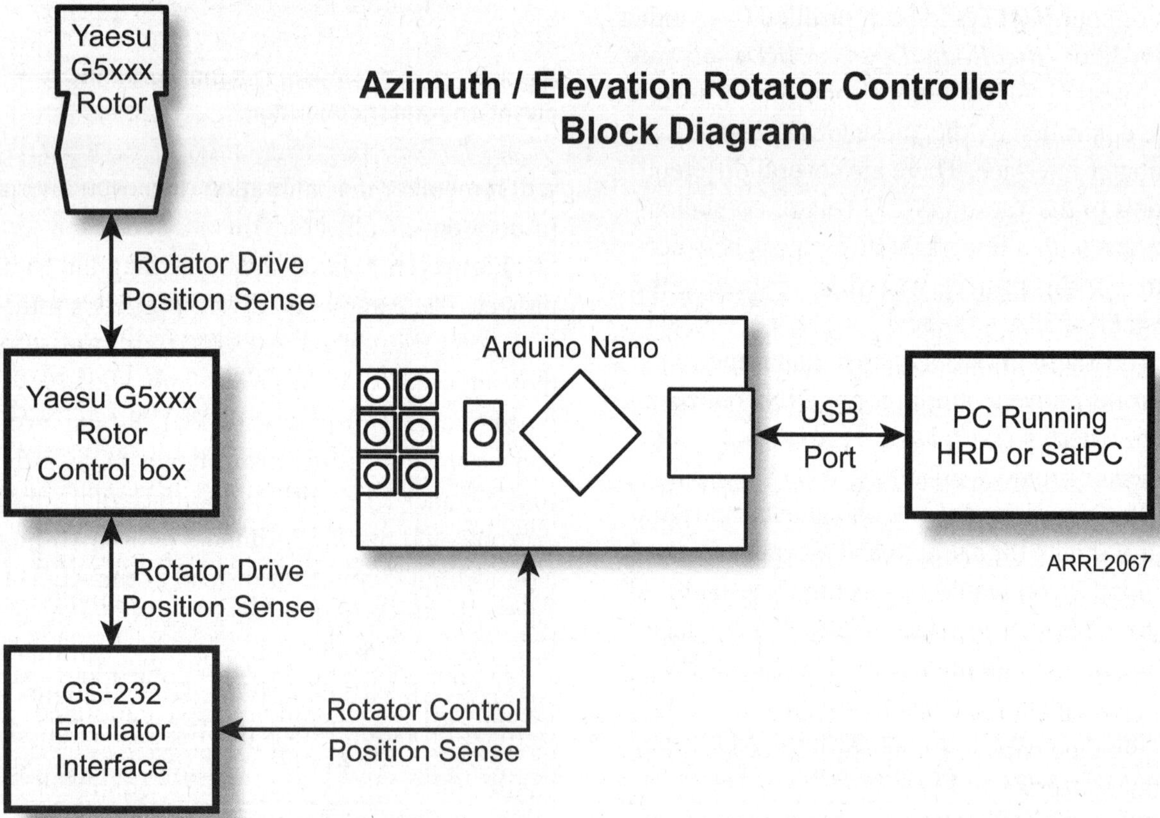

Figure 16.3 — Azimuth elevation rotator controller block diagram.

Figure 16.4 — The ADS1115 16-bit A/D module. ***(Ed. — This image has been enlarged. These modules are about the size of a quarter.)***

The Texas Instruments ADS1115 Analog to Digital Converter

The Texas Instruments ADS1115 (see **Figure 16.4**) is a 4 channel 16 bit analog-to-digital converter that communicates with the Arduino using the I²C bus. The analog inputs can be configured as 4 single-ended inputs or 2 differential inputs, with the input gain program-selectable in 6 steps from 256 mV all the way up to 6.144 V. It is important to note that the 6.144 V gain setting is only used as a range on the A/D converter. At no time should a voltage higher than VDD+0.3 V (5.3 V) ever be used as an analog input to the A/D, otherwise you may damage the chip. The sampling rate is program-selectable in 8 steps from 8 – 860 samples per second, along with a single-shot conversion mode. Because of the ADS1115's flexibility and ease of interfacing, it has become my personal analog-to-digital converter of choice when I need higher resolution than the Arduino's onboard 10-bit analog-to-digital converter.

Since the G-5400B rotator controller already has most of the interface circuitry we'll need built-in, constructing our portion of the interface on the breadboard should go quickly. It is important to note that unlike the other rotator controller projects, we don't need relays to drive this one. The Yaesu G-5400B rotator controller already has the control relays (with clamping diodes) built-in and connected to the external control connector, so all we need to do to drive the rotator is use a transistor for each rotator control pin. From your breadboard, you can build an 8-pin header to DIN plug adapter cable to connect to the external control socket on the rotator controller. When you move the project to the perfboard, you can use an 8-pin header on the board to connect the DIN plug adapter cable you've already built. We'll power the Azimuth-Elevation rotator controller from the Arduino's USB connector that attaches to a PC, so we won't need an external source of power for this project. **Figure 16.5**, which shows the schematic for the azimuth-elevation rotator controller, also includes the parts list.

Do not connect the rotator controller to your breadboard circuit until you have calibrated the output voltages on the controller's external control pins. First, use the rotator controller and manually position the antenna to full scale for azimuth and elevation. Using a voltmeter, measure the voltage between the ground pin (Pin 8) and

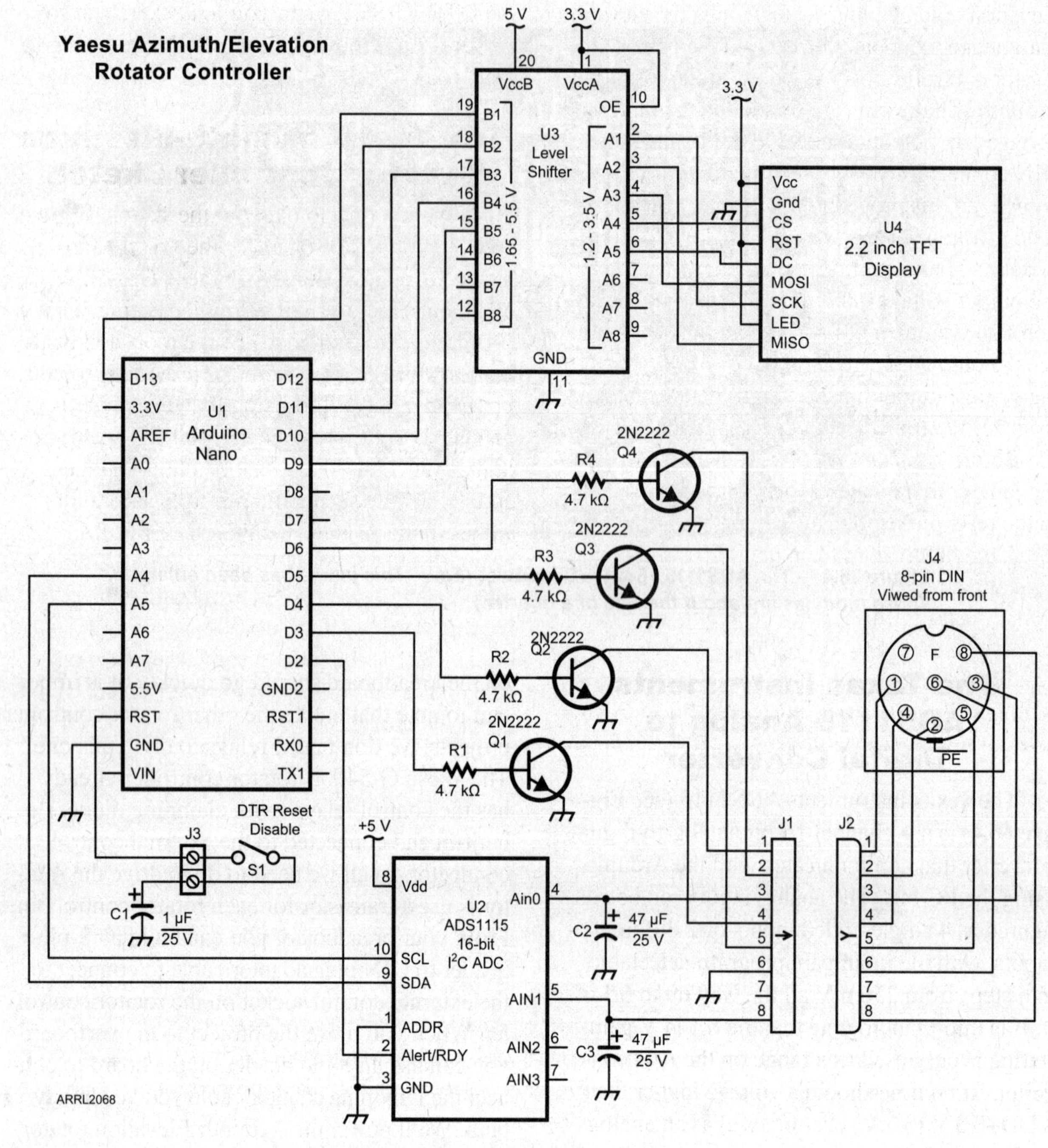

Figure 16.5 — Azimuth elevation rotator controller schematic.

Part	*Value*	*Part*	*Value*
C1	**1 µf capacitor, 25 V**	**U3**	**8-channel level shifter module (TXB0108 type)**
C2, C3	**47 µF capacitor, 25 V**	**U4**	**2.2-inch ILI9341-type color TFT LCD display**
J1, J2	**8-pin Dupont header/ socket connector**	**R1-R4**	**4.7 KΩ resistor**
J3	**2-pin Dupont header connector**	**Q1-Q4**	**2N2222A NPN transistor**
J4	**8-Pin DIN female connector**		
U1	**Arduino Nano**		
U2	**TI ADS1115 16-bit analog-to-digital converter**		

elevation voltage pin (Pin 1). Adjust the elevation out voltage adj. potentiometer on the back of the rotator controller for 5 V. Do the same for the azimuth voltage pin (Pin 6) and the azimuth out voltage adj. potentiometer. Do not connect the DIN connector to your breadboard until you have completed this procedure, otherwise you could send a voltage higher than 5 V to the A/D converter and damage it. When you have completed the output voltage calibration, you should be able to rotate your azimuth and elevation rotators and see the output change from 0 – 5 V on each of the voltage output pins as you rotate the antenna from zero to full scale on the front panel meters.

Before you can load your sketch, be sure that the jumper that connects the 1 µf capacitor (C1) to the reset pin is off. This capacitor is used to block the **RESET** command that *HRD* accidentally sends to the Arduino when it toggles the DTR signal line as it begins a connection attempt to the rotator controller. If you forget to short this jumper after loading your sketch, *HRD* will not be able to connect to the controller since it will be resetting. However, you do want to enable the DTR RESET when you load sketches, so be sure and remove the jumper before trying to load your sketch.

The Yaesu Azimuth-Elevation Rotator Controller Sketch

Next, it's time to plan out the sketch. **Figure 16.6** shows the flowchart for the sketch, and **Table 16.1** shows the actual Yaesu GS-232A commands that we will be implementing. This sketch will be using many of the tools and techniques we've learned from the previous projects. For example, we'll be using the Texas Instruments ADS1115 4-channel 16-bit I^2C analog-to-digital converter to read the position voltages, and we will be saving the position calibration values in the Arduino's onboard EEPROM. Using the Arduino's USB port, we will communicate with the satellite tracking software on the PC. To aid in troubleshooting, we will also implement a debug mode in the sketch will return diagnostic information to the Arduino IDE's serial monitor when set.

Table 16.1

Azimuth/Elevation Rotator Controller Commands (Subset of Yaesu GS-232A Commands)

Command	*Action*
A	Stop azimuth rotation
B	Return current elevation value in degrees (format +0eee)
C	Return current azimuth value in degrees (format +0aaa)
C2	Return current azimuth and elevation (format +0aaa+0eee)
D	Rotate elevation down
E	Stop elevation rotation
F	Set azimuth full scale calibration
F2	Set elevation full scale calibration
L	Rotate azimuth counterclockwise
Maaa	Rotate azimuth to aaa degrees
O	Set azimuth zero calibration
O2	Set elevation zero calibration
R	Rotate azimuth clockwise
S	Stop all rotation
U	Rotate elevation up
Waaa eee	Rotate azimuth to aaa degrees and elevation to eee degrees

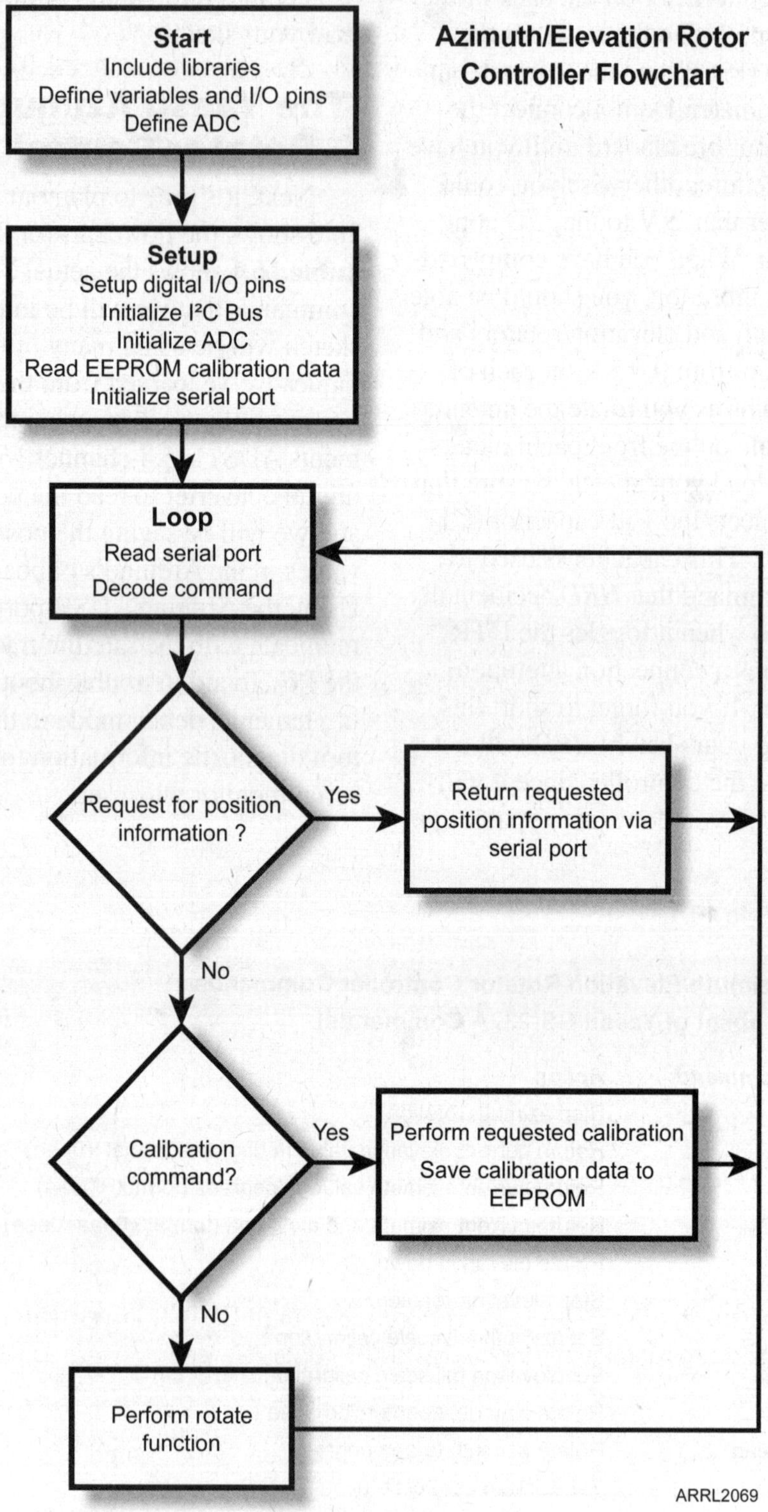

Figure 16.6 — The azimuth-elevation rotator controller flowchart.

From the flowchart you can see that in the `main loop()` we decode the command string from the tracking software on the PC into three basic groups of functions: status requests, calibration commands, and movement commands. By using this divide and conquer method, we break the sketch down into manageable chunks that can be independently created and debugged. The complete sketch and libraries can be found in online at **kw5gp.com**.

Starting out in the sketch, we'll define the libraries and the ADS1115 object.

```
#include <Adafruit_ADS1X15.h> // Include the adafruit ADS1X15 A/D library
#include <EEPROM.h> // Include the EEPROM library so we can save calibration
data
#include <SPI.h>
#include <Adafruit_GFX.h> // The Adafruit Graphics Library
#include <Adafruit_ILI9341.h> // The ILI9341 TFT Display Library

Adafruit_ADS1115 adc;  /* Use this for the 16-bit version */
```

Next, we'll define the Arduino digital I/O pins used to control the antenna rotation and the TFT display, along with the baud rate for the serial port used to interface with the tracking software on your PC. The baud rate in the sketch must match the baud rate you have set in the tracking software on your PC.

```
#define rotate_down 3
#define rotate_up 4
#define rotate left 5
#define rotate_right 6

#define TFT_CS 10 // Defines the CS pin for the TFT
#define TFT_DC 9  // Defines the DC pin for the TFT

#define BAUD_RATE 19200 // Max Baud Rate is 19200 if using Ham Radio Deluxe,
57600 if using SatPC32
```

Since we will be storing the positioning calibration data in EEPROM, we will need to define the format we will be using to read and store the EEPROM data.

```
#define EEPROM_ID_BYTE 1   // EEPROM ID to validate EEPROM data location
#define EEPROM_ID  55  // EEPROM ID Value
#define EEPROM_AZ_CAL_0 2    // Azimuth Zero Calibration EEPROM location
#define EEPROM_AZ_CAL_MAX 4  // Azimuth Max Calibration Data EEPROM location
#define EEPROM_EL_CAL_0 6    // Elevation Zero Calibration Data EEPROM location
#define EEPROM_EL_CAL_MAX 8  // Elevation Max Calibration Data EEPROM location
```

In case there is no calibration data stored in the EEPROM, default values are defined, along with a tolerance value for the rotator positioning. The tolerance value, which is measured in degrees, is used so that the controller can determine that the current rotator positioning is within the tolerance range to the desired position. The tolerance value is designed for use when you have excessive noise on the positioning signals and want to prevent rotation when the desired position and the actual position are calculated to be a few degrees apart from each other. In practical use, a tolerance of three works well in my shack.

```
#define AZ_CAL_0_DEFAULT 0
#define AZ_CAL_MAX_DEFAULT 27000
#define EL_CAL_0_DEFAULT 0
#define EL_CAL_MAX_DEFAULT 27000

#define AZ_Tolerance 3  // The G5400 controller needs some tolerance because
#define EL_Tolerance 3  // of the power supply drop during rotation
Next, we'll define all the variables used in the sketch:
//variables
byte inByte = 0;  // incoming serial byte
byte serial_buffer[50];  // incoming serial byte buffer
int serial_buffer_index = 0;
int set_AZ;  // Azimuth set value
int set_EL;  // Elevation set value
int current_AZ;  // Current Azimuth raw value
int current_EL;  // Current Elevation raw value
String Serial_Send_Data; // Data to send to Serial Port
int AZ_0;  // Azimuth Zero Value from EEPROM
int AZ_MAX; // Azimuth Max Value from EEPROM
int EL_0;  // Elevation 0 Value from EEPROM
int EL_MAX;  //Elevation Max Value from EEPROM
int AZ_Degrees; // mapped AZ ADC value to Degrees
int EL_Degrees; // mapped EL ADC value to Degrees
String Requested_AZ; // RS232 Requested Azimuth - M and short W command
String Requested_EL; //RS232 Requested Azimuth and Elevation - Full W command
int AZ_To; // Requested AZ Move
int EL_To; // Requested EL Move
int AZ_Distance; // Distance to move AZ
int EL_Distance; // Distance to move EL
int prev_AZ_Degrees = -1;  // The previous reading so we can update the TFT
int prev_EL_Degrees = -1; // Set to -1 so it will trigger an update

int background = tft_black; // Set the TFT background and text color
int text_color = tft_cyan;
```

Next, we define the default calibration values and the tolerance settings, followed by the sketch variables and finally, we'll instantiate (create) the TFT display object.

```
#define AZ_CAL_0_DEFAULT 0
#define AZ_CAL_MAX_DEFAULT 27000
#define EL_CAL_0_DEFAULT 0
#define EL_CAL_MAX_DEFAULT 27000

#define AZ_Tolerance 3  // The G5400 controller needs some tolerance because
#define EL_Tolerance 3  // of the power supply drop during rotation

//variables
byte inByte = 0;  // incoming serial byte
byte serial_buffer[50];  // incoming serial byte buffer
int serial_buffer_index = 0;
int set_AZ;  // Azimuth set value
int set_EL;  // Elevation set value
int current_AZ;  // Current Azimuth raw value
int current_EL;  // Current Elevation raw value
String Serial_Send_Data; // Data to send to Serial Port
int AZ_0;  // Azimuth Zero Value from EEPROM
int AZ_MAX; // Azimuth Max Value from EEPROM
int EL_0;  // Elevation 0 Value from EEPROM
int EL_MAX;  //Elevation Max Value from EEPROM
int AZ_Degrees; // mapped AZ ADC value to Degrees
int EL_Degrees; // mapped EL ADC value to Degrees
String Requested_AZ; // RS232 Requested Azimuth - M and short W command
String Requested_EL; //RS232 Requested Azimuth and Elevation - Full W command
int AZ_To; // Requested AZ Move
int EL_To; // Requested EL Move
int AZ_Distance; // Distance to move AZ
int EL_Distance; // Distance to move EL

int prev_AZ_Degrees = -1;  // The previous reading so we can update the TFT
int prev_EL_Degrees = -1; // Set to -1 so it will trigger an update

int background = tft_black; // Set the TFT background and text color
int text_color = tft_cyan;

// Use hardware SPI for TFT
Adafruit_ILI9341 tft = Adafruit_ILI9341(TFT_CS, TFT_DC);  // Create the TFT instance
```

In the `setup()` loop, we'll initialize the digital I/O pins and make sure that all the position control outputs are turned off.

```
void setup()
{
  pinMode(rotate_up, OUTPUT);
  pinMode(rotate_down, OUTPUT);
  pinMode(rotate_left, OUTPUT);
  pinMode(rotate_right, OUTPUT);

  digitalWrite(rotate_up, LOW);
  digitalWrite(rotate_down, LOW);
  digitalWrite(rotate_left, LOW);
  digitalWrite(rotate_right, LOW);
```

Then, we'll start the TFT display and display a brief startup message. Now, we'll initialize the serial port and the ADS1115 A/D converter. As the final step of initializing the A/D converter, we'll set it to read a range of 0 – 6.144 V, which works out to 0.1875 mV per A/D count.

```
  start_tft();   // Start the TFT display

  Serial.begin(BAUD_RATE); // initialize serial communication on the USB port

  adc.begin();   // Start the ADS1115 A/D

  adc.setGain(GAIN_TWOTHIRDS);   // 2/3x gain +/- 6.144V  1 bit = 3mV
0.1875mV (default)
```

As the last step in the `setup()` loop, the move flag variables are set to -1 to indicate that there is no active rotation command, and the calibration values (if any) are loaded from EEPROM and update the TFT display to show the current rotator positions.

```
  set_AZ = -1;
  set_EL = -1;

  read_eeprom_cal_data();

  update_tft();

  // End Setup
```

The `main loop()` for this sketch is rather unique. It consists of only three statements. The divide and conquer design strategy for this project resulted in much of the work being handled by functions. All the `main loop()` does is continually check for incoming commands on the serial port, decode and execute any commands it receives, and handle any move commands. Each decoded command will call a function to perform the action called for by the incoming command.

```
check_serial();
check_move();
update_tft();
```

Most of this project is controlled by the various functions. As part of the `setup()` loop, the `read_eeprom_cal_data()` function is called to load the saved calibration data into the sketch variables. This function checks to see if there is a valid EEPROM ID byte and loads the EEPROM data into the sketch variables if the ID is valid. If the EEPROM ID is not valid, the sketch calibration variables are loaded with the defined default values and saved to EEPROM. If the debug mode is defined, the IDE's serial monitor will display the values read from the EEPROM.

```
void read_eeprom_cal_data()
{
  if (EEPROM.read(EEPROM_ID_BYTE) == EEPROM_ID)
  {

#ifdef debug
    Serial.println("Read EEPROM Calibration Data Valid ID");
    Serial.println((EEPROM.read(EEPROM_AZ_CAL_0) * 256) + EEPROM.read(EEPROM_AZ_
CAL_0 + 1), DEC);
    Serial.println((EEPROM.read(EEPROM_AZ_CAL_MAX) * 256) + EEPROM.read(EEPROM_
AZ_CAL_MAX + 1), DEC);
    Serial.println((EEPROM.read(EEPROM_EL_CAL_0) * 256) + EEPROM.read(EEPROM_EL_
CAL_0 + 1 ), DEC);
    Serial.println((EEPROM.read(EEPROM_EL_CAL_MAX) * 256) + EEPROM.read(EEPROM_
EL_CAL_MAX + 1), DEC);
#endif

    AZ_0 = (EEPROM.read(EEPROM_AZ_CAL_0) * 256) + EEPROM.read(EEPROM_AZ_CAL_0 + 1);
    AZ_MAX = (EEPROM.read(EEPROM_AZ_CAL_MAX) * 256) + EEPROM.read(EEPROM_AZ_CAL_
MAX + 1);
    EL_0 = (EEPROM.read(EEPROM_EL_CAL_0) * 256) + EEPROM.read(EEPROM_EL_CAL_0 + 1);
    EL_MAX = (EEPROM.read(EEPROM_EL_CAL_MAX) * 256) + EEPROM.read(EEPROM_EL_CAL_
MAX + 1);
  } else
  { // initialize eeprom to default values
```

```
#ifdef debug
    Serial.println("Read EEPROM Calibration Data Invalid ID - setting to de-
faults");
#endif

    AZ_0 = AZ_CAL_0_DEFAULT;
    AZ_MAX = AZ_CAL_MAX_DEFAULT;
    EL_0 = EL_CAL_0_DEFAULT;
    EL_MAX = EL_CAL_MAX_DEFAULT;
    write_eeprom_cal_data();
  }
}
```

The `write_eeprom_cal_data()` function is called when data is to be saved to the EEPROM, either as part of the initial EEPROM setup, or when a calibration command is received via the USB serial port. The 16-bit integer values are broken into high and low bytes and written to the EEPROM one byte at a time.

```
void write_eeprom_cal_data()
{
#ifdef debug
  Serial.println("Writing EEPROM Calibration Data");
#endif

  EEPROM.write(EEPROM_ID_BYTE, EEPROM_ID);
  EEPROM.write(EEPROM_AZ_CAL_0, highByte(AZ_0));
  EEPROM.write(EEPROM_AZ_CAL_0 + 1, lowByte(AZ_0));
  EEPROM.write(EEPROM_AZ_CAL_MAX, highByte(AZ_MAX));
  EEPROM.write(EEPROM_AZ_CAL_MAX + 1, lowByte(AZ_MAX));
  EEPROM.write(EEPROM_EL_CAL_0, highByte(EL_0));
  EEPROM.write(EEPROM_EL_CAL_0 + 1, lowByte(EL_0));
  EEPROM.write(EEPROM_EL_CAL_MAX, highByte(EL_MAX));
  EEPROM.write(EEPROM_EL_CAL_MAX + 1, lowByte(EL_MAX));
}
```

The `check_serial()` function is where the serial commands from the PC tracking software are received and decoded. The function echoes all incoming characters back to the tracking software. This is required for the *HRD* rotator control program to function. As each character is received, it is added to the serial buffer until a carriage return is received. When a carriage return is received, the first character of the command is converted to uppercase as necessary, and the function is then decoded and executed.

```
void check_serial()
{
  if (Serial.available() > 0)
  {
    inByte = Serial.read();

#ifndef HRD // Echo the command if we're not using HRD
    Serial.print(char(inByte));
#endif

    if (inByte == 10)
    {
      return; // ignore Line Feeds
    }
    if (inByte != 13) { // Add to buffer if not CR
      serial_buffer[serial_buffer_index] = inByte;

#ifdef debug
      Serial.print(" Received = ");
      Serial.println(serial_buffer[serial_buffer_index]);
#endif

      serial_buffer_index++;

    } else
    {

      // It's a CR, execute command

      if ((serial_buffer[0] > 96) && (serial_buffer[0] < 123))
      { // it's lowercase, convert to uppercase
        serial_buffer[0] = serial_buffer[0] - 32;
      }
```

A `case()` statement is used to decode the various commands, and each command is executed by a function within it. If the debug mode is defined, additional diagnostic information will be sent to the Arduino IDE's serial monitor. The `serial_buffer_index` variable indicates how many characters were received for the decoded command and is used to decode commands that have multiple options.

```
switch (serial_buffer[0]) {  // Decode first character of command

      case 65:

#ifdef debug
        Serial.println("A Command Received");
#endif

        az_rotate_stop();
        break;

      case 66:

#ifdef debug
        Serial.println("B Command Received");
#endif
        send_current_el();
        break;

      case 67:      // C - return current azimuth

#ifdef debug
        Serial.println("C Command Received");
        Serial.println(serial_buffer_index);
#endif

        if ((serial_buffer_index == 2) & (serial_buffer[1] == 50)) {

#ifdef debug
          Serial.println("C2 Command Received");
#endif

          send_current_azel();
        } else {
          send_current_az();
        }
        break;

      case 68:

#ifdef debug
        Serial.println("D Command Received");
#endif

        rotate_el_down();
```

```
        break;

      case 69:

#ifdef debug
        Serial.println("E Command Received");
#endif
        el_rotate_stop();
        break;

      case 70:

#ifdef debug
        Serial.println("F Command Received");
        Serial.println(serial_buffer_index);
#endif

        if ((serial_buffer_index == 2) & (serial_buffer[1] == 50)) {

#ifdef debug
          Serial.println("F2 Command Received");
#endif

          set_max_el_cal();
        } else {
          set_max_az_cal();
        }
        break;
      case 76:

#ifdef debug
        Serial.println("L Command Received");
#endif
        rotate_az_ccw();
        break;

      case 77:

#ifdef debug
        Serial.println("M Command Received");
#endif

        rotate_to();
        break;
```

```
        case 79:

#ifdef debug
          Serial.println("O Command Received");
          Serial.println(serial_buffer_index);
#endif
          if ((serial_buffer_index == 2) & (serial_buffer[1] == 50))
          {
#ifdef debug
            Serial.println("O2 Command Received");
#endif

            set_0_el_cal();
          } else
          {
            set_0_az_cal();
          }
          break;

        case 82:

#ifdef debug
          Serial.println("R Command Received");
#endif

          rotate_az_cw();
          break;

        case 83:

#ifdef debug
          Serial.println("S Command Received");
#endif

          az_rotate_stop();
          el_rotate_stop();
          break;

        case 85:

#ifdef debug
          Serial.println("U Command Received");
#endif

          rotate_el_up();
          break;

        case 87:
```

```
#ifdef debug
        Serial.println("W Command Received");
#endif

        rotate_az_el_to();
        break;

    }
```

The **C** command sends the current azimuth in degrees back to the tracking software on the PC. However, this command may actually be a **C2** command, which needs to send the current azimuth and elevation back to the tracking software, so be careful in checking. The length of the command string as determined by the `serial_buffer_index` variable is used to decide what information needs to be sent back to the tracking software.

The **D** command manually rotates the elevation down by calling the `rotate_el_down()` function.

The **E** command stops all elevation rotation by calling the `el_rotate_stop()` function. It is important to note that this command will not affect any active azimuth commands.

The **F** command is another command that can have a secondary function. The **F** command by itself reads the current azimuth position and saves it as the azimuth full scale calibration value, while the **F2** command reads the current elevation position and saves it as the elevation full scale value.

The **L** command manually rotates the azimuth rotator counter-clockwise by calling the `rotate_az_ccw()` function.

The **M** command manually rotates the azimuth rotator to a set point by calling the `rotate_to()` function.

The **O** command sets the zero calibration settings. An **O** command by itself will save the current position as the azimuth zero calibration value, while an **O2** command saves the current position as the elevation zero calibration value.

The **R** command manually rotates the azimuth rotator clockwise by calling the `rotate_az_cw()` function.

The **S** command stops all rotation by calling the `az_rotate_stop()` and `el_rotate_stop()` functions.

The **U** command manually rotates the elevation up by calling the `rotate_el_up()` function.

The **W** command manually rotates the azimuth and elevation rotators to a set point by calling the `rotate_az_el_to()` function.

When the command has been decoded and executed, the incoming serial character buffer is cleared, and the buffer index is set to zero. The function will then return to the `main loop()` to wait for the next command.

The `send_current_az()` function is used to build and send the response to the **C** command sent by the tracking software on the PC. The tracking software expects the response to be in the format of "+0aaa," where aaa is the azimuth value in degrees. This function will read the A/D converter and map the value to degrees based on the current azimuth calibration values. A string is then built using the Yaesu GS-232 format that the tracking software is expecting, and then sent back to the tracking software via the USB serial port. Because the G-5400B rotator has 0°/360° at midscale, the data must be adjusted to return the correct azimuth value.

```
void send_current_az()
{
  read_adc();
  // Map it to degrees

#ifdef debug
  Serial.println(current_AZ);
#endif

  AZ_Degrees = map(current_AZ, AZ_0, AZ_MAX, 0, 360);
  if (AZ_Degrees > 180) {
    AZ_Degrees = AZ_Degrees - 180;
  } else
  {
    AZ_Degrees = AZ_Degrees + 180;
  }

#ifdef debug
  Serial.println(AZ_Degrees);
#endif

  // Send it back via serial
  Serial_Send_Data = "";
  if (AZ_Degrees < 100)
  { // pad with 0's if needed
    Serial_Send_Data = "0";
  }
  if (AZ_Degrees < 10)
  {
    Serial_Send_Data = "00";
  }
  Serial_Send_Data = "+0" + Serial_Send_Data + String(AZ_Degrees);
  Serial.println(Serial_Send_Data);   // Return value via RS-232 port

}
```

The `send_current_azel()` function is used to build and send the response to the **C2** command. The tracking software expects the response to be in the format of "+0aaa+0eee," where aaa is the azimuth value, and eee is the elevation value. This function will read the A/D converter and map the values to degrees based on the current azimuth and elevation calibration values. A string is then built using the Yaesu GS-232 format that the tracking software is expecting, and then sent back to the tracking software via the USB serial port. As with the **C** command, the azimuth data has to be adjusted to return the correct azimuth value.

```
void send_current_azel()
{
  read_adc();
  // Map AZ to degrees

#ifdef debug
  Serial.println(current_AZ);
#endif

  AZ_Degrees = map(current_AZ, AZ_0, AZ_MAX, 0, 360);
  if (AZ_Degrees > 180)
  {
    AZ_Degrees = AZ_Degrees - 180;
  } else {
    AZ_Degrees = AZ_Degrees + 180;
  }

  // Map EL to degrees
  EL_Degrees = map(current_EL, EL_0, EL_MAX, 0, 180);
#ifdef debug
  Serial.println(current_EL);
  Serial.println(EL_Degrees);
  Serial.println(AZ_Degrees);
#endif

  // Send it back via serial
  Serial_Send_Data = "";

  if (AZ_Degrees < 100)
  { // pad with 0's if needed
    Serial_Send_Data = "0";
  }
  if (AZ_Degrees < 10)
  {
    Serial_Send_Data = "00";
  }
  Serial_Send_Data = "+0" + Serial_Send_Data + String(AZ_Degrees) + "+0";

#ifdef HRD
  if (EL_Degrees > 90) // Limit elevation response to 90 degrees for HRD
  {
    EL_Degrees = 90;
  }
#endif
```

```
  if (EL_Degrees < 100)
  { // pad with 0's if needed
    Serial_Send_Data = Serial_Send_Data + "0";
  }

  if (EL_Degrees < 10)
  {
    Serial_Send_Data = Serial_Send_Data + "0";
  }

  Serial_Send_Data = Serial_Send_Data + String(EL_Degrees);
  Serial.println(Serial_Send_Data);  // Return value via RS-232 port
}
```

The `set_max_az_cal()` function saves the azimuth full scale value to the Arduino's onboard EEPROM.

The `set_max_el_cal()` function saves the elevation full scale value to the Arduino's onboard EEPROM.

```
void set_max_az_cal() {

#ifdef debug
  Serial.println("Calibrate Max AZ Function");
#endif

  read_adc();
  // save current az and el values to EEPROM - Zero Calibration

#ifdef debug
  Serial.println(current_AZ);
#endif

  AZ_MAX = current_AZ;
  write_eeprom_cal_data();
  Serial.println("Max Azimuth Calibration Complete");
}

void set_max_el_cal() {

#ifdef debug
  Serial.println("Calibrate EL Max Function");
#endif
```

```
  read_adc();
  // save current az and el values to EEPROM - Zero Calibration

#ifdef debug
  Serial.println(current_EL);
#endif

  EL_MAX = current_EL;
  write_eeprom_cal_data();
  Serial.println("Max Elevation Calibration Complete");
}
```

The `rotate_az_ccw()`, `rotate_az_cw()`, `rotate_el_up()`, and `rotate_el_down()` functions control the relays inside the Yaesu G-5400B rotator controller to perform the actual rotation functions.

```
void rotate_az_ccw()
{
  digitalWrite(rotate_left, HIGH);
  digitalWrite(rotate_right, LOW);
}

void rotate_az_cw()
{
  digitalWrite(rotate_right, HIGH);
  digitalWrite(rotate_left, LOW);
}

void rotate_el_up()
{
  digitalWrite(rotate_up, HIGH);
  digitalWrite(rotate_down, LOW);
}

void rotate_el_down()
{
  digitalWrite(rotate_down, HIGH);
  digitalWrite(rotate_up, LOW);
}
```

The `az_rotate_stop()` and `el_rotate_stop()` are used to turn off the relays and stop rotation.

```
void az_rotate_stop()
{
  digitalWrite(rotate_right, LOW);
  digitalWrite(rotate_left, LOW);
}

void el_rotate_stop()
{
  digitalWrite(rotate_up, LOW);
  digitalWrite(rotate_down, LOW);
}
```

The `rotate_to()` function is used to implement the **M** command. This command is in the format of "Maaa," where aaa is the desired azimuth value in degrees. This function will read the desired azimuth from the **M** command, compare the requested value to the reading from the A/D, and rotate the azimuth rotator in the proper direction. While the rotator is moving to the desired position, this function will allow processing to return to the `main loop()` and check for additional commands, such as a stop command. The status and control of this move command is handled in the `main loop()` by the `check_move()` function.

```
void rotate_to()
{

  // Decode Command - Format Mxxx - xxx = Degrees to Move to


#ifdef debug
  Serial.println("M Command -  Rotate Azimuth To Function");
  Serial.println(serial_buffer_index);
#endif

  if (serial_buffer_index == 4)
  {

#ifdef debug
    Serial.println("Value in [1] to [3]?");
#endif

    Requested_AZ = (String(char(serial_buffer[1])) + String(char(serial_buf-
```

```
fer[2])) + String(char(serial_buffer[3]))) ;
    AZ_To = (Requested_AZ.toInt()); // AZ Degrees to Move to as integer

    if (AZ_To < 0)
    {
      AZ_To = 0;
    }

    if (AZ_To > 360)
    {
      AZ_To = 360;
    }
    if (AZ_To > 180)
    {
      AZ_To = AZ_To - 180;
    } else
    {
      AZ_To = AZ_To + 180;
    }

#ifdef debug
    Serial.println(Requested_AZ);
    Serial.println(AZ_To);
#endif

    // set the move flag and start
    read_adc();
    // Map it to degrees
    AZ_Degrees = map(current_AZ, AZ_0, AZ_MAX, 0, 360);

#ifdef debug
    Serial.println(current_AZ);
    Serial.println(AZ_Degrees);
#endif

    AZ_Distance = AZ_To - AZ_Degrees;
    set_AZ = AZ_To;
    if (abs(AZ_Distance) <= AZ_Tolerance) { // No move needed
      az_rotate_stop();
      set_AZ = -1;
    } else
    { // Move Azimuth - figure out which way
```

```
      if (AZ_Distance > 0)
      { //We need to move CCW
        rotate_az_ccw();
      } else {
        rotate_az_cw();
      }
    }
  }
}
```

The send_current_el() function prepares and returns the response to the **B** command. The response is in the format "+0eee," where eee is the current elevation value in degrees.

```
void send_current_el()
{
  read_adc();
  // Map it to degrees
  EL_Degrees = map(current_EL, EL_0, EL_MAX, 0, 180);

#ifdef debug
  Serial.println(current_EL);
  Serial.println(EL_Degrees);
  Serial.println(AZ_Degrees);
#endif

  // Send it back via serial
  Serial_Send_Data = "";

  if (EL_Degrees < 100)
  { // pad with 0's if needed
    Serial_Send_Data = "0";
  }

  if (EL_Degrees < 10)
  {
    Serial_Send_Data = "00";
  }

  Serial_Send_Data = "+0" + Serial_Send_Data + String(EL_Degrees);
  Serial.println(Serial_Send_Data);  // Return value via RS-232 port
}
```

The `rotate_az_el_to()` function is similar to the `rotate_to()` function, except it performs azimuth and elevation rotation simultaneously in response to the **W** command. The **W** command is in the format "Waaa eee," where Waaa is the desired azimuth, and eee is the desired elevation. This function will read the desired azimuth and elevation from the **W** command, compare the requested values to the readings from the A/D, and rotate the azimuth and elevation rotators in the proper direction. While the rotators are moving to the desired position, this function will allow processing to return to the `main loop()` and check for additional commands, such as a stop command. The status and control of this move command is handled in the `main loop()` by the `check_move()` function. Since the azimuth and elevation rotations are handled separately, both axes of rotation can occur simultaneously, and each will stop independently when the desired position is reached.

```
void rotate_az_el_to()
{
  // Decode Command - Format Wxxx yyy where xxx = Azimuth to Move to yyy = El-
evation to Move to

#ifdef debug
  Serial.println("W Command -  Rotate Azimuth and Elevation To Function");
  Serial.println(serial_buffer_index);
#endif

  if (serial_buffer_index == 8)
  {

#ifdef debug
    Serial.println("Value in [1] to [3]?");
#endif

    Requested_AZ = (String(char(serial_buffer[1])) + String(char(serial_buf-
fer[2])) + String(char(serial_buffer[3]))) ;
    AZ_To = (Requested_AZ.toInt()); // AZ Degrees to Move to as integer
    if (AZ_To < 0)
    {
      AZ_To = 0;
    }

    if (AZ_To > 360)
    {
      AZ_To = 360;
    }
```

```
    if (AZ_To > 180)
    {
      AZ_To = AZ_To - 180;
    } else
    {
      AZ_To = AZ_To + 180;
    }
#ifdef debug
    Serial.println(Requested_AZ);
    Serial.println(AZ_To);
    Serial.println("Value in [5] to [7]?");
#endif

    Requested_EL = (String(char(serial_buffer[5])) + String(char(serial_buf-
fer[6])) + String(char(serial_buffer[7]))) ;
    EL_To = (Requested_EL.toInt()); // EL Degrees to Move to as integer
    if (EL_To < 0)
    {
      EL_To = 0;
    }

    if (EL_To > 180)
    {
      EL_To = 180;
    }

#ifdef debug
    Serial.println(Requested_EL);
    Serial.println(EL_To);
#endif

    // set the move flag and start
    read_adc();
    // Map it to degrees
    AZ_Degrees = map(current_AZ, AZ_0, AZ_MAX, 0, 360);

#ifdef debug
    Serial.println(current_AZ);
    Serial.println(AZ_Degrees);
#endif
```

```
    AZ_Distance = AZ_To - AZ_Degrees;
    set_AZ = AZ_To;
    EL_Degrees = map(current_EL, EL_0, EL_MAX, 0, 180);

#ifdef debug
    Serial.println(EL_Degrees);
#endif

    EL_Distance = EL_To - EL_Degrees;
    set_EL = EL_To;
    // Set Azimuth
    if (abs(AZ_Distance) <= AZ_Tolerance)
    { // No move needed
      az_rotate_stop();
      set_AZ = -1;
    } else
    { // Move Azimuth - figure out which way
      if (AZ_Distance > 0) {  //We need to move CW
        rotate_az_ccw();
      } else {
        rotate_az_cw();
      }
    }
    // Set Elevation
    if (abs(EL_Distance) <= EL_Tolerance)
    { // No move needed
      el_rotate_stop();
      set_EL = -1;
    } else
    { // Move Azimuth - figure out which way
      if (EL_Distance > 0)
      { //We need to move CW
        rotate_el_up();
      } else
      {
        rotate_el_down();
      }
    }
  }
}
```

The `set_0_az_cal()` function implements the **O** command and saves the azimuth zero calibration value to the Arduino's EEPROM.

The `set_0_el_cal()` function implements the **O2** command and saves the elevation zero calibration value to the Arduino's EEPROM.

```
void set_0_az_cal()
{
  read_adc();
  // save current az and el values to EEPROM - Zero Calibration

#ifdef debug
  Serial.println("Calibrate Zero Function");
  Serial.println(current_EL);
  Serial.println(current_AZ);
#endif

  AZ_0 = current_AZ;
  write_eeprom_cal_data();
  Serial.println("Zero Azimuth Calibration Complete");
}

void set_0_el_cal()
{
#ifdef debug
  Serial.println("Calibrate Zero Function");
#endif

  read_adc();
  // save current az and el values to EEPROM - Zero Calibration

#ifdef debug
  Serial.println(current_EL);
  Serial.println(current_AZ);
#endif

  EL_0 = current_EL;
  write_eeprom_cal_data();
  Serial.println("Zero Elevation Calibration Complete");
}
```

The read_adc() function is used to read the analog azimuth and elevation voltages coming from the rotator controller, and converts those values to raw, uncalibrated 16-bit digital values using the ADS1115 A/D. The values are stored in the global variables current_EL and current_AZ and used by the other functions to determine the current azimuth and elevation of the rotators.

```
void read_adc()
{
#ifdef debug
  Serial.println("Read ADC Function  ");
#endif

  int RotorValue;
  adc.setGain(GAIN_TWOTHIRDS);
  delay(10); // adc settling delay
  current_EL = adc.readADC_SingleEnded(0);

#ifdef debug
  Serial.println(current_EL);
#endif

  delay(10); // adc settling delay
  current_AZ = adc.readADC_SingleEnded(1);

#ifdef debug
  Serial.println(current_AZ);
#endif

}
```

The next main function in our sketch is the check_move() function. When we receive a movement command, the command is decoded and the relays on the rotator controller are energized to move the antenna to the requested position. This function is called from the main loop() and continually checks the status of the rotator positioning, then stops the rotation when the rotators reach the requested position.

```
void check_move()
{
  if (set_AZ != -1 || set_EL != -1) {   // We're moving - check and stop as
needed
    read_adc();
```

```
    // Map AZ to degrees

#ifdef debug
    Serial.println(current_AZ);
#endif
    AZ_Degrees = map(current_AZ, AZ_0, AZ_MAX, 0, 360);
    // Map EL to degrees

#ifdef debug
    Serial.println(current_EL);
#endif

    EL_Degrees = map(current_EL, EL_0, EL_MAX, 0, 180);

#ifdef debug
    Serial.println(EL_Degrees);
    Serial.println(AZ_Degrees);
#endif

    if (set_AZ != -1)
    {
      AZ_Distance = set_AZ - AZ_Degrees;
      if (abs(AZ_Distance) <= AZ_Tolerance) { // No move needed
        az_rotate_stop();
        set_AZ = -1;
      } else
      { // Move Azimuth - figure out which way
        if (AZ_Distance > 0) {  //We need to move CW
          rotate_az_cw();
        } else
        {
          rotate_az_ccw();
        }
      }
    }

    if (set_EL != -1)
    {
```

```
    EL_Distance = set_EL - EL_Degrees;

    if (abs(EL_Distance) <= EL_Tolerance)
    {
      // No move needed
      el_rotate_stop();
      set_EL = -1;
    } else
    { // Move Azimuth - figure out which way
      if (EL_Distance > 0)
      { //We need to move CW
        rotate_el_up();
      } else {
        rotate_el_down();
      }
    }
  }
 }
}
```

And finally, the last main function in our three function `loop()` is the `update_tft()` function. This function will update the position information on the TFT display if any changes have occurred.

```
void update_tft() // Updates the current Azimuth and Elevation on the TFT display
{
  int Azimuth, Elevation;
  String s_Azimuth, s_Elevation;

  read_adc();
  // Map Azimuth to degrees
  Azimuth = map(current_AZ, AZ_0, AZ_MAX, 0, 360); // Map the AZ to 0 - 360
  // Adjust for North centered display
  if (Azimuth > 180)
  {
    Azimuth = Azimuth - 180;
  } else {
    Azimuth = Azimuth + 180;
  }
```

```
  // Map elevation to degrees
  Elevation = map(current_EL, EL_0, EL_MAX, 0, 180); // Map the EL to 0 - 180

  // Check to see if it's changed
  if (Elevation != prev_EL_Degrees || Azimuth != prev_AZ_Degrees)
  {
    // Position has changed - update TFT

    // Adjust to 3 characters each
    s_Azimuth = String(Azimuth);
    s_Elevation = String(Elevation);

    while (s_Azimuth.length() <3)
    {
      s_Azimuth = "0" + s_Azimuth;
    }

    while (s_Elevation.length() <3)
    {
      s_Elevation = "0" + s_Elevation;
    }

    tft.setTextSize(status_font_size);
    tft.setCursor(30, 150);
    tft.print("Elevation");
    tft.setCursor(210, 150);
    tft.print("Azimuth");

    tft.setTextSize(font_size);
    tft.setCursor(35, 90);
    tft.print(s_Elevation);
    tft.setCursor(200, 90);
    tft.print(s_Azimuth);


#ifdef debug1
    Serial.print("Azimuth: ");
```

```
    Serial.print(Azimuth);
    Serial.print("   Elevation:");
    Serial.print(Elevation);

    Serial.print("  AZ length ");
    Serial.print(s_Azimuth.length());
    Serial.print("   EZ length ");
    Serial.println(s_Elevation.length());
    delay(5000);
#endif

    prev_EL_Degrees = Elevation;
    prev_AZ_Degrees = Azimuth;

  }
}
```

Looking back, this might seem like a big sketch, but breaking it down into small blocks helps to keep it manageable. This sketch does a lot of things, but since each section is a modular block of code, you can build it piece by piece and test it as you go using the debug mode method. When you're finished, undefine the debug mode and upload the sketch with the debug mode changes. This turns off the diagnostic messages, and you're ready to go.

Calibrating the Yaesu Azimuth-Elevation Rotator Controller

To test the interface and set the azimuth and elevation calibration values, you can use the Arduino IDE's serial monitor to send the GS-232A commands directly to the Arduino. Manually rotate the azimuth and elevation rotator to zero (both meters at minimum deflection), and use the **O** and **O2** commands to set the zero calibration points. Then, manually rotate the azimuth and elevation rotators to full scale and use the **F** and **F2** commands to set the azimuth and elevation full scale values. Once you have done that, your controller is calibrated, and the calibration values are stored in the Arduino's onboard EEPROM. You can recalibrate your controller at any time by repeating the calibration process. Now, you can send the various GS-232A controller commands listed in Table 16.1 to test all the functions on your controller. When you're finished, undefine the debug mode and upload the sketch with the debug mode changes. This turns off the diagnostic messages, and you're ready to go.

Operating the Yaesu Azimuth-Elevation Rotator Controller

Now that your controller is calibrated, you're ready to use it with the SatPC32 or *HRD* satellite tracking software on your PC. Don't forget to short the jumper on capacitor C1 to disable the DTR reset, and make sure that you have the sketch loaded with the debug undefined. Configure your tracking software to use the Arduino's COM port and select the Yaesu GS-232A AZ/El rotator controller in your tracking software using the baud rate you selected in the sketch.

The default in the sketch is set to 19,200. Now your G-5400B rotator should move and track the selected satellite.

Once you have everything working on the breadboard, you can build the circuit on a 6 × 8 cm piece of copper-clad perfboard using the schematic in Figure 16.5. For the upgraded version of this project, I mounted the perfboard in a Solarbotics Mega S.A.F.E. enclosure with the TFT display attached to the top of the enclosure, as shown in **Figure 16.7**. I mounted an 8-pin DIN female connector in the side of the enclosure to allow an 8-pin DIN to DIN patch cable to interface to the Yaesu G-5400B rotator controller. We won't need to connect any source of external power to the Arduino since we'll be using the power provided by the PC and the Arduino's USB connector.

You can also use this project to control the Yaesu G-5500 series of rotator controllers, however, the G-5500 azimuth meter starts at 0° and wraps around 360° to 90° for a full 450° of rotation. You'll need to modify the azimuth functions to allow for this difference.

Since the Yaesu GS-232 Rotator Computer Interface is also used by the Yaesu DXA-series of azimuth-only rotators, you can use this controller to drive those rotators as well. *HRD* has an option to use the Yaesu GS-232A for azimuth control only. Instead of the 8-pin DIN connector, the Yaesu DXA azimuth-only rotators use a 6-pin DIN connector since they don't need the two pins used for elevation in the G-5400B. You can even omit the elevation relay transistors used for elevation control if you're going to use this project to control an azimuth-only Yaesu DXA-series rotator.

Figure 16.7 — The azimuth-elevation rotator controller display.

The complete sketch and libraries for this project can be found in online at **kw5gp.com**.

17 The CDE/Hy-Gain Keypad Rotator Controller

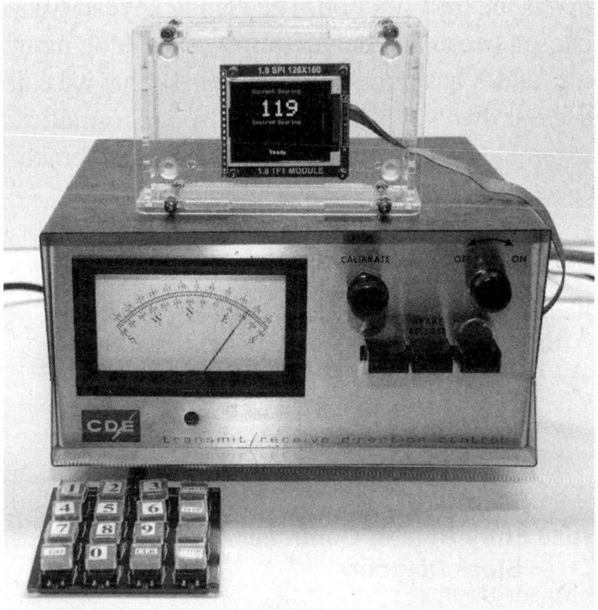

Figure 17.1 — The CDE/Hy-Gain Keypad Rotator Controller.

The 4 × 4 matrix keypad has been used with the Arduino since the early days and is included in just about every Arduino starter kit. But, I'm not a huge fan of the thin, flexible keypads, and they also tie up eight digital I/O lines to decode the key presses. Recently, several new types of the 4 × 4 keypads have become available. Adafruit has the NeoTrellis keypad (see **Figure 17.2**), which is a nice elastomer 4 × 4 keypad with addressable RGB or monochrome LED backlighting that uses the I^{2}C bus. RobotDyn has an inexpensive analog version of the keypad, that uses resistors to select the output voltage based on the key pressed, similar to the Yaesu FH-2 keypad. The nice thing about the RobotDyn keypad (see **Figure 17.3**) is that it has removable clear keycaps so that you can create your own key legends.

One of things that has always been an issue with computer-controlled rotators is that you need an attached workstation running software, such as *Ham Radio Deluxe (HRD)*, to manage the rotation. Another common issue while using the meter on the CDE/Hy-Gain rotator is having to mentally convert the meter reading to a bearing in your head to make sure you point the antennas in the right direction. To expedite this process, I thought it would be cool to create a build where you could use a 4 × 4 keypad to enter the bearing, press go, and have the Arduino take care of the rotating while the bearings show on a TFT display. This project will allow you to directly enter a bearing using an analog 4 × 4 keypad and have the Arduino manage the rest of the rotation for you as shown in **Figure 17.1**.

The Block Diagram

Based on the rotator projects in the earlier chapters, we have all the pieces we need to make this project work. **Figure 17.4** shows the block diagram for the project. We'll be using many of our favorite components from previous projects, including an Adafruit 1.8-inch ST7735 color graphic TFT display, the ADS1115 16-bit programmable gain analog-to-digital converter, and

Figure 17.2 — The Adafruit NeoTrellis keypad.

Figure 17.3 — The RobotDyn analog keypad.

an analog 4 × 4 keypad. It is important to note that you should use the Adafruit ST7735 display, as it runs on 3.3 – 5 V and does not need level shifters. If you choose to build this chapter with a different display, don't forget to add a level shifter and power the display with 3.3 V. Due to the length of the ribbon cable to the display, I've found that the Adafruit ST7735 running at 5 V is the best solution. The analog keypad was chosen because it only requires one analog input pin, and it has removable keycaps so that we can use our own key labels. We'll also use a small external relay control board to control the brake release and rotation relays.

Figure 17.5, which shows the schematic for the keypad entry rotator controller project, also includes the parts list. The plan is to mount the Arduino, prototyping shield, and relay board

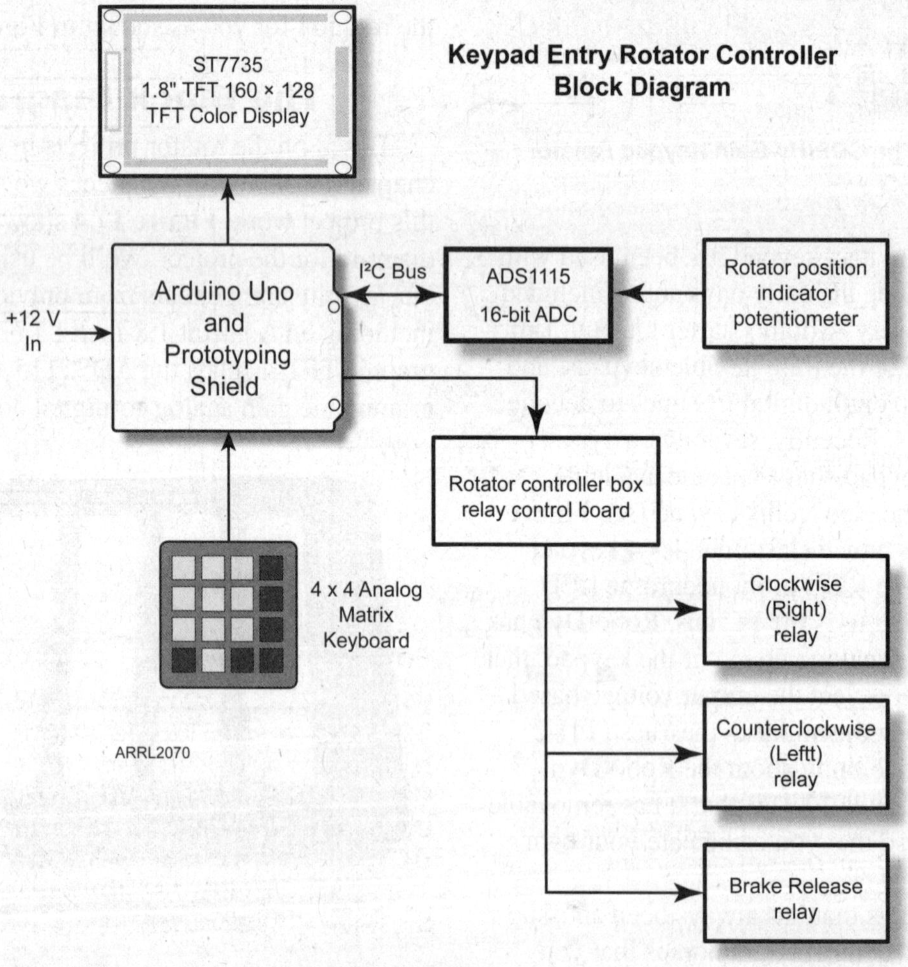

Figure 17.4 — The keypad rotator controller block diagram.

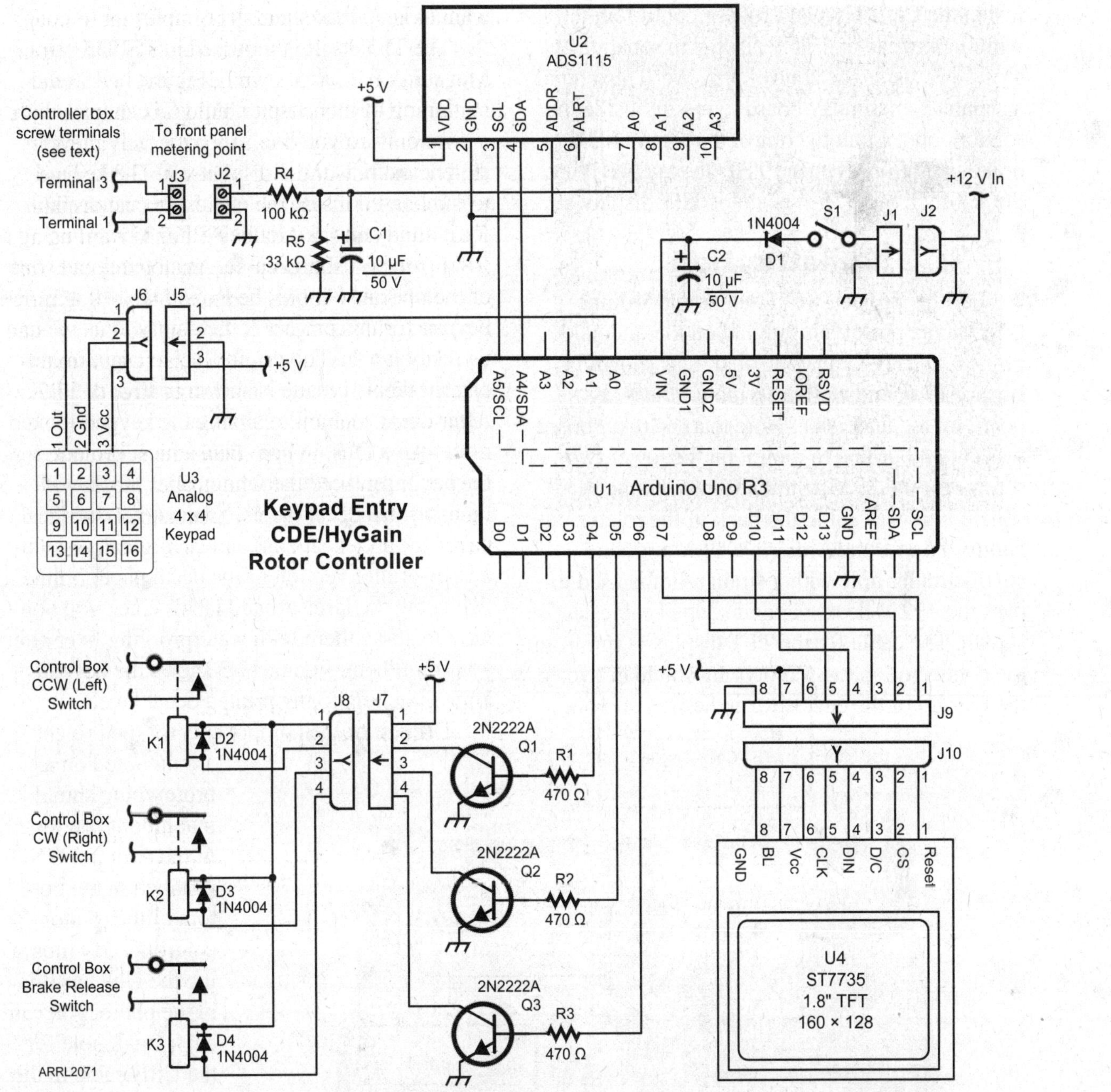

Figure 17.5 — The keypad rotator controller schematic.

Part	*Value*	*Part*	*Value*
C1, C2	10 µF capacitor, 50 V	R5	33 kΩ resistor, 1/8 W
D1-D4	1N4004 diode	S1	SPST toggle switch
J1-J4	2-pin Dupont male header	U1	Arduino Uno
J5, J6	3-pin Dupont male header	U2	ADS1115 A/D converter
J7, J8	4-pin Dupont male header	U3	Analog 4 × 4 keypad
J9, J10	8-pin Dupont male header	U4	ST7735-type 1.8-inch color TFT display
K1-K3	5 V SPST relay		
Q1-Q3	2N2222A NPN transistor	Enclosure	Solarbotics Mega S.A.F.E.
R1-R3	470 Ω resistor, 1/8 W		
R4	100 kΩ resistor, 1/8 W		

inside the CDE/Hy-Gain rotator control unit, with the keypad and TFT display in separate enclosures outside the control box. We'll also build a separate small relay board to mount inside the rotator control unit to control the rotation and brake operations. For the TFT display, we'll use the 1.8-inch ST7735-type color TFT display.

Construction

For this project, we'll mount the Arduino UNO and a prototyping shield on the underside of the CDE/Hy-Gain control unit, as shown in **Figure 17.6**, and we'll also mount the relay board on the underside. Note that the relay board was cut and drilled to mount on the lower two screws for the 28 V ac transformer mount. A square hole was cut in the back of the rotator control box for the Arduino's USB port, and a small circular hole with a strain relief is used to pass the +12 V dc power and keypad cables as shown. The cable for the TFT display is run from the connector on the prototyping shield between the rear lip of the enclosure and cover, so you want to keep it as short as possible.

The TFT itself is mounted in a Solarbotics Mega S.A.F.E., as shown in Figure 17.1 at the beginning of this chapter, but you can use whatever enclosure you like, since the only thing in the enclosure is the display itself. The keypad was mounted inside the enclosure custom built for it and connected to the Arduino board using a short piece of ribbon cable. As stated above, one of the reasons I chose the RobotDyn 4 × 4 analog keypad for this project is the ability to use your own key labels. The original set of plain inkjet-printed labels I made is shown in **Figure 17.7**. After doing that and realizing the keypad looked more like a ransom note than a nice, project keypad, I replaced the white paper labels with laser-printed decal labels. You can get the decal paper for inkjets at craft stores. You can also buy a laser-printer version of the decal paper online. I like using the laser printed labels since you don't have to spray them with waterproofing fixer after you print them. Figure 17.3 shows the keypad after switching to the printed decal labels.

The rest of the components for the project are mounted on a prototyping shield that mounts on top of the Arduino UNO installed in the bottom of the rotator controller. By mounting the UNO as it is in the photo, you can cut a small hole for the USB cable in the rear of the controller. This will allow you to access the UNO's USB port and modify the sketch or use the serial monitor without having to open the controller unit. On the prototyping shield, we'll mount the ADS1115 16-bit programmable A/D

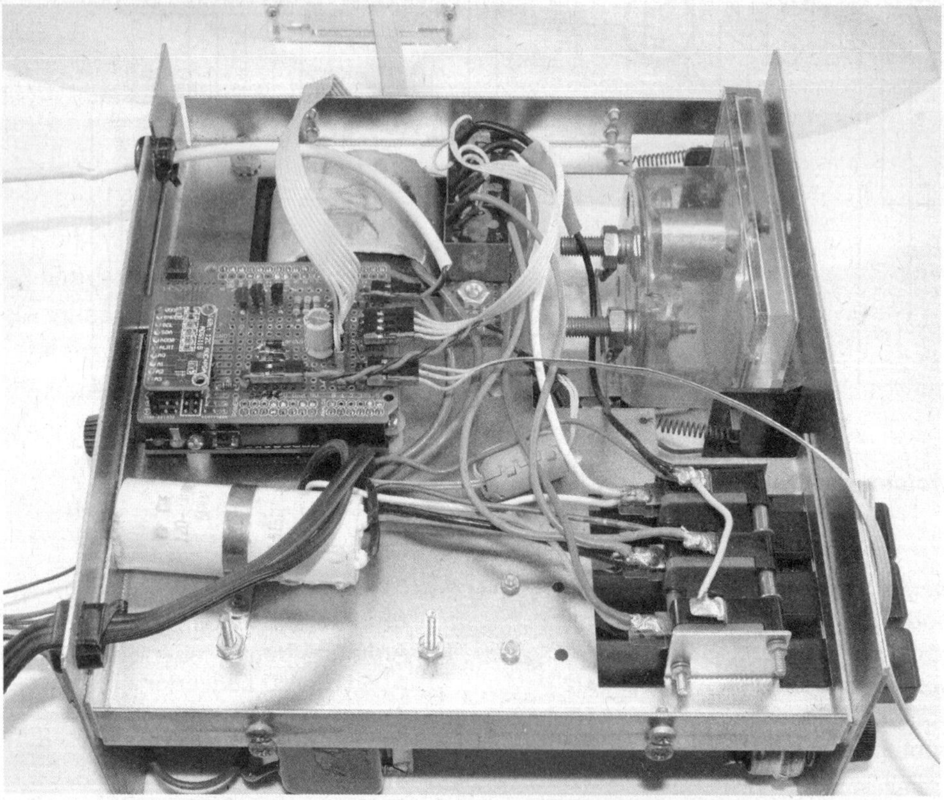

Figure 17.6 — The keypad rotator controller mounted in the bottom of the CDE/Hy-Gain rotator control unit.

Figure 17.7 — The "ransom note" keypad labels.

converter. The module shown in the photo is the most recent ADS1115 module available, although any style ADS1115 module will work fine.

The ADS1115 A/D is used to read the CDE/Hy-Gain's rotator position potentiometer. For this project, we'll be using a 100 kΩ/33 kΩ resistor divider to reduce the 0 – 13.5 V dc position voltage to a safe level of 0 – 3.5 V dc. The reason for these high resistor values is so that we don't interact with the rotator controller's meter. The basic rule with resistors is that if the value you're using is 10 times or greater than the resistance of the circuit (known as the 10 to 1 rule), the added resistance has virtually no effect. By using a total of 133 kΩ worth of resistance, the position signal seen by the A/D converter is reduced to a safe value for the ADS1115 A/D, but also has virtually no impact on the existing meter in the controller.

While we're talking about the CDE/Hy-Gain direction indication meter, I found out something that I think we all knew, but never really considered. During my testing, I discovered that the accuracy could vary by as much as 5°. Just adding the 10 μF capacitor on the rotator position potentiometer signal smoothed the signal out and helped a lot to improve the accuracy. As you'll find with the keypad and computer controller rotator projects that use the 16-bit ADS1115 A/D converter, accuracy down to 1° is actually possible. At the end of the day, though, there's no reason to obsess over a few degrees since there's really no antenna that will have a 1° beamwidth.

Two Flowcharts

The Keypad Calibration Flowchart

Since we're dealing with an inexpensive analog keypad comprised of resistors it's likely that no two keypads will have the exact same analog values for the various keys. Theoretically, the resistance spacing between the keys is enough not to be a concern, but it never hurts to calibrate things just to be sure. To help with this, I found a library for the analog keypad that included a calibration routine in its examples. While that library wasn't suitable to be used in this project, I was able to adapt the calibration sketch to provide the values we'll need to calibrate the keypad for the main project.

Odds are that you can just build the project and use the project sketch without making any changes, as the analog keypad values are not that critical for proper key decoding. However, on the off chance that you're having key decode errors, you can use the keypad calibration sketch to get the analog values for your keypad to insert into

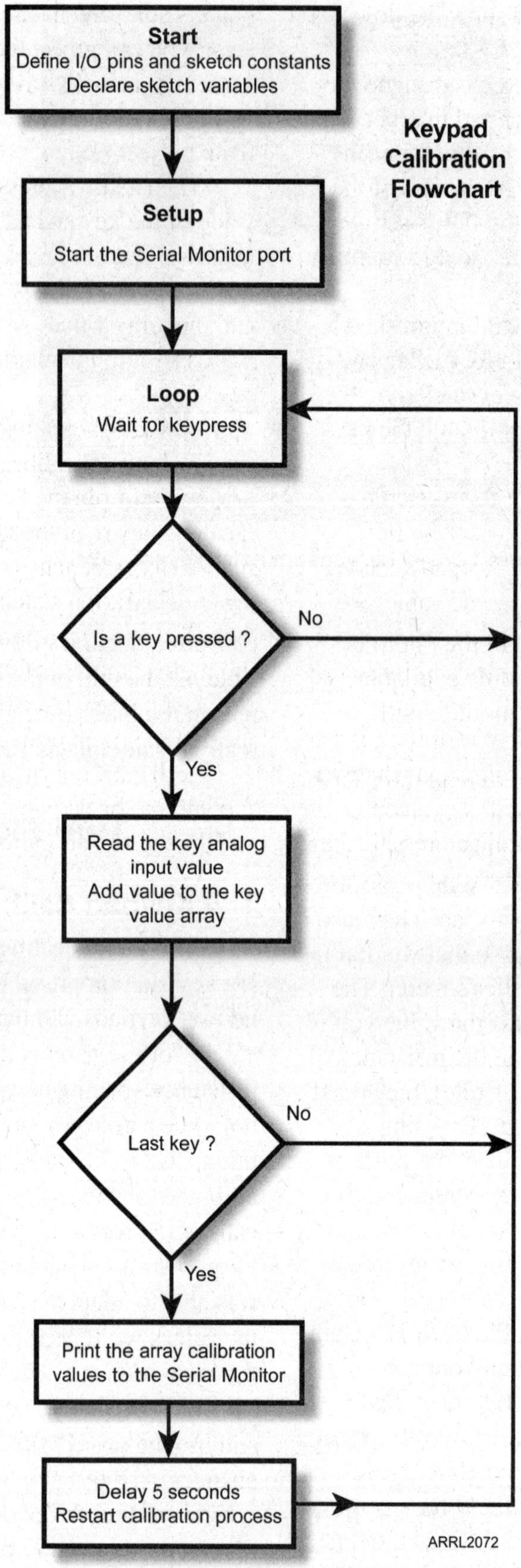

Figure 17.8 — Keypad calibration sketch flowchart.

the project sketch. **Figure 17.8** show the flowchart for the keypad calibration sketch.

The keypad calibration sketch is designed to have you press every key on the pad in order. The sketch will then read the analog value for each key and save them in an array. At the end of the sketch, the array is printed on the IDE's serial monitor for you to cut and paste the information into the main project sketch.

We'll start out the keypad calibration sketch by defining the analog input pin for the keypad and declaring the sketch variables, including the 16-element array we'll be using to hold the key-press analog values.

The `setup()` function for this sketch is about as simple as it gets. We'll start the IDE serial monitor port and read the keypad's analog input pin to get the "no key pressed" value. We'll print the basic instructions for the calibration process to the serial monitor, and then start the actual calibration process.

In the `main loop()`, we'll wait for a key to press. Note, if you press the wrong key, whatever key you pressed will be assigned the value that the A/D converter reads. It's best to just press the keys sequentially as they are numbered on the keypad board's silk screen lettering. If you do make an error, you can repeat the calibration process, as no values are actually saved, just printed to the serial monitor so they can be copied and pasted into the main project sketch.

When a key is pressed, we'll read the analog value of the key and add that value to the key value array. Once we read the last key on the keypad, we'll print the full text of the command statement and the array values so that you can just copy and paste this into the main project sketch.

The Keypad Calibration Sketch

The keypad calibration sketch will read the analog input pin the keypad is connected to and create an array of the key values for use by the main project sketch. The sketch will output an array declaration statement that you can copy from IDE's serial monitor display and paste into the main project sketch. The initial comments for the sketch show the default values for the keypad used in the main project sketch.

We'll start out by defining the analog input pin and the basic parameters and variables that we'll need for the calibration sketch.

```
#define keypad_pin  A0
#define no_key_pressed 0
#define samples 5 // The number of times to read the key
#define read_delay 10  // Delay between data samples
#define variance 10

int key_value;
int keypad_map[16]; // The number of keys on the keypad
int key;
int zero_value;
int analog_value;
```

In the `setup()` function, we'll start the IDE's serial monitor and read the analog value when no key is pressed. You can transfer this value into the main sketch for the `no_key_pressed` variable as needed. Then you'll be instructed to press one key at a time until all the keys have been read and their value stored in the array. You will notice on the keypad circuit board silk screen that the key switches are numbered from 1 – 16. You'll need to press the keys in the order of their silk screen label on the keypad circuit board to complete the calibration sequence.

```
void setup()
{
  Serial.begin(115200);

  Serial.print("No key pressed value defined as: ");
  Serial.print(no_key_pressed);

  read_key(); // Read key 5 times and average - put value in key_value
  Serial.print("   Actual Value:");
  Serial.println(key_value);
  zero_value = key_value;
  Serial.println("Keypad Calibration Sequence Initialized");
  Serial.println();
  Serial.println("Press one button at a time, hold it until instructed to re-
lease");
  Serial.println();
}
```

In the main loop(), we'll set up a For... and a Do... loop to read the analog value for each keypress. We'll then save the keypress value in an array.

```
void loop()
{
  for (key = 0 ; key <= 15; key++)
  {
    Serial.print("Press Key ");
    Serial.println(key + 1);

    do
    {
      read_key();
    } while (key_value < (zero_value + variance));

    read_key();
    Serial.print("Release Key ");
    Serial.println(key + 1);

#ifdef debug
    Serial.print("  Key Value = ");
    Serial.println(key_value);
#endif

    keypad_map[key] = key_value;
```

```
    do
    {
      read_key();
    } while (key_value > (zero_value + variance));


    if (key != 16)
    {
      Serial.println("Delaying for next key");
      delay(read_delay);
    }
  }
```

Once we have all the key values in the array, we'll format it to look like the array declaration statement you'll need to plug into the main project sketch. We'll then print the generated command and you can use the copy and paste function in the IDE to copy it into the main project sketch.

```
// All key data acquired - Print the keypad map array

  Serial.println("Copy the following line into your sketch");
  Serial.println();

  Serial.print("const int keypad_map[] = {");
  for (key = 0; key <= 15; key++)
  {

    Serial.print(keypad_map[key]);
    if (key != 15)
    {
      Serial.print(",");
    } else
    {
      Serial.println("};");
      Serial.println();
    }
  }

  Serial.println("Keypad calibration complete");
```

The `read_key()` function is used to read the analog value of the pressed key. When a key is pressed, we'll read the Arduino's A/D converter several times and use the average of the key values as the calibration value for that key.

```
void read_key()
{
#ifdef debug
  Serial.println();
  Serial.print("Averaging ");
#endif

  int average = 0;
  int count = samples;
  for (count = 0; count < samples + 1 ; count++)
  {
    analog_value = analogRead(keypad_pin);
    delay (read_delay);
    if (count != 0)
    {
      average = average + analog_value;

#ifdef debug
      Serial.print(" ");
      Serial.print(count);
      Serial.print(": ");
      Serial.print(average / count);
#endif

    }
    key_value = average / samples;
  }
}
```

The Keypad Entry Rotator Controller Flowchart

Figure 17.9 shows the keypad entry rotator controller flowchart. This sketch is very similar to the computer-controlled rotator control sketches, except the initial rotation is commanded by the keypad and stopped when either the destination is reached, or the stop key is pressed.

We'll start out by including the Adafruit GFX, Adafruit ST7735, and the SPI library for the TFT display. We'll also include the Adafruit ADS1015 and wire libraries for the ADS1115 A/D converter on the I^2C bus.

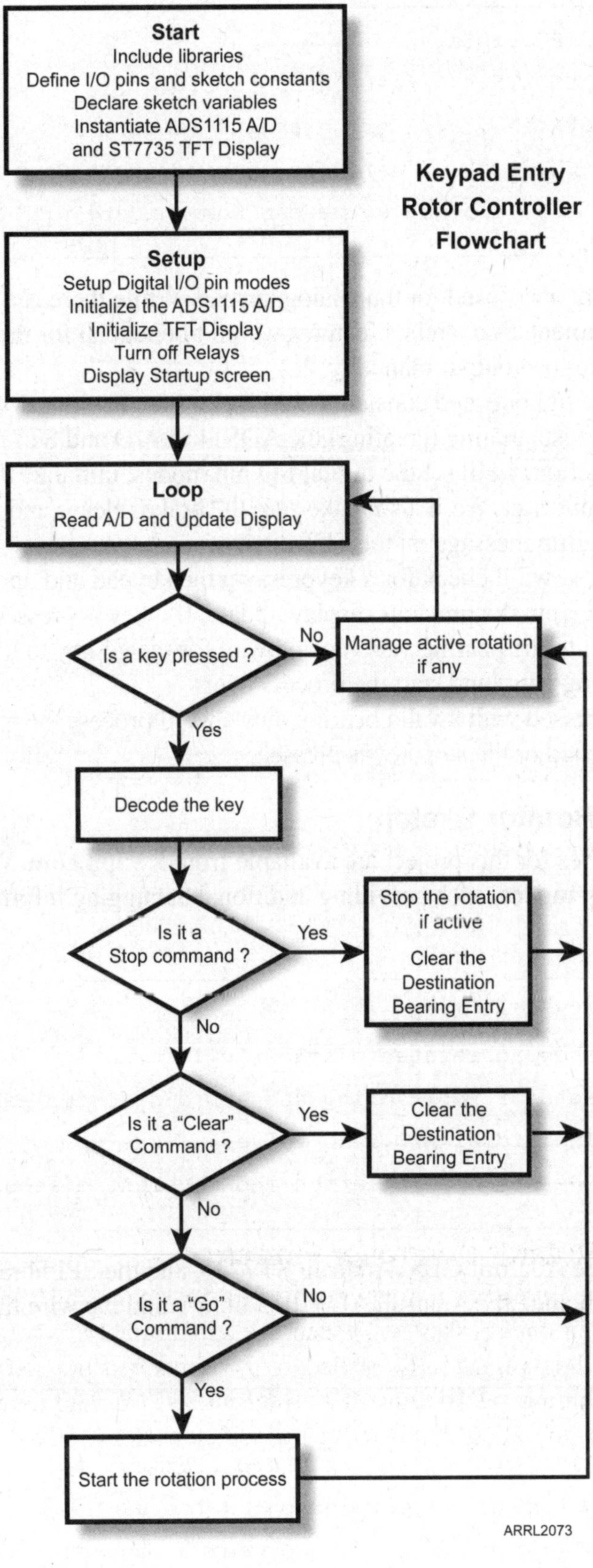

Figure 17.9 — The keypad rotator controller flowchart.

```
#include <Wire.h> //I2C Library
#include <Adafruit_ADS1X15.h> // ADS1x15 A/D Library
#include <Adafruit_GFX.h>     // Core graphics library
#include <Adafruit_ST7735.h> // Hardware-specific library
#include <SPI.h>  // SPI library so we can communicate with the TFT display
```

You'll note that no library is used for the analog keypad. While there is a library available, I was not happy with the implementation, and its features weren't beneficial for this particular project, so we just handled the analog keypad output manually.

Next, we'll define the I/O pins and constants used in the sketch, along with the keypad configuration array. Then we'll be instantiating (creating) the ADS1115 A/D and ST7735 TFT display objects.

In the `setup()` function, we'll set the digital I/O pin modes, initialize the ST7735 TFT display, and the ADS1115 A/D converter. We'll also make sure the brake release and rotation relays are turned off and display a brief startup message on the TFT display.

In the `main loop()`, we'll check for a keypress on the keypad and update the rotator bearing on the TFT if it has changed from the previous display update. If a key is pressed, we'll decode it, and if it's a number, we'll add it to the bearing variable and wait for the go key to be pressed. If it's the clear key, we'll erase the bearing entry and start the process over.

Once the go key is pressed with a valid bearing entry, we'll process the rotation until the specified destination has been reached or the stop key is pressed.

The Keypad Entry Rotator Sketch

The sketch and libraries for this project are available from **kw5gp.com**. We'll start out the sketch by defining several debug modes, each providing additional debugging information about specific portions of the sketch.

```
//#define debug   // Enables diagnostic information
//#define debug1  // Keypad related diagnostic information
//#define debug2 // Rotator Positioning and A/D diagnostic information
//#define debug3 // Individual A/D read values
//#define debug4  // Rotation calculation and movement diagnostic information
```

Next, we'll include the Adafruit GFX, Adafruit ST7735, and the SPI libraries for the ST7735 color TFT display. We'll also include the Adafruit ADS1015 library and the wire library for the ADS1115 A/D converter.


```
#include <Wire.h> //I2C Library
#include <Adafruit_ADS1X15.h> // ADS1x15 A/D Library
#include <Adafruit_GFX.h>     // Core graphics library
#include <Adafruit_ST7735.h> // Hardware-specific library
#include <SPI.h>  // SPI library so we can communicate with the TFT display
```

Next, we'll define the calibration settings for the rotator position potentiometer. Note that unlike the other rotator control projects, this project does not dynamically calibrate the rotator position potentiometer and save the calibration settings to EEPROM. You can easily add this back in yourself as an enhancement to the project.

We'll also define the keypad calibration values and map the keys on the keypad to the text that we will use to match with them. This is the part of the sketch where you would replace the `keypad_map[]` array with the values from the keypad calibration sketch if you need to. We'll also create a string with the numbers 0 – 9 that we will use to determine if a keypress is a number or a command key.

```
#define default_cal_0 150  // The default ADC Zero (-180 degrees) value
#define default_cal_max 26895   // The default ADC Max (+180 degrees) value

// Map the calibrated analog keypad values to an array
const int keypad_map[] = {1023, 931, 853, 787, 675, 633, 596, 563, 503, 480,
458, 438, 401, 322, 269, 231};
// Map the keypad keys to their matching text
const String key_text[] = {"1", "2", "3", "A", "4", "5", "6", "B", "7", "8",
"9", "C", "D",  "0", "E", "F"};
const String numbers = "1234567890";  // Creates a string used to compare if a
pressed key is a number
```

Next, we'll define the I/O pins and other definitions we'll need in the sketch. The CDE/Hy-Gain HAM-series rotators do not have as much of a positioning error issue as the AR-40, but we'll still want the ability to trim the position just a bit. The trim settings are in this section as well. Note, when adjusting the trim settings, be sure to turn off all debug information so that the sketch runs at its normal operating speed to track properly with the rotation speed.

```
#define comm_speed 115200 // Define the Arduino Serial Monitor baud rate
#define keypad_pin  A0  // Define the keypad analog input pin
#define no_key_pressed 0  // The A/D value for when no key is pressed
#define samples 5 // The number of times to read the key when pressed
#define read_delay 10  // Delay between keypad data samples
#define variance 10 // The A/D count tolerance for the calibrated key value ver-
sus the actual value
#define update_time 1000  // The display update time in milliseconds
#define degree_tolerance 1  // The rotator error tolerance in degrees
#define brake_time 3000 // Define the Brake delay in milliseconds
#define release_delay 500 // Define the delay between the brake release and rota-
tion start
#define left_rotate_trim -3 // Define the rotate error fine tuning value for left
(counterclockwise) rotation
```

```
#define right_rotate_trim 3 // Define the rotate error fine tuning value for right
(clockwise) rotation

#define TFT_CS      10  // Assign the TFT CS to pin 10
#define TFT_RST     7  // Assign the TFT RST to pin 7
#define TFT_DC      8  // Assign the TFT DC to pin 6
#define tft_delay 10  // set the TFT command delay to 10ms

#define left_relay 4  // Define the I/O pin for the rotate left relay
#define right_relay 5 // Define the I/O pin for the rotate right relay
#define brake_relay 6 // Define the I/O pin for the brake release relay
```

Then, we'll declare the remainder of the variables we'll need for the sketch and instantiate (create) the ADS1115 A/D and ST7735 TFT display objects.

```
int key_value;  // Variable to hold the averaged A/D value of the pressed key
int analog_value; // Variable to hold the converted analog value of the pressed
key
int zero_point = no_key_pressed + variance;
int key_index =  -1;  // The index point to the location of the pressed key in
the keypad_map array
String test_key;  // Variable to hold the text value of the pressed key
String move_to = "";  // Variable for the text of the entered bearing
bool moving = false;  // Flag to indicate if the rotator is turning
bool data_changed = true; // Flag to indicate that the display data has changed
int entered_bearing = -1; // Variable to hold the entered destination
int current_bearing;  // Variable to hold the current bearing
int previous_bearing = -1;  // Variable to hold the previous bearing
int actual_bearing; // Variable to hold the raw bearing value (-180 to +180)
boolean first_pass = true;  // Variable used to determine if first pass through
the main loop
int adc_value;  // Variable to hold the ADS1115 A/D Value
int adc_cal_0 = default_cal_0;  // Variable to hold the ADS1115 ADC zero cali-
bration point
int adc_cal_max = default_cal_max;  // Variable to hold the ADS1115 ADC max
calibration point
int destination_bearing;  // Variable to hold the destination bearing
String rotate_direction;  // Variable to indicate if the rotation is to the left
(counterclockwise) or right (clockwise)
unsigned long next_update_time = 0; // Variable to hold the next display update
```

```
time in millis()

Adafruit_ADS1115 ads;  // Create an instance for the ADS1115 16-bit A/D

Adafruit_ST7735 tft = Adafruit_ST7735(TFT_CS,  TFT_DC, TFT_RST);  // Create the
TFT display instance
```

In the `setup()` function, we'll set the digital I/O pin modes and verify that the brake release and rotation control relays are turned off. We'll also start the serial monitor port if debugging is enabled.

```
void setup()
{
  pinMode(left_relay, OUTPUT);  // Set the rotate left relay pin to output
  pinMode(right_relay, OUTPUT);  // Set the rotate right relay pin to output
  pinMode(brake_relay, OUTPUT);  // Set the brake release relay pin to output

  digitalWrite(left_relay, LOW);  // Turn off the rotate left relay
  digitalWrite(right_relay, LOW);  // Turn off the rotate right relay
  digitalWrite(brake_relay, LOW);  // Turn off the brake release relay

#ifdef debug
  Serial.begin(comm_speed); // Set the Serial Monitor port speed
  Serial.println("Starting");
#endif
```

Next, we'll set the ADS1115 A/D gain for a range of 0 – 4.096 V and start the A/D converter.

```
  ads.setGain(GAIN_ONE);         // Set the ADC Gain  to 1x gain   +/- 4.096V  1
bit = 0.125mV

  ads.begin();// Start the A/D Converter
```

Then we'll start the ST7735 TFT display and display a brief startup message.

```
  tft.initR(INITR_18BLACKTAB);   // initialize a 1.8" TFT with ST7735S chip,
black tab
  delay(tft_delay);
  tft.fillScreen(ST7735_BLACK); // Clear the display - fill with BLACK background
  delay(tft_delay);
  tft.setRotation(1); // Set the screen rotation
```

```
delay(tft_delay);
tft.setTextWrap(false); // Turn off Text Wrap
delay(tft_delay);
tft.setTextSize(3); // Set the Font Size
delay(tft_delay);
tft.setTextColor(ST7735_BLUE); //Set the Text Color
delay(tft_delay);
tft.setCursor(40, 10);  //Set the Cursor and display the startup screen
delay(tft_delay);
tft.print("KW5GP");
delay(tft_delay);
tft.setTextSize(2);
delay(tft_delay);
tft.setCursor(10, 60);
delay(tft_delay);
tft.print("Keypad Entry");
delay(tft_delay);
tft.setCursor(45, 80);
delay(tft_delay);
tft.print("Rotator");
delay(tft_delay);
tft.setCursor(25, 100);
delay(tft_delay);
tft.print("Controller" );

delay(5000);  //Wait 5 seconds then clear the startup message
```

After the startup message, we'll clear the display, read the A/D converter, and update the TFT display with the current rotator position information.

```
tft.fillScreen(ST7735_BLACK); // Clear the display - set to use a black back-
ground
delay(tft_delay);

read_adc(); // Read the ADC to get the initial values

update_display(); // update the display
```

In the `main loop()`, we'll first call the `read_key()` function to read the current value of the keypad analog input pin. Next, we'll call the `update_display()` function that will update the bearing on the display if the update timer has timed out and if the bearing has changed.

```
void loop()
{
  read_key(); // Check the keypad to see if a key is pressed
  if ((millis() > next_update_time && data_changed) || first_pass)  // Check to
see if we need to update the TFT display
  {
    update_display(); // Update the data on the TFT display
  }
```

Then we'll check to see if a valid key has been pressed, and if so, we'll get the text associated with the pressed key. Next, we'll check to see if the pressed key is a number. If the rotator is not currently rotating, this number will be added to the bearing input string.

```
  if (key_index != -1)  // Check to see if there is a valid key pressed
  {
    test_key = key_text[key_index]; // Get the text for the pressed key
    // We have a valid key

#ifdef debug1
    Serial.print("Valid Key: ");
    Serial.println(key_index);
#endif

    // Key has been decoded - determine action
    //If it's a number, check that's it's a valid number
    // Key codes 0-2, 4-6, 8-10 and 13 are digit entries

#ifdef debug1
    Serial.print("Key Text: ");
    Serial.print(test_key);
    Serial.print("   Checking numbers string: ");
    Serial.println(numbers.indexOf(test_key));
#endif

    if (numbers.indexOf(test_key) != -1)  // Check to see if the test key is in
the numbers string (it's a number key)
```

```
    {
      if (!moving) // Allow bearing entry when not moving
      {
        // It's a valid number - add it to direction string

        move_to = move_to + key_text[key_index];  // Add the key entry to the
desired bearing
        entered_bearing = move_to.toInt();  // Convert the desired bearing to a
string

#ifdef debug
        Serial.print("Valid Number: ");
        Serial.println(key_text[key_index]);
        Serial.print("Bearing: ");
        Serial.println(entered_bearing);
#endif
```

Now, we'll check to be sure that the entered bearing hasn't gone over 359°. If it has, we'll clear the bearing entry and restart the `loop()`.

```
        if (entered_bearing >= 360)  // Clear the entered bearing if it's over
359
        {
#ifdef debug
          Serial.println("Entry Error - Bearing Cleared");
#endif
          move_to = "";
          entered_bearing = -1;
        }
        data_changed = true;  // Flag to indicate the display needs to be up-
dated
      }
    }
```

If the key pressed is not a number, we'll check to see if it is a command key.

First, we'll check to see if it is an E key, indicating that we want to clear the destination bearing entry.

```
    // It's a command key
#ifdef debug
      Serial.println("Command Key Pressed");
#endif
      if (test_key == "E")  //Clear Bearing Entry
      {
        if (!moving)  // Don't execute if moving
        {
          move_to = ""; // Reset the bearing entry string
          entered_bearing = -1; // Reset the entered bearing
          data_changed = true;// Flag to indicate the display needs to be up-
dated

#ifdef debug
          Serial.println("Clear Key Pressed - Bearing cleared");
#endif
        }
      }
```

Next, we'll check to see if it is the D key, which performs the role of the go key. If a valid bearing has been entered and we're not already rotating, we'll call the `start_rotate()` function to begin the rotation process.

```
if (test_key == "D")  // Go Key
        {
          if (entered_bearing != -1 && !moving) // Only execute if not moving
and a valid destination has been entered
          {

#ifdef debug
            Serial.print("Go Key Pressed - Destination Bearing: ");
            Serial.println(entered_bearing);
#endif

            start_rotate(); // Process the rotation

          }
        }
```

Lastly, we'll check to see if it is the F key, indicating that it's the stop key. The stop key will immediately stop all rotation, clear the destination bearing entry, and update the display.

```
if (test_key == "F")   // Stop Key
        {

#ifdef debug
            Serial.print("Stop Key Pressed");
#endif

            if (moving) // If rotating, stop rotation
            {
#ifdef debug
              Serial.println(" - Rotation stopped");
#endif

              stop_all(); // Stop the rotation
            }
            if (entered_bearing != -1)  // Clear the bearing entry if it's a
valid destination
            {
#ifdef debug
              Serial.println(" - Bearing cleared");
#endif
            }
            move_to = ""; //  Reset the bearing string
            entered_bearing = -1; // Reset the bearing entry
            data_changed = true;  // Indicate the display needs updating
        }
      }
    }
  }
}
```

We'll then call the `read_adc()` function to read the ADS1115 16-bit programmable gain A/D converter. If we're rotating, we'll call the `check_move()` function to manage any active rotation process.

```
  read_adc(); // Read the ADS1115 A/D

  if (moving) // If we're moving - check where we are and stop if we're at the
destination
  {
    check_move(); // Calls the function to process the rotation
  }
```

The read_key() function will read the analog input pin for the 4 × 4 analog keypad. The keypad signal will be read five times and the results averaged to get us the analog value of the pressed key.

```
void read_key() // read_key function - Reads the 4x4 analog keypad
{
  int average = 0;  // Local variable to hold the average analog value
  int count = samples;  // Local variable to hold the number of data samples to
get
  key_value = 0;  // Clear the incoming key value
  key_index = -1; // Clear the keypad_map array pointer

  if (analogRead(keypad_pin) > zero_point)  // Valid Key Pressed - otherwise
exit
  {
    for (count = 0; count < samples + 1 ; count++)  // Get the desired number of
keypad data samples
    {
      analog_value = analogRead(keypad_pin);  // Read the keypad analog input pin
      delay (read_delay); // Delay between samplings
      if (count != 0) // Only if the number of requested samples is not zero
      {
        average = average + analog_value; // Add the analog value to the average
      }
      key_value = average / samples;  // Calcluate the average value
    }

#ifdef debug
    if (key_value > zero_point) // If the key analog value is not at the zero
voltage point
    {
#ifdef debug1
```

```
      Serial.print("Key Value: ");
      Serial.println(key_value);
#endif

    }

#endif

    do
    {
      // Wait until the key is released
    } while (analogRead(keypad_pin) > zero_point);  // Stay in the loop as long
as the keypad is pressed

    decode_key(); // Decode the key press
  }
}
```

We'll wait for the key to be released and then call the `decode_key()` function, which will take the value of the pressed key and scan the contents of the `keypad_map[]` array, looking for a matching value.

```
void decode_key() // Function to translate key code to its position number
{
  key_index = 1;  // Local variable for the pointer to the keypad_map array

#ifdef debug1
  Serial.println("Scanning Array");
#endif

  for (int x = 0; x <= 15; x++)  // Scan the keypad_map array looking for clos-
est valid match
  {
#ifdef debug1
    Serial.print("Index: ");
    Serial.print (x);
#endif

    if (abs(key_value - keypad_map[x]) < variance)  // If the data matches with-
in the specified tolerance
```

```
    {

#ifdef debug1
      Serial.print("    Match");
      Serial.print("    Key Char: ");
      Serial.println(key_text[x]);
#endif

      key_index = x;  // Set the key pointer to the matching key

      break;  // Exit as soon as we have a valid match
    }

#ifdef debug1
     Serial.println();
#endif

   }
#ifdef debug1
  Serial.print("Key Index: ");
  Serial.println(key_index);
#endif
}
```

The `read_adc()` function will first read the rotator position potentiometer on input 0 of the ADS1115 A/D converter. If the value has changed since the last A/D read, we'll then convert and map the analog value to a degree value from –180° to +180° and flag the TFT display to update the rotation bearing value.

```
void read_adc() // Function to read the ADS 1115 A/D Converter
{
  adc_value = ads.readADC_SingleEnded(0); // Get the raw ADC value

#ifdef debug3
  Serial.print("Current ADC Value: ");
  Serial.println(adc_value);
#endif

  // Now map the value to degrees (-180 to +180)
  actual_bearing = map(adc_value, adc_cal_0, adc_cal_max, -180, 180); // This is
```

```
the true bearing (-180 to + 180 degrees)
  // Only process the data if it's outside of the degree tolerance value or if
it's the first time through this code
  if ((abs(actual_bearing - previous_bearing) > degree_tolerance) || previous_
bearing == -1)
  {
    // Convert the bearing from -180 to +180 into two zones of 180 - 359 and 0 -
180 for proper calculation and display
    if (actual_bearing < -180)  // Set the lower limit to -180
    {
      actual_bearing = -180;
    }

    if (actual_bearing > 180) // Set the upper limit to +180
    {
      actual_bearing = 180;
    }

#ifdef debug2
    Serial.print("Actual Bearing: ");
    Serial.print(actual_bearing);
#endif

    if (actual_bearing < 0)    // Now map it to the display value
    {
      current_bearing = actual_bearing + 360;
    } else {
      current_bearing = actual_bearing;
    }

#ifdef debug2
    Serial.print("   Current Bearing: ");
    Serial.println(current_bearing);
#endif

    previous_bearing = actual_bearing;  // Update the previous bearing value
    data_changed = true;  // Indicate the display needs to be updated
  }
}
```

The first time the `update_display()` function is called, it will clear the display and then display a basic template of display data that doesn't change during sketch operation. This is done to reduce the amount of information that has to be erased and rewritten on every update that can make the TFT display flicker.

```
void update_display() // Updates the TFT display
{
  if (first_pass) // Only do this part the first time the function is called
  {
    // Clear the screen and display normal operation
    tft.fillScreen(ST7735_BLACK); // Clear the display
    delay(tft_delay);
    tft.setTextSize(1); // Set the text size to 1
    delay(tft_delay);
    tft.setTextColor(ST7735_BLUE); // Set the text color to BLUE
    delay(tft_delay);
    tft.setCursor(35, 5);
    delay(tft_delay);
    tft.print("Current Bearing");  // Display Current Rotator Bearing
    tft.setCursor(35, 60);
    delay(tft_delay);
    tft.print("Desired Bearing");  // Display Entered Rotator Bearing
    delay(tft_delay);
    tft.setCursor(67, 115);
    delay(tft_delay);
    tft.setTextColor(ST7735_GREEN); // Set the text color to GREEN
    delay(tft_delay);
    tft.print("Ready");
    first_pass = false; // Turn off the first pass flag
  }
```

If this is not the first time the function has been called, we'll call the `clear_data()` function to clear the data display areas and display the current bearing. If the rotator is rotating, we'll display this in yellow, otherwise it will be displayed in green.

```
    clear_data();  // Clear the value display area

    tft.setTextSize(4); // Set the text size to 4
    delay(tft_delay);
    if (moving) // If the rotator is turning set the current bearing color to
```

```
yellow - otherwise set it to green
      {
        tft.setTextColor(ST7735_YELLOW); // Set the text color to GREEN
        delay(tft_delay);
      } else
      {
        tft.setTextColor(ST7735_GREEN); // Set the text color to GREEN
        delay(tft_delay);
```

If there is a valid destination bearing entered, the display will be updated with the currently entered destination bearing and the next display update time is calculated.

```
      convert_bearing(current_bearing); // convert the current bearing to a 3
digit formatted string
      tft.setTextColor(ST7735_GREEN); // Set the text color to GREEN
      delay(tft_delay);
      if (entered_bearing != -1)  // If the entered bearing is a valid entry
      {
        tft.setCursor(50, 80);
        delay(tft_delay);
        convert_bearing(entered_bearing);// Convert the entered bearing to a 3
digit formatted string
      }
      data_changed = false; // The display is updated - Turn off the display up-
date flag
    }

    next_update_time = millis() + update_time;  // Calculate the time for the next
update
```

The `clear_data()` function is used to clear specific areas of the display. This is done to reduce the amount of erasing and rewriting of display data that causes flickering. We'll only clear the areas where the current bearing and the entered destination bearing are displayed.

```
void clear_data() //Clears the data area of the display
{
  tft.fillRect(45, 23, 78, 33, ST7735_BLACK); // Clear the display data areas
  delay(tft_delay);
  tft.fillRect(45, 78, 78, 33, ST7735_BLACK); // Clear the display data areas
  delay(tft_delay);
}
```

The `clear_status()` function is similar to the `clear_data()` function, except it will clear the bottom line of the TFT display that provides status information.

```
void clear_status() // Clears the status line of the display
{
  tft.fillRect(0, 112, 159, 15, ST7735_BLACK); // Clear the display data areas
  delay(tft_delay);
}
```

The `start_rotate()` function is used to set up and start the rotation process. It will use the difference between the current bearing and the destination bearing to determine the proper direction to rotate. Note that rotation will not occur if the entered destination bearing is within the tolerance range of the current rotator position.

```
void start_rotate() // Handles the rotate process
{
  // sets up and starts the rotation process

  // Check to be sure it's a valid move (greater than degree tolerance) Move has
already been checked for being a valid entry

  // We use actual bearing value since we may cross the 359 to 0 degree boundary
  // We already have actual - we need to calculate it for the desired destina-
tion

  if (entered_bearing > 180)  // Convert the destination to the -180 to +180 ac-
tual degree format
  {
    // We're in 181 to 359 area - subtract 180 degrees
    destination_bearing = entered_bearing - 360;
  } else
  {
    // We're in 0 to 180 - don't do anything with it
    destination_bearing = entered_bearing;
  }

#ifdef debug4
  Serial.print("Current = ");
  Serial.print(actual_bearing);
  Serial.print("   Start Moving - Destination = ");
```

```
  Serial.print(entered_bearing);
  Serial.print("  Actual Destination Bearing = ");
  Serial.print(destination_bearing);
#endif

  // Only start if the desired destination is outside of the degree tolerance
  if (abs(actual_bearing - destination_bearing) > degree_tolerance)
  {
```

Next, we'll check to see if the entered value is 180°. On the CDE/Hy-Gain HAM-series rotators, you have two 180° points on the meter, one on the far left and one on the far right. If we're commanded to go to 180°, we'll rotate to the closest one.

```
    if (destination_bearing == 180) // If desired bearing is 180 degrees, turn
towards the closest 180 degree point
    {
      // Find the nearest one and go to it
      if (actual_bearing < 0) // Convert the destination bearing to the actual
bearing format
      {
        // Turn Left
        destination_bearing = -180;

#ifdef debug4
        Serial.println("  Turn left");
#endif

        rotate_direction = "L"; // The rotation direction to the nearest 180
should be Left
        rotate(); // Start the rotation
      } else
      {
        // Turn Right

#ifdef debug4
        Serial.println("  Turn right");
#endif
        rotate_direction = "R"; // The rotation direction to the nearest 180
should be Right
        rotate(); // Start the rotation
      }
```

If the desired bearing is not 180°, we'll determine which direction to rotate and call the `rotate()` function to begin the rotation process. If the desired rotation is within the tolerance range, we won't execute the move and will instead exit the start rotation process.

```
      // Do we turn right (clockwise) or left (counterclockwise)?
      if (actual_bearing > destination_bearing) //  If the current bearing is
greater than the destination - turn left
      {
        // Turn left

#ifdef debug4
        Serial.println("  Turn left");
#endif

        rotate_direction = "L"; //  The rotation direction should be Left
        rotate(); // Start the rotation

      } else
      {
        // Turn right
#ifdef debug4
        Serial.println("  Turn right");
#endif

        rotate_direction = "R"; //  The rotation direction should be Right
        rotate(); // Start the rotation
      }

      moving = true;  // Set the rotating flag
    }
  } else
  {
    // Don't execute the move
    moving = false; // Turn off the moving flag
    move_to = ""; // Reset the desired bearing string
    entered_bearing = -1; // Reset the entered bearing value
    data_changed = true;  // Indicate that the display needs to be updated
  }
}
```

The `convert_bearing()` function will receive an integer bearing value, convert it to a three-character string, and display it on the TFT.

```
void convert_bearing(int calc_bearing)  // Function to convert int bearing to
String
{
  // Convert the bearing to a 3 character string and pad with zeroes
  String display_bearing = String(calc_bearing);
  while (display_bearing.length() < 3)    //  Do this while the string length is
less than 3 digits
  {
    // While the string length is less than 3 digits
    display_bearing = "0" + display_bearing;  // Pad with leading zeroes until
we have a 3 digit value
  }
  tft.print(display_bearing); // Display the padded bearing value
}
```

The `stop_all()` function stops all rotation. It will first clear the TFT status line, update it with a red braking message, and de-energize the brake release and rotation control relays. After a three second delay to allow rotation to stop, the braking message will be cleared.

```
void stop_all() // Function to stop all rotation
{
  // Stop all rotation and trigger braking delay
#ifdef debug4
  Serial.println("Stop all Rotation - Initiate Braking delay");
#endif

  clear_status(); // Clear the display status line
  tft.setTextSize(1); // Set the text size to 1
  tft.setTextColor(ST7735_RED); // Set the text color to RED
  delay(tft_delay);
  tft.setCursor(65, 115);
  delay(tft_delay);
  tft.print("Braking"); // Indicate we are braking
  digitalWrite(left_relay, LOW);
  digitalWrite(right_relay, LOW);
  delay(brake_time);
  digitalWrite(brake_relay, LOW);
```

```
  clear_status();
  tft.setTextColor(ST7735_GREEN); // Set the text color to GREEN
  delay(tft_delay);
  tft.setCursor(67, 115);
  delay(tft_delay);
  tft.print("Ready");

  moving = false; // Turn off the moving flag
  move_to = ""; // Reset the desired bearing string
  entered_bearing = -1; // Reset the entered bearing value
  data_changed = true;  // Indicate that the display needs to be updated
}
```

The `rotate()` function starts the rotation operation. It will first change the color of the current bearing display to yellow to indicate rotation is occurring, then it energizes the brake release relay along with the proper rotation direction control relay.

```
void rotate() // Function to perform rotation
{
  clear_status(); // Clear the display status line
  tft.setTextSize(1); // Set the text size to 1
  delay(tft_delay);
  tft.setTextColor(ST7735_YELLOW); // Set the text color to YELLOW
  delay(tft_delay);
  tft.setCursor(60, 115);
  delay(tft_delay);
  tft.print("Rotating"); // Indicate we are rotating
  if (rotate_direction == "L")  // Turn on the left rotation relay and release 
the brake
  {
    digitalWrite(brake_relay, HIGH);  // Release the brake
    delay(release_delay);
    digitalWrite(left_relay, HIGH); // Enable the left relay
  } else
  {
    // Turn on the left rotation relay and release the brake
    digitalWrite(brake_relay, HIGH);  // Release the brake
    delay(release_delay);
    digitalWrite(right_relay, HIGH); // Enable the right relay
  }
}
```

The `check_move()` function manages an active rotation process. This function will check to see if we've reached the destination bearing. If so, we'll call the `stop_all()` function to end the rotation process.

```
void check_move()
{
  // We only get here if it's already moving
  // Checks to see if we've reached the destination bearing
  // We've just read the ADC so no need to read it again
  if (rotate_direction == "L")  // destination is less than current position
  {
    if (actual_bearing <= (destination_bearing + left_rotate_trim))  // Stop
when we get there - include left rotate trim
    {
      stop_all(); // Stop all rotation
    }
  } else
  {
    // We're turning right
    if (actual_bearing >= (destination_bearing + right_rotate_trim))   // Stop
when we get there - include right rotate trim
    {
      stop_all(); // Stop all rotation
    }
  }
}
```

Figure 17.10 shows the CDE/Hy-Gain keypad rotator controller TFT display.

The sketch and libraries for this project are available from **kw5gp.com**.

Figure 17.10 — The CDE/Hy-Gain keypad rotator controller TFT display.

18 Yaesu CAT Display

Many of today's transceivers can be controlled via computer using computer aided transceiver control, also known as CAT. Many CAT interfaces on various radios are based on RS-232, or some other serial-based communication method such as Icom's CI-V interface. Over time, the CAT control functions have expanded so much that you can practically do anything over the CAT interface that you can with the radio's physical knobs and buttons. In addition, the advancements in radio technology offer more features than can be viewed or managed using only the rig's front panel display and control. The days of moving more and more features into a menu system, no matter how well the system is done, are nearly behind us, as it becomes a massive undertaking to remember which menu has the feature you're wanting to view or change. For this book, we've updated this project to use an Arduino Nano, and added a TXB0108 level converter to the ST7735-type TFT display (see **Figure 18.1**).

One of the first things I do when I get a new radio, or more specifically, a radio that is new to me, is go through the manual to make myself familiar with the menu system. Then, I'll skim through the CAT commands, and realize that I'd probably need a workstation connected to the rig with some sort of CAT control program to access some of the functions on the radio. There are times though that I don't want to carry a workstation with me out into the field on a portable operation, but still would like to have access to some of the CAT functions, what should I do? Here's where my "wouldn't it be cool…" tendency kicks in; wouldn't it be cool if I could have an Arduino accessing some of the CAT functions on the rig and displaying them on a color TFT graphic display?

Figure 18.1 — The Yaesu CAT display.

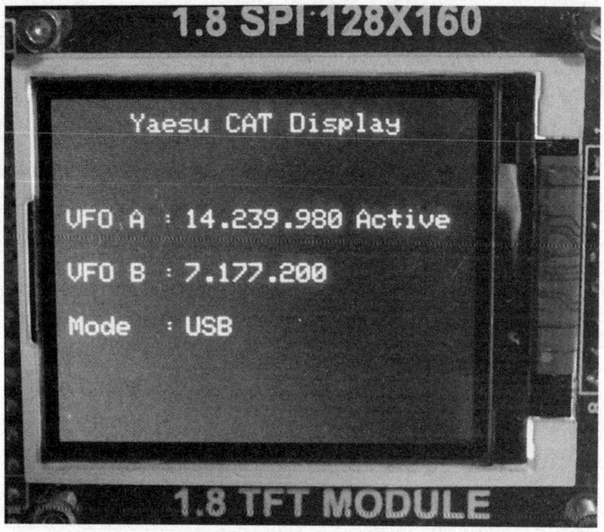

Figure 18.2 — Yaesu CAT display output.

For this project, we'll build a basic Arduino-powered CAT interface and display (see **Figure 18.2**) and set it up to use with my Yaesu FT-450 Go-Box that I use for portable events such as ARRL Field Day and the Mississippi QSO party, two of my favorite events. The plan is to remotely display some of the basic operating parameters so that visitors, my logger, etc. can see what frequency we're operating on, for example, without having to lean over and see the actual rig's display. Since the Yaesu FT-450 (see **Figure 18.3**)

Figure 18.3 — The Yaesu FT-450.

shares the same RS-232 interface, and many of the same CAT commands, with the FT-950 and FT-1200, this project can be used with those transceivers as well. With minor changes to the sketch, this project could be used with virtually any transceiver that has CAT control capability and a serial interface.

The parameters I have chosen to display for this project are the VFO-A and VFO-B frequencies, an indicator as to which VFO is currently active, and the current operating mode. There are dozens of choices for what to display available through the CAT menu, feel free to pick the ones you would like. The Yaesu CAT command structure is very straightforward and easy to work with as you'll see when we get to the sketch for this project.

The Block Diagram

Figure 18.4 shows the block diagram for the Yaesu CAT display project. As you can see, the hardware for this project is little more than the Arduino UNO, TFT display, and the MAX232 RS232 driver/receiver chip we use to interface

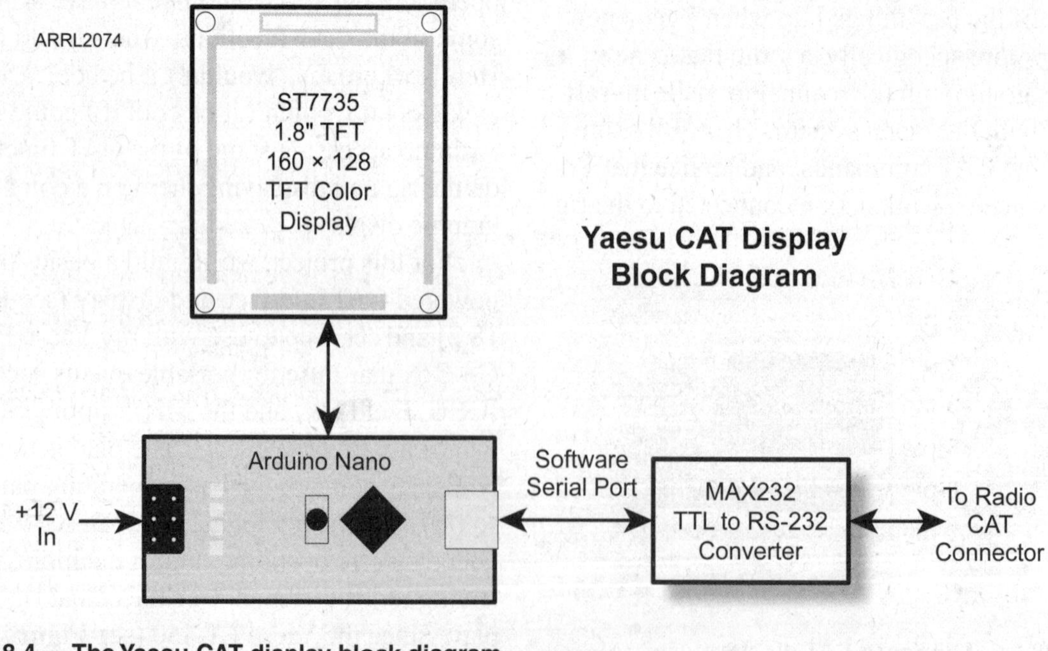

Figure 18.4 — The Yaesu CAT display block diagram.

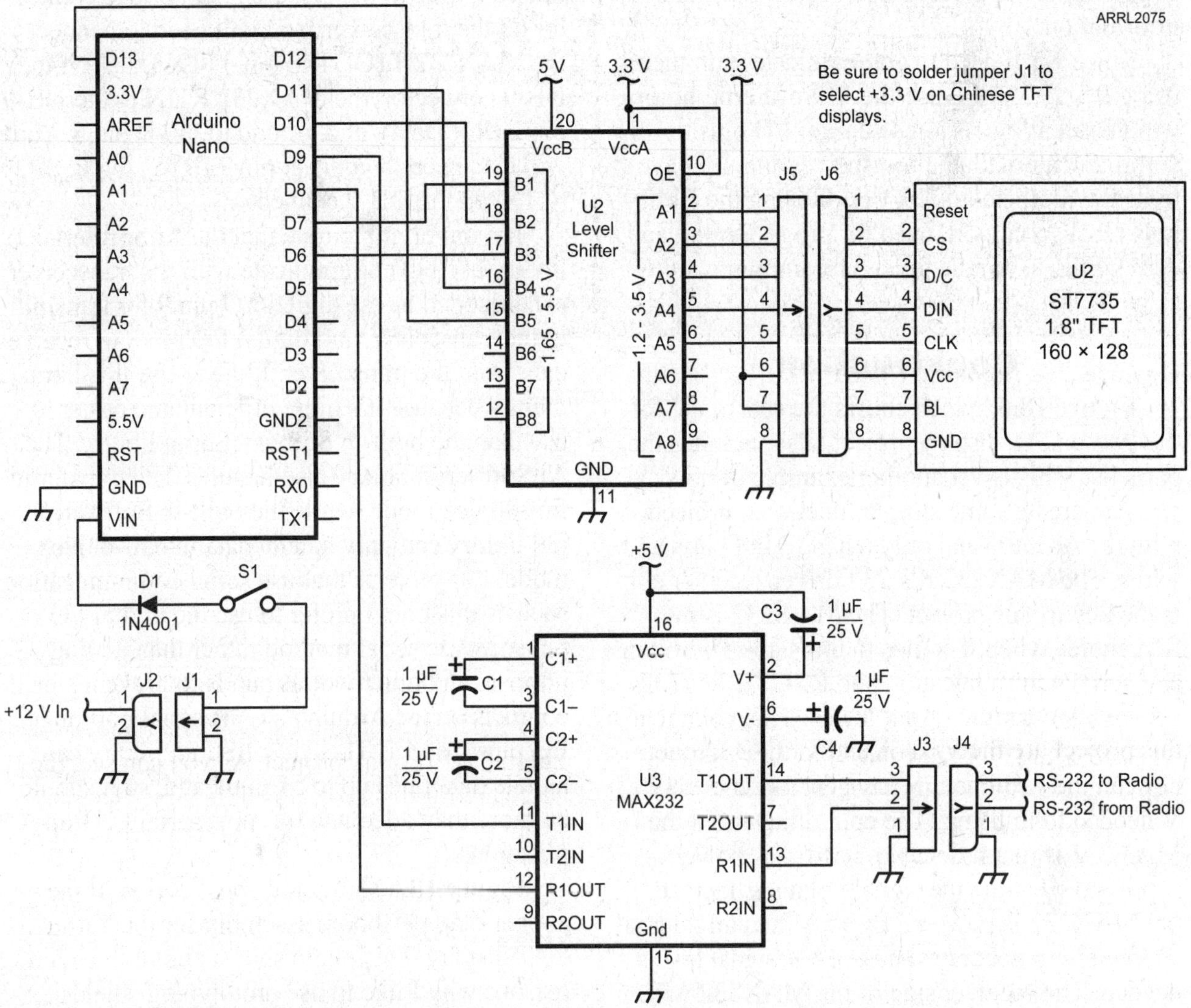

Figure 18.5 — The Yaesu CAT display schematic.

Part	*Value*
C1-C4	1 µF 25 V capacitor
D1	1N4004 diode
J1	2-pin Dupont male header
J2	2-pin Dupont female header
J3	3-pin Dupont male header
J4	3-pin Dupont female header
J5	8-pin Dupont male header
J6	8-pin Dupont female header
S1	SPST toggle switch
U1	Arduino Nano
U2	ST7735-type 1.8-inch color TFT display
U3	MAX232 RS-232 driver/receiver
U4	8-channel level shifter module (TXB0108 type)
Enclosure	Solarbotics Mega S.A.F.E.

to the transceiver. We'll use an ST7735-type 1.8-inch color graphic 160 × 128 TFT display, although there's no reason you couldn't use a larger display to give yourself room for more information.

Since we'll need to communicate with the rig using RS-232, we'll use the AltSoftSerial library, which will allow us to use digital I/O pins to communicate with the transceiver and not have to deal with the complexities of using the Arduino's USB serial port, used for programming and debugging, to handle the CAT communications as well.

Construction

Figure 18.5, which shows the schematic for the Yaesu CAT display project, also includes the parts list. This is yet another example of how you can create some simple functional projects with the Arduino and only a handful of parts and wires. The MAX232 RS-232 driver/receiver chip is the key to this project. The MAX232 is my first choice when it comes to translating between 5 V serial communication and RS-232. RS-232 uses ± 12 V for the signal levels. This voltage is far too high to directly connect to the Arduino, nor can the Arduino directly generate the ± 12 V needed to transmit. The cool thing about the MAX232 is that it does not need a ± power supply to generate the signal voltages. Instead, the MAX232 is powered by +5 V, and an internal charge pump generates the ± 12 V signal levels needed. The receiver side of the MAX232 will convert the ± 12 V signal levels into standard 5 V TTL logic levels that can be directly connected to the Arduino I/O pins. To connect to the Yaesu FT-450/FT-950/FT-1200 transceivers, you will need to build a cable to connect from the MAX232 to the DB-9 CAT interface connector on the transceiver. You will need to connect the MAX232 T1OUT to pin 3 (TX Data) on the DB-9 connector, the MAX232 R1IN pin to DB-9 pin 2 (RX Data), and ground to DB-9 pin 5. You will also need to connect pin 7 (RTS) to pin 8 (CTS) on the DB-9 connector.

Figure 18.6 — The Yaesu CAT display copper-clad perfboard assembly.

It's important to note that the AltSoftSerial library used to communicate with the transceiver is hardcoded to use digital I/O pin 9 for transmit to the transceiver, and digital I/O pin 8 to receive data from the transceiver. I like to use this library mainly because it's more efficient and easier to use than the built-in SoftwareSerial library. The AltSoftSerial library also handles data transfer in full-duplex mode, while the built-in SoftwareSerial library can only handle data in half-duplex mode. For projects that use serial communication such as this one, I prefer to use the digital I/O pin software serial method rather than adding jumpers and other workarounds to prevent signal conflicts on the Arduino's USB/serial port on I/O pins 0 and 1. The AltSoftSerial library can handle data rates up to 57,600 baud, so it should be more than adequate for most serial CAT applications.

Figure 18.6 shows the construction of the copper-clad perfboard assembly for the Yaesu CAT display. This photo shows one of the main reasons why I like to use prototyping shields along with the 2.54 mm Dupont-style headers and sockets. This arrangement allows me to quickly disconnect everything and remove the board assembly to correct wiring errors or make modifications to the project. I like to use 2 mm hardware to mount the TFT display, as seen in Figure 18.1, to the top cover of the enclosure. You'll also notice a pair of PowerPole connectors is used in a chassis mount to supply the 12 V DC for the project. Since this project is made to work in conjunction with a 12 V DC-powered transceiver, you can easily add an Anderson Powerpole cable from the power supply to it.

Figure 18.7 shows the flowchart for the Yaesu CAT Display. You can download the sketch and libraries for the project at **kw5gp.com**. For

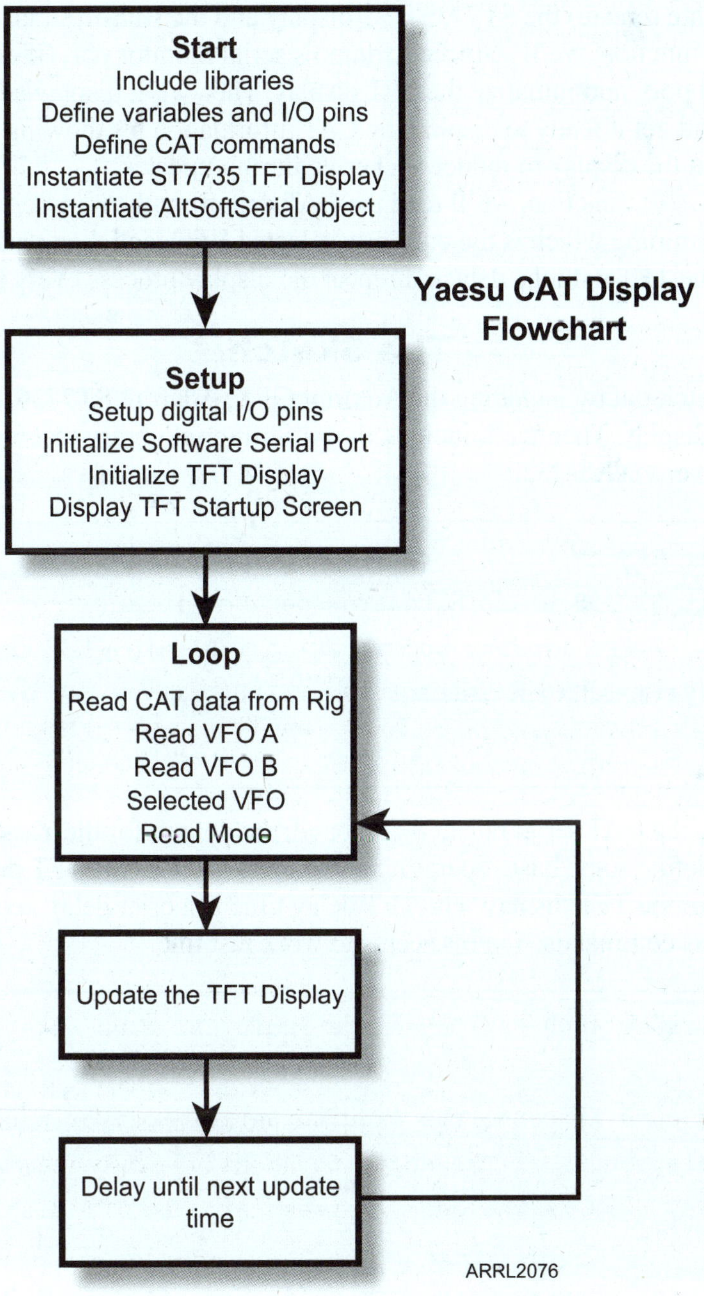

Figure 18.7 — The Yaesu CAT display flowchart.

this sketch, we'll need to include the Adafruit GFX, Adafruit ST7735, and SPI libraries for the TFT display. We'll also need to include the AltSoftSerial library to handle serial communication with the transceiver's CAT interface.

Next, we'll define the digital I/O pins for the TFT and data rate for the CAT interface. The CAT interface baud rate defaults to 4,800 baud for the FT-450, FT-950, and FT-1200 transceivers. We'll also define the CAT commands used to have the transceiver send us the information we want to display. Finally, we'll instantiate (create) the ST7735 TFT display and the AltSoftSerial objects.

In the `setup()` function, we'll start the Arduino's serial monitor port (if debugging is enabled), start the AltSoftSerial port, and initialize the TFT display. Then we'll display a brief startup message on the TFT display and get it ready to display the CAT information by showing a template of the data that doesn't change on the display to reduce flickering during updates.

In the `main loop()` function, we'll read the CAT data from the transceiver. First reading VFO-A, then VFO-B, determining which is the currently selected VFO, and then reading the operating mode. We'll update the TFT with the data and repeat the display process every five seconds.

The Sketch

We'll start the sketch out by including the Adafruit GFX, Adafruit ST7735, and the SPI libraries for the ST7735 TFT display. Then we'll include the AltSoftSerial library that we'll use to communicate with the transceiver via RS-232.

```
#include <Adafruit_GFX.h>     // Core graphics library
#include <Adafruit_ST7735.h> // Hardware-specific library
#include <SPI.h>  // SPI library so we can communicate with the TFT display
#include <AltSoftSerial.h>  // AltSoft Serial Library - better than the Softwa-
reSerial library
```

Next, we'll define the CAT serial rate at 4,800 baud, the default for the Yaesu FT-450, FT-950, and FT-1200. We'll also define some basic operating parameters, such as the TFT display update delay and the I/O pins needed for the TFT display. The TFT delay time is a brief delay needed to allow the TFT display time to process commands. Ten ms seems to work just fine.

```
#define CAT_serial_rate 4800  // Define the radio CAT baud rate - Default is 4800
baud
#define comm_speed 9600 // Define the Arduino Serial Monitor baud rate
#define timeout_delay 5000  // The delay to signify a data timeout from the radio
#define update_delay 5000 // The delay between Arduino TFT display updates

#define TFT_CS      10  // Assign the TFT CS to pin 10
#define TFT_RST     7  // Assign the TFT RST to pin 7
#define TFT_DC      6  // Assign the TFT DC to pin 6

#define tft_delay 10  // set the TFT command delay to 10ms
```

Next, we'll define the CAT command strings we'll need to send to the transceiver. For this project we'll get the VFO-A and VFO-B frequencies, along with which of the two VFO's is currently active, and the current operating mode. You can customize this display and get different CAT information simply by changing or adding CAT commands and modifying the display.

```
#define get_VFO_A "FA;" // Get VFO-A command
#define get_VFO_B "FB;" // Get VFO-B command
#define get_vfo "VS;" // Get which VFO is selected
#define get_mode "MD0;"   // Get Mode command
```

Next, we'll declare and initialize the remaining variables needed in the sketch. Then, we'll instantiate (create) the ST7735 TFT display and AltSoftSerial objects.

```
boolean first_pass = true;  // Variable used to determine if first pass through
the main loop
char rx_char; // Variable for each character received from the radio
unsigned long timeout, current_millis;  // Variables to handle communication
timeout errors
bool receiving, timeout_error;  // Variables for the receive radio data process
String CAT_buffer;  // String to hold the received CAT data
String VFO_A, VFO_B, vfo, mode; // Variables to hold the parsed and formatted
CAT data

Adafruit_ST7735 tft = Adafruit_ST7735(TFT_CS,  TFT_DC, TFT_RST);  // Initialize
the TFT display instance

AltSoftSerial altSerial;  // Initialize the software serial instance
```

In the `setup()` function, we'll first start the Arduino IDE serial monitor port if debugging is enabled, then we'll start the software serial port on digital I/O pins 8 and 9 with a rate of 4,800 baud, the default rate.

```
#ifdef debug
  Serial.begin(comm_speed); // Set the Arduino Serial Monitor baud rate
  while (!Serial) ; // wait for Arduino Serial Monitor to open
  Serial.println("FT-450/900/1200 CAT Display Begin");
#endif

  //Start the Software Serial port
  altSerial.begin(CAT_serial_rate); // Start the software serial port to the
transceiver
```

Next, we'll initialize the ST7735 TFT display and show a brief startup message. Finally, we'll clear the TFT display and display a template of the TFT data that doesn't change between updates to save time and reduce display flickering.

```
  tft.initR(INITR_18BLACKTAB);   // initialize a 1.8" TFT with ST7735S chip,
black tab
  delay(tft_delay);
  tft.fillScreen(ST7735_BLACK); // Clear the display - fill with BLACK background
  delay(tft_delay);
  tft.setRotation(1); // Set the screen rotation
  delay(tft_delay);
  tft.setTextWrap(false); // Turn off Text Wrap
  delay(tft_delay);
  tft.setTextSize(3); // Set the Font Size
  delay(tft_delay);
  tft.setTextColor(ST7735_BLUE); //Set the Text Color
  delay(tft_delay);
  tft.setCursor(40, 10);  //Set the Cursor and display the startup screen
  delay(tft_delay);
  tft.print("KW5GP");
  delay(tft_delay);
  tft.setTextSize(2);
  delay(tft_delay);
  tft.setCursor(50, 60);
  delay(tft_delay);
  tft.print("Yaesu");
  delay(tft_delay);
  tft.setCursor(15, 80);
  delay(tft_delay);
  tft.print("Transceiver");
  delay(tft_delay);
  tft.setCursor(15, 100);
  delay(tft_delay);
  tft.print("CAT Display" );

  delay(5000);  //Wait 5 seconds then clear the startup message
  tft.fillScreen(ST7735_BLACK); // Clear the display
  delay(tft_delay);

  update_display(); // update the display
  first_pass = false; // turn off the update display first pass flag
}
```

In the main loop(), we'll start by calling the send_command() function to get the frequency of VFO-A. Then we'll call the get_response() function to receive the CAT response from the transceiver, and then the format_string() function to format the CAT response for the TFT display. This is the basic sequence needed to request and receive CAT data for all the Yaesu CAT commands.

```
void loop()
```

```
{
  send_command(get_VFO_A);  // Send the Get VFO-A Frequency Command

  get_response();  // Receive the VFO-A frequency response

  format_string();  // Format the VFO Response data for display
  VFO_A = CAT_buffer;

#ifdef debug
  Serial.print("VFO-A: ");
  Serial.println(VFO_A);
#endif
```

We'll then repeat the same sequence for VFO-B.

```
  send_command(get_VFO_B);  // Send the Get VFO-B Frequency Command


  get_response();  // Receive the VFO-B frequency response

  format_string();  // Format the VFO Response data for display
  VFO_B = CAT_buffer;

#ifdef debug
  Serial.print("VFO-B: ");
  Serial.println(VFO_B);
#endif
```

Next, we'll get the CAT information for which VFO is currently selected. Since we're only extracting a single character from the response we don't need to format the string for the display, as this information will be formatted by the display_update() function.

```
send_command(get_vfo);     // Send the VFO Selected command

  get_response();    // Receive the VFO Selected response
  vfo = CAT_buffer.substring(2);

#ifdef debug
  Serial.print("VFO: ");
  Serial.println(vfo);
#endif
```

Finally, we'll get the CAT information for the current operating mode. We'll then extract the number of the mode from the CAT response and call the `convert_mode()` function to translate the CAT response into text for the display.

```
  send_command(get_mode);    // Send the Get Mode Command

  get_response();    // Receive the Get Mode response
  mode = CAT_buffer.substring(3);

  convert_mode(); // Convert the CAT Mode response into the corresponding text

#ifdef debug
  Serial.print("Mode: ");
  Serial.println(mode);
#endif
```

Then we'll call the `update_display()` function to update the information on the TFT display, delay five seconds, then repeat the process.

```
  update_display(); // Update the TFT display

  delay(update_delay);  // Delay until the next display update
```

The `get_response()` function receives the CAT command response from the transceiver. The first thing this function does is set up a timeout timer so that the sketch doesn't hang waiting for a response from the transceiver that for some reason failed or was garbled.

```
void get_response() // Receives the CAT command response from the radio
{
#ifdef debug
  Serial.print("    Start RX   ");
```

```
#endif

  // Set a timeout value
  current_millis = millis();  // Get the current time
  timeout = current_millis + timeout_delay; // Calculate the timeout time
  // Check for millis() rollover condition - the Arduino millis() counter rolls
over about every 47 days

  if (timeout < current_millis) // We've calculated the timeout during a mil-
lis() rollover event
  {
    timeout = timeout_delay; // Go ahead and calculate as if we've rolled over
already (adds a few millis to the timeout delay)
  }
```

You'll note that the timeout calculation is a bit more complex than you'd think, and you'd be correct. The `millis()` timer in the Arduino returns an unsigned long integer, which ranges from zero to 4,294,867,295. This means that the `millis()` timer will roll over every 49.7 days, which we will need to take into account when dealing with timeout calculations. Since it is possible for this project to be left running for extended periods of time, we need to take into account that the `millis()` timer may roll over. That's why the timeout calculation formula used in this sketch is a bit more involved than usual.

Next, we'll get ready to receive a CAT response from the radio. This of course assumes that we've already sent a CAT command to the radio and are expecting a CAT response. We'll keep looping, waiting for received characters until the loop either times out or receives a semicolon (;) indicating an end of CAT message from the radio.

```
  // Get ready to receive CAT response from the radio
  receiving = true;
  timeout_error = false;
  CAT_buffer = "";
  do
  {
    if (millis() > timeout) // We've exceeded the timeout delay
    {
      // We timed out - exit
      receiving = false;
      timeout_error = true;

      CAT_buffer = "";
      break;
```

```
    }
    if (altSerial.available() && receiving) // If there's a character in the rx
buffer and we're ok to receive
    {
      rx_char = altSerial.read(); // Get the incoming character
      if (rx_char == ';') // ";" indicates the end of the response from the ra-
dio
      {
        receiving = false;  // Turn off the ok to receive flag
      }  else
      {
        CAT_buffer = CAT_buffer + rx_char;  // Add the received character to the
CAT rx string
      }

    }
  } while (receiving);  // Keep looping while we're ok to receive data from the
radio

#ifdef debug
  if (timeout_error)
  {
    Serial.println("   Timeout ");
  } else
  {
    Serial.print("   Received complete: ");
    Serial.println(CAT_buffer);

  }
#endif
}
```

The send_command() function sends a CAT command string to the radio. The CAT command to send is passed as a parameter to the function so that this function can be reused to send any desired CAT command.

```
void send_command(String CAT_command) // Sends the CAT command string to the ra-
dio
{
  // Send the CAT Command
```

```
#ifdef debug
  Serial.print("Sending: ");
  Serial.print(CAT_command);
#endif

  altSerial.print(CAT_command); // Sends the CAT command to the radio

#ifdef debug
  Serial.print("  Send Complete  ");
#endif
}
```

The `convert_mode()` function converts the single character operating mode CAT response to a text string that can be displayed on the TFT display. We'll use a Switch…Case statement to handle the mode character to text conversion.

```
void convert_mode()  // Convert the mode to the mode name
{
  // Mode is a single character string

  switch (mode[0])// convert the single character mode string to a char variable
  {
    case '1': // LSB
      mode = "LSB";
      break;

    case '2': // USB
      mode = "USB";
      break;

    case '3': // CW
      mode = "CW";
      break;

    case '4': // FM
      mode = "FM";
      break;

    case '5': // AM
      mode = "AM";
      break;
```

```
    case '6': // DATA (RTTY-LSB)
      mode = "DATA (RTTY-LSB)";
      break;

    case '7': // CW-R
      mode = "CW-R";
      break;

    case '8': // USER-L
      mode = "USER-L";
      break;

    case '9': // DATA (RTTY-USB)
      mode = "DATA (RTTY-USB)";
      break;

    case 'B': // FM-N
      mode = "FM-N";
      break;

    case 'C': // USER-U
      mode = "USER-U";
      break;

    default: // Unknown mode
      mode = " ";
      break;

#ifdef debug
      Serial.print("Mode String: ");
      Serial.println(mode);
#endif
  }
}
```

The update_display() function will update the CAT information displayed on the TFT display. The first time this function is called it will display a template of the display information that doesn't change. This is done to reduce flickering from clearing and redisplaying the template information on every update. Since this information doesn't change, don't waste time updating it every five seconds.

```
void update_display() // Updates the TFT display
{
  if (first_pass) // Only do this part the first time the function is called
  {
    // Clear the screen and display normal operation
    tft.fillScreen(ST7735_BLACK); // Clear the display
    delay(tft_delay);
    tft.setTextSize(1); // Set the text size to 1
    delay(tft_delay);
    tft.setTextColor(ST7735_BLUE); // Set the text color to BLUE
    delay(tft_delay);
    tft.setTextSize(1); // Set the text size to 2
    delay(tft_delay);
    tft.setCursor(30, 5);
    delay(tft_delay);
    tft.print("Yaesu CAT Display");  // Display screen title
    tft.setCursor(5, 40);
    delay(tft_delay);
    tft.print("VFO A :");  // Display VFO A
    tft.setCursor(5, 60);
    delay(tft_delay);
    tft.print("VFO B :");  // Display VFO B
    delay(tft_delay);
    tft.setCursor(5, 80);
    delay(tft_delay);
    tft.print("Mode  :");  // Display Mode
    delay(tft_delay);
  }
```

If this is not the first time through the operation, we'll update the CAT information on the display. The `clear_data()` function is used to clear the data area of the TFT display.

```
    clear_data();  // Clear the value display area

    tft.setCursor(50, 40);
    delay(tft_delay);
    tft.print(VFO_A); // Display VFO-A Value

    tft.setCursor(50, 60);
    delay(tft_delay);
```

```
    tft.print(VFO_B); // Display VFO-B Value

    tft.setCursor(50, 80);
    delay(tft_delay);
    tft.print(mode);  // Display mode

    if (vfo == "0")
    {
      // VFO-A is active
      tft.setCursor(115, 40); // Select the active VFO line
    }

    if (vfo == "1")
    {
      // VFO-B is active
      tft.setCursor(115, 60); // Select the active VFO line
    }

    tft.print("Active"); // Indicate the active VFO

  }
}
```

Due to how TFT displays work, we can't just type over the existing data, so we must erase it first. With a TFT display, the way to erase an area is to fill it with the background color.

```
void clear_data() //Clears the data area of the display
{
  tft.fillRect(50, 40, 110, 50, ST7735_BLACK); // Clear the CAT data area
  delay(tft_delay);
}
```

You can download the sketch and libraries for the project at **kw5gp.com**.

19 USB CW Keyboard

Figure 19.1 — The USB CW keyboard.

One of the most popular projects in my first Arduino book was the PS/2-type CW Keyboard (see **Figure 19.1**). That was actually one of my first Arduino projects, and I still use it in my own shack whenever I work CW. I keep saying I'll go back to using a straight key or a set of paddles, but it hasn't happened yet. It turns out I'm much better at typing than I am at keying. However, as computer technology advances, the 1980's era PS/2 style keyboard has all but been replaced by USB or wireless keyboards, and it's getting harder and harder to find the old PS/2 type keyboards.

For my third Arduino book, I wanted to include some USB and Bluetooth projects, so I did a lot of research into those technologies on the Arduino, hoping to come up with some next-generation Arduino projects. As it turns out, the Arduino doesn't directly support the human interface device (HID) USB role very well at all. Virtually, yes, all Arduinos can connect to a workstation via USB, but only as a device, not as a USB host. Since you can't connect a USB device directly to another USB device, you must have a USB host to communicate with the device. This means I couldn't just take a USB keyboard and connect it to an Arduino. So much for the easy way.

As for the Bluetooth projects, it's pretty much the same story. The standard Bluetooth modules for the Arduino don't support the HID role. I still haven't given up on the Bluetooth projects but figuring that out would have taken far more research time than was available. Rest assured, the lab workbench is piled high with Bluetooth-related components just waiting for met to finish this book.

To solve the USB issue, several vendors created a USB host shield in the earlier days of the Arduino to allow it to be a USB host. I will not talk about plugging shields in and writing a few lines of code in this book, though, so I kept researching and thinking up ideas to try and get around the USB host shield issue to create a ham radio-related Arduino project.

Recently, I came across the mini–USB host shield/module with an SPI interface. Technically, it's still a shield in the general sense of the word, but now I could fit it all in an enclosure with an Arduino Nano, and give it a much better, larger 2.2-inch color TFT display compared to the LCD used in the original PS/2 keyboard project. As you can see from the photo in **Figure 19.2**, though, the mini-USB host shield/module is not a shield in the traditional sense, it is more like a module than anything else. As it turns out, this module was designed as a shield for the Arduino Pro-Mini. Since I prefer to use Arduino Nanos,

Figure 19.2 — The USB mini host module.

for this project we'll just treat it like any other module and go from there.

A Brief Look at USB

USB is short for Universal Serial Bus. Created in 1996, the USB standard was designed to be a standard for cables, connectors, and protocols to interface between computers, peripherals, and even other computers. As part of the standard, USB also provides power to the attached devices. USB was designed as a far simpler method of connecting devices to a computer than the large multiconductor parallel and RS-232 serial cables used prior. A major advantage of USB was the higher data transfer speeds of 12 Mb/s with USB 1.0. USB 2.0 gave us speeds up to 480 Mb/s, and now with USB 3.0 the speeds are up to 5 Gb/s, and are headed even higher with USB 3.1, 3.2, and USB 4.

One nice thing about USB is that the cables are the same for USB 1.0 and 2.0. USB 3.0 introduced a new device-end connector, but for all intents, a USB host port is compatible with older USB devices running a lesser version of the USB protocol. USB cables are also hot-swappable, meaning they can be plugged and unplugged without the need to turn off either the host or device.

USB is also highly versatile. The USB interface on a computer is self-configuring, eliminating the need for manually configuring device settings. Most computers will automatically install any needed drivers for a new USB device without any action needed on the part of the user. However, herein lies the issue with the Arduino when acting as a USB Host. There are literally thousands of USB devices, each with their own specific drivers, that can be connected to a USB host port. These devices can include keyboards, mice, headphones, microphones, webcams, external hard drives, thumb drives, and even transceivers such as the Yaesu FT-991A and others. The Arduino would need a driver for each of these for it to be able to properly communicate with every USB device.

The USB protocol breaks the functionality of various devices into classes. Each USB device will send its class code to the USB host when it is first connected so that the host can determine what type of device it is. The host will then query the new device and gather more information to load the correct device drivers and complete the auto-configuration process. Since the basic Arduino project doesn't have enough memory or storage to store every possible USB

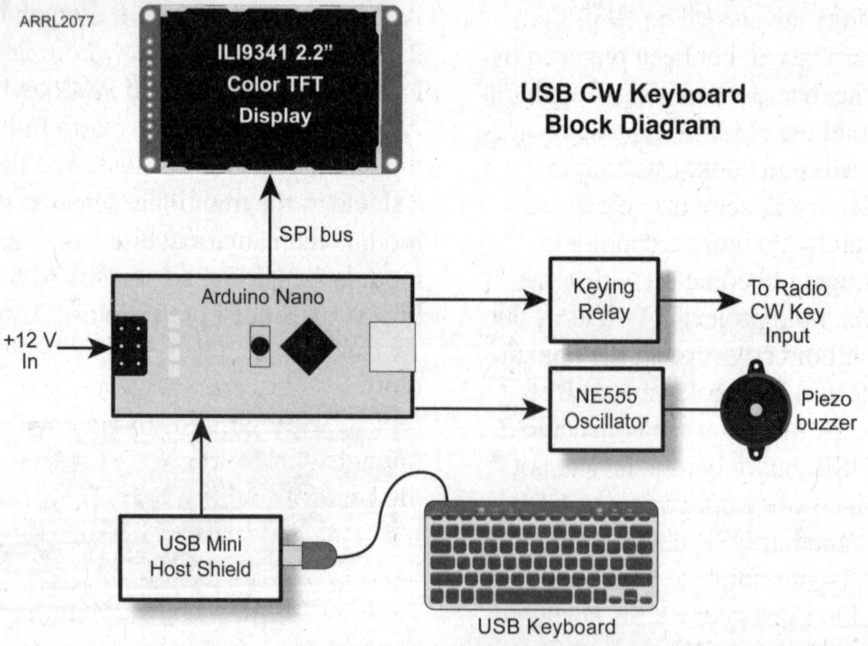

Figure 19.3 — The USB CW keyboard block diagram.

Figure 19.4 — The USB CW keyboard 2.2-inch TFT display.

device configuration, nor does it have the ability to automatically download and install any missing drivers, it can only support a subset of the USB devices available. Fortunately for us, things like keyboards and mice are well supported in the Arduino USB host library. I was even able to use a wireless keyboard with a USB dongle with the Arduino USB host library. Even better, this library is fully compatible with the mini-USB host module that we'll be using in this project. For those of you who wish to know more about the USB protocol, I found an excellent explanation on USB at **beyondlogic.org/usbnutshell/usb1.shtml**.

The Block Diagram

So now that we have the biggest piece of the Arduino and USB keyboard connection puzzle solved, we can go on with the process of designing the USB CW Keyboard. **Figure 19.3** shows the block diagram for the USB CW keyboard project.

The goal for this project is to duplicate the functionality of the original PS/2 CW keyboard, but also give it a bit of an upgrade. We'll be using an Arduino Nano and the mini-USB host shield/module we talked about earlier. In addition, we'll be upgrading from the original 16-character by 2-line LCD display to a 2.2-inch ILI-9341-type 240 × 320-pixel color graphic TFT display (see **Figure 19.4**) and using a TXB0108 level converter for the TFT display. Due to some differences between the original PS/2 keyboard library and the USB host library, which we'll discuss in detail later, we won't be able to use the original Morse library that was used in the PS/2 version. The new interrupt-driven Morse code library does not have the ability to send a CW tone, so we'll need to add an external NE555 timer-based oscillator driving a piezo buzzer to provide the CW tone. As with the PS/2 version, this version will also drive a reed relay used to key the rig itself.

Building the USB CW Keyboard

Figure 19.5, which shows the schematic for the USB CW keyboard project, also includes the parts list. The project is built on a piece of copper-clad perfboard cut to fit within a Solarbotics Mega S.A.F.E. enclosure. This will allow us to house the electronics in the enclosure and mount the 2.2-inch TFT display on the inside of the top cover using 2.2 mm hardware as shown in **Figure 19.6**. I've found that when working with the Arduino and the various modules, 2 mm screws, washers, and nuts are nearly a perfect fit for the mounting holes in most modules. I prefer the hex head version of the screws and recommend getting them in various lengths.

When wiring up the USB host module, there are a few things to be aware of. The mini-USB host module is a 3.3 V device, and it will be damaged if you use 5 V to power it. However, most USB keyboards don't run at 3.3 V, so to power the attached devices with 5 V we'll need to cut a trace on the USB host module and apply 5 V to the VBUS pin as shown in **Figure 19.7**. Cutting the trace will isolate the USB power pin from the rest of the USB module and only supply 5 V to the attached USB device. We'll also use a TXB0108-type level converter module between the Nano, the USB host module, and TFT display for the SPI bus signals to reduce those voltages to a safe level. You could also replace the mini-USB module with a regular Arduino UNO shield and use an UNO in place of the Nano. The full-size USB host shield for the UNO is completely 5 V powered, and some even have a 3.3 V/5 V option.

Some of the inexpensive mini-USB host modules you get from places such as eBay may have an error on the silk-screen labeling on the

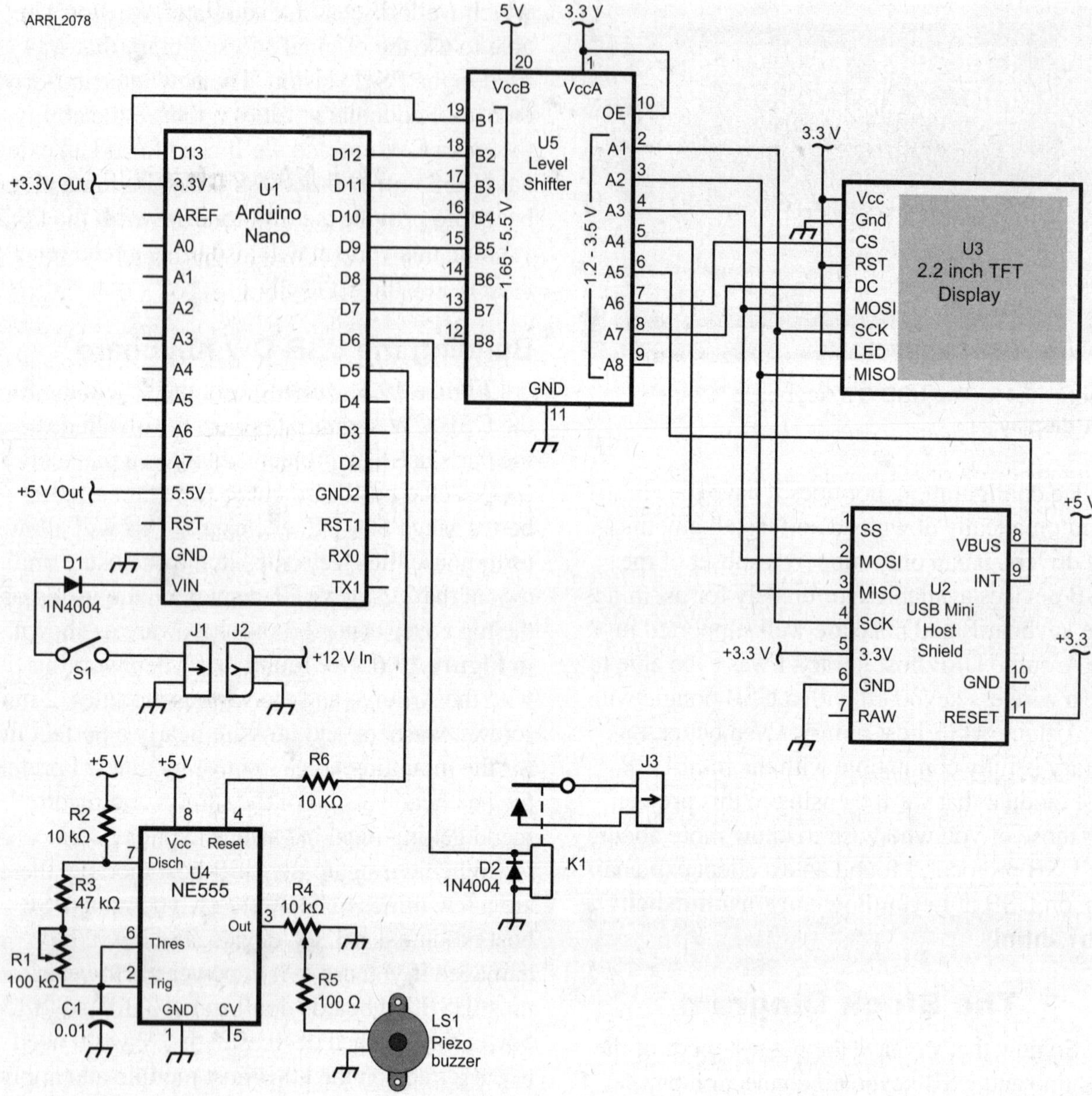

Figure 19.5 — The USB CW keyboard schematic.

Part	*Value*	*Part*	*Value*
C1	.01 µF capacitor, 25 V	R4	10 kΩ 10-turn potentiometer
D1, D2	1N4004 diode	R5	100 Ω resistor ¼ W
J1, J3	2-pin 2.54 mm Dupont male header	S1	SPST toggle switch
J2	2-pin 2.54 mm Dupont female header	U1	Arduino Nano
		U2	ILI-9341 type 2.2" color TFT display
K1	5 V SPST DIP reed relay	U3	Mini USB host shield/module
LS1	piezo buzzer	U4	NE555 timer IC
R1	100 kΩ 10-turn potentiometer	U5	8-channel level shifter module (TXB0108 type)
R2, R6	10 kΩ resistor 1/8 W		
R3	47 kΩ resistor 1/8 W	Enclosure	Solarbotics Mega S.A.F.E.

Figure 19.6 — The USB keyboard project in the enclosure.

board. These boards have the MOSI and SCK pins mislabeled. The correct labeling is shown in Figure 19.7.

The ILI9341-type 2.2-inch color TFT display has somewhat similar power issues, except you can supply 5 V to the display for power without damaging the device. However, the signal lines must all connect to the Nano using level converters or voltage dividers. I have found that the inexpensive Chinese versions of the TFT display are a bit fussy on the voltage divider resistor values, which is why I use a 1 kΩ/1.5 kΩ resistor voltage divider for those projects when I don't use a level converter module. However, using a TXB0108-type level converter module is my preferred method.

The NE555 timer circuit is a simple astable oscillator that is triggered from the Arduino Nano. You can control both the frequency and the volume of the piezo buzzer on the output of the NE555 circuit. For the NE555 oscillator circuit to trigger properly from the Arduino I/O pin, we'll need to add a 10 kΩ pulldown resistor on the I/O pin.

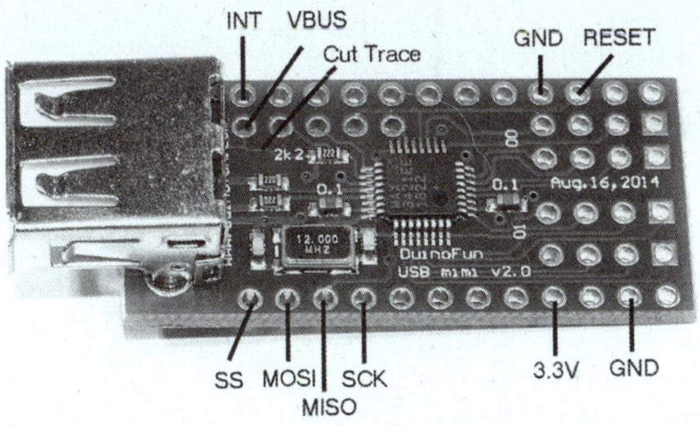

Figure 19.7 — The USB host module pinout and trace cut location.

The completed USB CW keyboard perfboard assembly is shown in **Figure 19.8**.

The Flowchart

Figure 19.9 shows the flowchart for the USB CW keyboard. What makes this sketch work is the power in all the libraries we'll be using. For the IL9341 color TFT display, we'll need to include the Adafruit GFX, Adafruit ILI9341, and SPI libraries. The USB host library has several sub-libraries, but for this project, we'll include the hidboot library, which is for human interface devices (HID) in boot protocol mode, a simple USB HID-device communication method that is supported by all USB keyboards. We'll also include the Lewis Morse code library. Due to the way that the USB host library handles communication with the keyboard, we'll need to use an interrupt-driven library to send the Morse code, as the USB keyboard implementation doesn't work well with anything that spends a lot of time in delays such as the CW timing between elements. Without this interrupt-driven Morse code library, I'm not sure this project would have been viable at all. And finally, we'll need to include the TimerOne library to set timer interrupts for the Morse library to check for outbound CW data.

Next, we'll define the variables and I/O pins we'll need in the sketch. It is important to note that the SS and INT pins on the USB host module must be connected to digital I/O pins 10 and 9 respectively. These I/O pins are hardcoded in

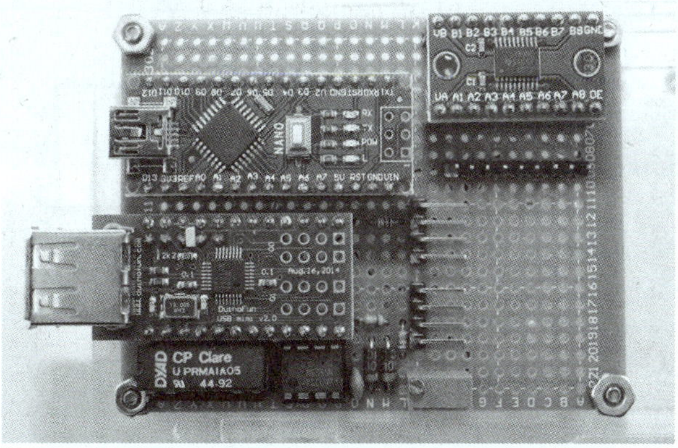

Figure 19.8 — The USB CW keyboard assembly.

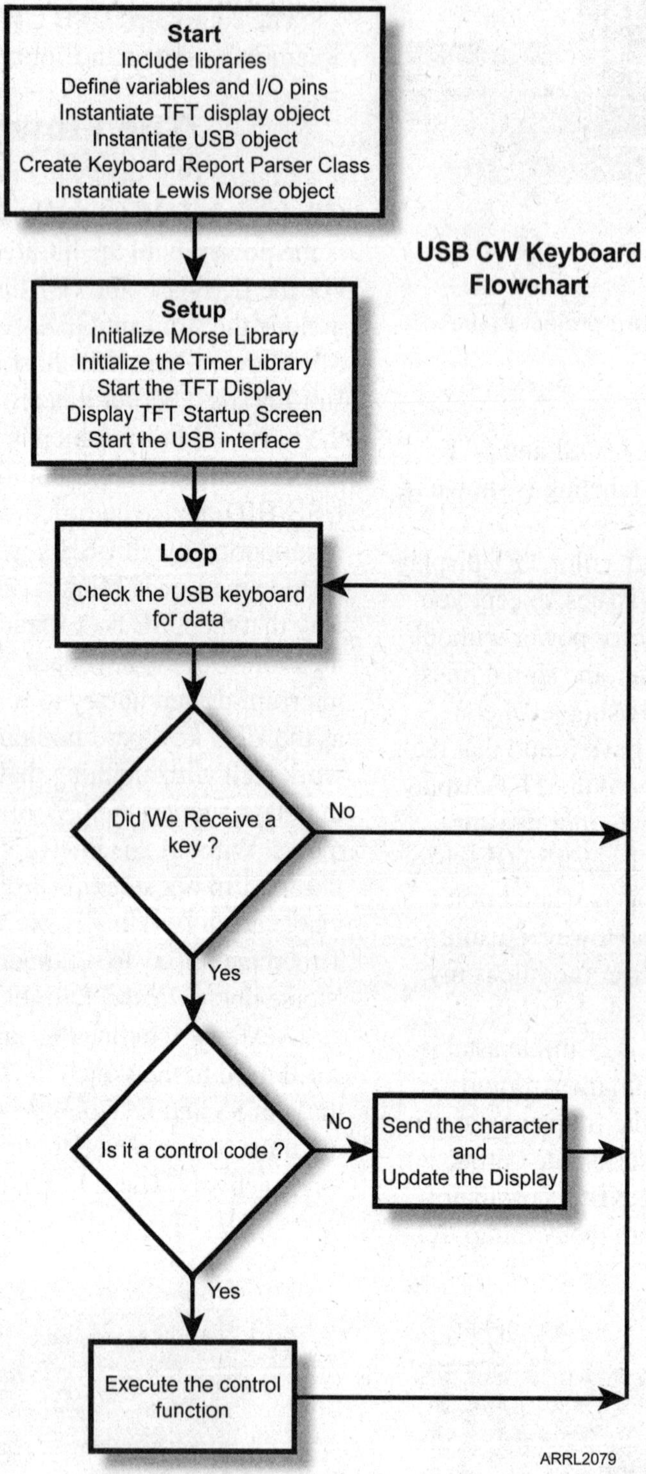

Figure 19.9 — The USB CW keyboard flowchart.

the USB host library, and while they theoretically can be changed by manually editing the pin definitions in the library, avoid that trouble and just use the pins they are configured for instead.

We'll need to instantiate (create) the USB host and TFT display objects, as well as the Lewis library Morse object. One special initialization task for this project is defining a keyboard report parser class and object for the USB keyboard data. Unlike a PS/2 keyboard that will send one character at a time, the USB keyboard will follow the standard USB protocol for transferring data and send the data over in blocks known as "reports." The USB host will need to decode these reports that are based on the USB device type and extract the keyboard data that way. Because of the way the keyboard data is transferred, and the way it is handled by the USB host library, the typeahead buffer for the USB keyboard is only ten characters long. A fast typist could easily overrun the keyboard buffer, so you will need to be mindful of not typing too fast. This will especially be noticeable at the slower CW speeds. This could easily be corrected with the implementation of a keyboard buffer routine in the sketch that could add new characters to the tail end of a larger buffer to keep from overrunning the CW sending side of things. But remember, one of the goals of the projects in this book is to provide you with a working and functional platform that you can improve on and add your own personal touches to, so I'll leave the larger type-ahead buffer up to you as an enhancement challenge.

In the `setup()` function, we'll initialize the Lewis Morse code, and the TimerOne libraries. We'll also start the TFT display and the USB host interface. Then we'll display a brief startup message on the TFT display.

The `loop()` function for this project is relatively straightforward. Unlike the previous two projects where we had the sketch broken out into multiple functions with a smaller `loop()` function, in this project we'll have the majority of the sketch in the `main loop()`. The reason for this is that this sketch performs basically a single operation, you wait for a keystroke on the keyboard, and either send it or perform the selected command operation.

The first thing we'll do is check the keyboard for data. Once we receive a keystroke, we'll check to see if it is either a valid Morse code character or a command code. The command codes in this sketch will allow you to adjust the keying speed from 5 – 30 words per minute and enable either the beep or keying modes.

If the received key is a morse code character, we'll send it to either the NE555 tone circuit or the keying relay. If it's a command code, we'll execute the specified command. We'll then go back to the top of the loop and wait for the next keystroke.

The Sketch

The libraries and sketch for this project can be downloaded from **kw5gp.com**. The USB CW keyboard has multiple levels of debugging information. Uncommenting the debug #define statements as needed when troubleshooting the sketch will provide a wealth of information on the Arduino IDE's serial monitor as the sketch runs. Next, we'll include all the libraries needed for the sketch. The USB host library has several individual USB host-related sub-libraries to support a number of USB host functions. For this sketch, we'll be using the hidboot library to support the USB keyboard in boot protocol mode. For the TFT display, we'll include the Adafruit GFX, Adafruit ILI9341, and the SPI library. Finally, we'll include the Lewis Morse code library, along with the TimerOne library to support time-based interrupt operations.

```
#include <hidboot.h>  // USB Host shield library for Human Interface Devices in
boot protocol mode
#include <SPI.h>  // Library to drive SPI devices
#include <Adafruit_GFX.h> // The Adafruit Graphics Library
#include <Adafruit_ILI9341.h> // The ILI9341 TFT Display Library
```

```
#include <Lewis.h>  // The Lewis Interrupt driven Morse Code Library
#include <TimerOne.h> // Library to set timer interrupts
```

Next, we'll define the digital I/O pins and the various messages and parameters needed in the sketch.

```
#define TFT_DC 7  // Define the TFT DC pin
#define TFT_CS 8  // Define the TFT CS pin
#define beep_pin 5  // Define the pin for the CW tone
#define key_pin 6 // Define the relay keying pin

#define cw_start_speed 10 // Set the keyer starting speed to 10 wpm

#define osc_fail "USB OSC Fail" // Define the USB oscillator failure error text
#define usb_ok "USB OK" // Define the USB oscillator ok status text
#define repeat_delay 100  // Set repeating key delay to 100ms

#define cw_font_size 4  // The CW text display font size
#define status_font_size 2  // The font size for the status line

String usb_status_msg;  // Variable for the current USB status message to dis-
play
String cw_data = "";  // String variable to hold cw output text data
String key_ascii = "";  // Single character string variable to hold converted
key value
bool system_status = true;  // Flag indicating current system status - true =
ok, false = bad
bool repeat_enabled = false;  // Flag indicating if we've met the requirements
to repeat keys
int first_pass_delay = 10000; // Number of milliseconds to wait for USB Host to
identify and begin communication with HID Keyboard
int repeat_timeout = 500; // Number of milliseconds a key must remain pressed to
initiate repeat
int key_code; // Variable for the USB non-ASCII key code
int ascii_key_code; // Variable for the ASCII key code
int prev_key_code;  // Variable for the previous USB non-ASCII key code
int prev_ascii_key_code;  // Variable for the previous USB ASCII key code
unsigned long repeat_timer; // Timeout timer variable
int key_speed = cw_start_speed; // Define the initial keyer speed and set it to
the defined start_speed
int last_speed = 0; // Variable for the previous keyer speed
bool key_mode = false;  // Used to select beep or key 0=beep 1=key
bool last_mode = true; // Set the previous mode opposite to the starting key
mode so it will update the tft
```

Then we'll create the ILI9341 TFT display instance and do something that we haven't done before, create a new C++ class that will be used to parse the keyboard report data.

```
// Use hardware SPI (on Uno, #13, #12, #11) and the above for CS/DC
Adafruit_ILI9341 tft = Adafruit_ILI9341(TFT_CS, TFT_DC);  // Create the TFT in-
stance

class KbdRptParser : public KeyboardReportParser    // Define a class to parse
the incoming keyboard data
{
    void PrintKey(uint8_t mod, uint8_t key);

  protected:
    void OnControlKeysChanged(uint8_t before, uint8_t after);

    void OnKeyDown  (uint8_t mod, uint8_t key);
    void OnKeyUp  (uint8_t mod, uint8_t key);
    void OnKeyPressed(uint8_t key);
};
```

Finally, we'll create a USB instance and start the USB Keyboard in HID boot protocol mode. Along with that, we'll create a keyboard report parser instance and a Lewis Morse code instance.

```
USB     Usb;  // Create a USB instance

HIDBoot<USB_HID_PROTOCOL_KEYBOARD>    HidKeyboard(&Usb);  // Start keyboard in
HID Boot protocol mode

KbdRptParser Prs; // Create a Keyboard Report Parser Instance

Lewis Morse;  // Create a Morse Library Instance
```

In the `setup()` function, we'll start the Morse code instance in beep mode at ten words per minute, using interrupts. The ability to use interrupts is the key for this sketch and project to work properly. This allows the sketch to process other operations, such as receiving the next keystroke and implementing command code functions, while it is also sending the CW characters. Next, we'll initialize the timer interrupt to have the morse process check every ten milliseconds for a character to send, and attach the `myISR()` interrupt service function that will handle the actual interrupt processing. Next, we'll start the TFT display, clear it, then display a brief startup message.

```
Morse.begin(-1, beep_pin, 10, true); // Morse.begin(rx_pin, tx_pin, words_per_
minute, use_interrupts)
```

```
  Timer1.initialize(10000); // Initialize the timer interrupt to 10 milliseconds
  Timer1.attachInterrupt(myISR);  // Attach the Interrupt Service Routine

  tft.begin();  // Start the TFT display
```

We'll then start the USB host port. Note that if for some reason the USB port fails to start, the sketch will loop endlessly at this point until the issue is resolved and the Nano is reset. Then we'll set up the HID keyboard report parser, which will be used to extract the keystrokes from the keyboard.

```
  if (Usb.Init() == -1) // Check for USB initialization failure
  {
    usb_status_msg = osc_fail;  // Set the status message to oscillator fail
    system_status = false;  // Indicate a system error
    update_status();  // Update the status line on the bottom of the TFT display

#ifdef debug
    Serial.println(osc_fail);
#endif

    do  // Loop endlessly, no point in continuing
    {
      // do nothing further until reset
    } while (true);
  } else
  {
    usb_status_msg = usb_ok;  // Indicate USB started ok
    update_status();  // Update the status line on the bottom of the TFT display
  }

  delay( 200 ); // Wait for USB to come online

  HidKeyboard.SetReportParser(0, &Prs); // Set up the HID report parser
#ifdef debug
  Serial.println("USB Ready");
  Serial.println("");
#endif
```

Finally, we'll get things ready for receiving keystrokes, move the TFT display cursor to the top line, and set the CW keying mode to beep.

```
  key_code = 0; // Clear the incoming key code
  ascii_key_code = 0; // Clear the incoming ASCII key code
```

```
  repeat_timer = millis() + first_pass_delay; // Set the repeat key timer to wait
a bit on startup

  tft.setCursor(0, 10); // Set the TFT cursor to the top line

  mode_set();  // Set the CW keying mode (beep or key)
}
```

In the main loop(), we'll start out by clearing the incoming key variables and set up to allow us to repeat keystrokes when the key is held down. That's another issue with the USB host library as it relates to keyboards; there's no repeat function. So, we'll implement one manually in the sketch.

```
void loop()
{
  key_code = 0; // Clear the incoming key code
  ascii_key_code = 0; // Clear the incoming ASCII key code

  if (!repeat_enabled && (millis() > first_pass_delay)) // Check to see if we can
repeat the key and it's past the startup delay time
  {
    repeat_timer = millis() + repeat_timeout; // Set the repeat timer timeout
time
  }
```

Then, we'll check the USB host port for keyboard data. Next, we'll check to see if the key has been held long enough to trigger the repeat key operation and prepare to repeat the key. Then we'll check the received key code to verify that it's either a valid international Morse code character, or one of the command codes we use for setting the speed or changing the beep/key mode.

```
Usb.Task(); // Check the USB keyboard for data

  if (millis() >= repeat_timer) // if the key is still down and we've exceeded
the repeat timeout
  {
    delay(repeat_delay);  // Delay the repeat delay
    key_code = prev_key_code; // Set the key code to last received key code
    ascii_key_code = prev_ascii_key_code; // Set the ASICC key code to last re-
ceived ASCII key code
  }

  if (key_code != 0)  // If the key code is not zero, we have received a key
  {
```

```
#ifdef debug1
    Serial.print("Key code: ");
    Serial.print(key_code);
#endif
    if (ascii_key_code != 0)  // If the ASCII key code is not zero, the key is
an ASCII key
    {
      // It's an ASCII character

#ifdef debug1
      Serial.print("   ASCII code: ");
      Serial.print(ascii_key_code);
      Serial.print("   ASCII Character: ");
      Serial.println(char(ascii_key_code));
#endif

      prev_key_code = key_code; // Set the previous key code to current received
key code
      prev_ascii_key_code = ascii_key_code; // Set the previous ASCII key code
to current ASCII received key code
```

First, we'll check for invalid ASCII values in the received key codes and ignore them. If it is a valid ASCII key code, we'll update the TFT display with the new character and send it using the Lewis morse code library.

```
   // Verify the ASCII character is a valid International Morse Code character,
display and send it

      switch (ascii_key_code) // Check for invalid keys and ignore
      {
        case 19:  // <Enter>
          break;

        case 35:  // #
          break;

        case 37:  // %
          break;

        case 42:  // *
          break;
```

```
        case 91:  // - [
          break;

        case 92:  // - \
          break;

        case 93:  // - ]
          break;

        case 94:  // ^
          break;

        case 96:  // `
          break;

        case 123:  // {
          break;

        case 124:  // |
          break;

        case 125:  // - }
          break;

        case 126:  // ~
          break;

        default:  // It's a valid character - send it
          update_cw_text(); // Update the TFT display with the new key

          send_cw();  // Send the CW character
          break;
      }
    }
```

If the received key code is not a valid ASCII code, we'll next check to see if it's a command code. For this project, we use the up and down arrow keys on the keyboard to increase and decrease the CW sending speed. We'll use the left arrow key to enable the beep mode, and the right arrow key to enable the key mode. We'll also configure the del (delete) key to send three dits to indicate that we made a typo and let the receiving station know it's an error. You could change it and have it send all eight dits, but three dits, space like individual letter E is often used to indicate an error for the sake of convenience. That completes the `main loop()` for this project. As you can see, it's relatively straightforward.

```
 //It's a control code
#ifdef debug
      Serial.print(" - Control Code");
#endif
      switch (key_code)
      {
        case 82:
          // Up Arrow - Increase speed
#ifdef debug
          Serial.println("  Up Arrow");
#endif

          if (key_speed < 30) // Limit max speed to 30 wpm
          {
            key_speed = key_speed + 1;


          }
          mode_set();
          update_status();
          break;

        case 81:
          // Down Arrow - Decrease speed
#ifdef debug
          Serial.println("  Down Arrow");
#endif

          if (key_speed > 5)
          {
            key_speed = key_speed - 1;
          }
          mode_set();
          update_status();
          break;

        case 80:
          // Left Arrow - Set beep mode
#ifdef debug
```

```
          Serial.println("  Left Arrow");
#endif
          key_mode = false;  // Enable Beep mode
          mode_set();
          update_status();
          break;

        case 79:
          // Right Arrow - Set Key Mode
#ifdef debug
          Serial.println("  Right Arrow");
#endif

          key_mode = true; // Enable key mode
          mode_set();
          update_status();
          break;

        case 76:
          // Delete Key - Send 3 dits (EEE)
#ifdef debug
          Serial.println("  Delete/Error");
#endif

          for (int i = 1; i <= 3; i++)
          {
            ascii_key_code = 69;  // The ASCII code for "E"
            key_ascii = "E";

            tft.setTextColor(ILI9341_RED, ILI9341_BLACK);  // Red on black -
background color required to "clear" previous text
            update_cw_text();
            tft.setTextColor(ILI9341_GREEN, ILI9341_BLACK);  // Red on black -
background color required to "clear" previous text

            send_cw();  // Send the CW character
          }
          break;
```

```
        default:
          // If no match - ignore the key
#ifdef debug
          Serial.println();
#endif
          break;
      }
    }
  }
}
```

Now, let's dig into the functions and see what they do. The first function is the `clear_display()` function, which fills the TFT display with the background color. This is how we erase data on a TFT display. There is no way to type over existing information on the screen and have it erase the previous information unless you first fill that area with the background color.

```
void clear_display()  // Clears the TFT display
{
  tft.fillScreen(ILI9341_BLACK);  // Fill the screen with the background color
}
```

The next function is the `update_status()` function. The bottom line of the TFT display reports status information on the USB host port, the CW mode, and speed.

```
void update_status() // Prints status messages on bottom line of TFT display in
green
{
  if (system_status)  // True if good, false if bad
  {
    //We're all good - update status in green
    tft.setTextColor(ILI9341_GREEN, ILI9341_BLACK);  // Green on black back-
ground color required to "clear" previous text
  } else
  {
    // There is a system error
    tft.setTextColor(ILI9341_RED, ILI9341_BLACK);  // Red on black background
color required to "clear" previous text
  }
  tft.setTextSize(status_font_size);  // Set the font size to the status line
font size
```

```
  tft.setCursor(10, 220); // Move to the bottom line of the display

  tft.print(usb_status_msg);  // Display the USB status

  if (key_mode != last_mode)  // Check to see if the keyer mode has changed -
and if so, update the display
  {
    tft.setCursor(130, 220);
    if (key_mode)
    {
      tft.print("Key ");
    } else
    {
      tft.print("Beep");
    }
    last_mode = key_mode;
  }

  if (key_speed != last_speed)  // Check to see if the keyer speed has changed -
and if so, update the display
  {
    tft.setCursor(230, 220);
    tft.print("WPM: ");
    tft.print(key_speed);
    tft.print(" "); // Clear the second digit if less than 10 wpm
    last_speed = key_speed;
  }
  tft.setTextSize(cw_font_size);  // Set the display font size back to the CW
text font size

  // Return the cursor to where we were
  // Set the cursor to line 0 and reprint the cw_data string - neat trick..it
overwrites what is there without blinking and sets
  // the cursor back to where we were
  tft.setCursor(0, 10);
  tft.print(cw_data);
}
```

The `update_cw_text()` function is used to update the text on the TFT display. We use a larger font to display the CW characters, and we'll also scroll a line up at a time when we reach the bottom of the text area on the display. Since there is no scrolling function for the TFT display, we'll have to handle that manually. For this reason, you'll need to use the Adafruit_ILI9341 TFT library, or a library that's as fast as that one, as several of the other TFT libraries I tested were too slow to handle the scrolling operation without causing the display to irritatingly flicker.

```
void update_cw_text() // Updates the cw text on the display
{
  tft.setTextSize(cw_font_size);   // Set the cw text font size

  key_ascii = String(char(ascii_key_code)); // Convert the ascii key code to a
string
  key_ascii.toUpperCase();   // Make all the characters uppercase

  // Scroll bottom line up a line if at end of line 4
  if (cw_data.length() >= 52)
  {
    cw_data = cw_data.substring(13);   // Remove the first 13 characters
    tft.setCursor(0, 10); // Move the cursor to the top line
    tft.print(cw_data); // Print the cw data string variable
    for (int i = 1; i <= 13; i++) // Print a space 13 times to scroll the data
up a line
    {
    tft.print(" ");
    }
    tft.setCursor(0, 107);   // Set the cursor to the start of line 4
  }

  cw_data = cw_data + key_ascii;   // Add the character to the cw data string

  tft.print(key_ascii); // display the received ASCII character

#ifdef debug
  Serial.print("cw_data string length: ");
  Serial.print(cw_data.length());
#endif

}
```

The send_cw() function sends the CW key using the currently selected beep or key mode.

```
void send_cw()  // Sends the CW character
{
#ifdef debug
  Serial.print(" Sending: ");
  Serial.println(key_ascii);
#endif

  Morse.print(key_ascii); // Send the character to the Morse library
}
```

The mode_set() function is called when a left or right arrow keystroke is detected. In key mode, the Lewis Morse library will send the CW to the relay on digital I/O pin 6 that is used to key the rig. In beep mode, the library will send the CW using digital I/O pin 5 that is used to trigger the NE555 oscillator. The function will also update the TFT display to indicate the new status information.

```
void mode_set()   // Function to Set the mode to beep or keying and/or set key-
ing speed
{
  if (key_mode) // Keying mode True = Key False = Beep
  {
    Morse.begin(-1, key_pin, key_speed, true); // Key on pin 6
  } else
  {
    Morse.begin(-1, beep_pin, key_speed, true); // Key on pin 5
  }

#ifdef debug
  Serial.print("Key Mode: ");
  Serial.print(key_mode);
  Serial.print("  WPM: ");
  Serial.println(key_speed);
#endif

  update_status();  // Update the information on the status line

}
```

The myISR() function comes from the Morse library example code to process the interrupts for the Lewis Morse library. This allows the sketch to do other operations while sending the CW data simultaneously.

```
// myISR function
void myISR()  // Function to handle interrupts for the Morse library
{
  Morse.timerISR();
}
```

The USB host library functions come directly from the HIDboot keyboard example sketch. Since USB devices don't send data on a per-character basis, and instead use the USB data report mechanism, we'll need these functions to parse the received keyboard data for us. This sketch is a perfect example of how to use the example code that is often provided with Arduino libraries. Even with all the research I have done regarding USB operations, I doubt very seriously if I would even know where to start with getting the data from the keyboard. Fortunately, the example code that came with the USB host library has everything I need to get the job done. I may not fully understand how and what these functions are doing, but when it comes to things like the Arduino, sometimes you just trust the library, and copy and paste the pieces you need from the example code. That's what the Arduino and open-source communities are all about.

The first USB host function is the KbdRptParser::Printkey() function, which is used to display information about the pressed key on the Arduino IDE's serial monitor if the usb_debug information is enabled at the beginning of the sketch.

```
void KbdRptParser::PrintKey(uint8_t m, uint8_t key)
{
  MODIFIERKEYS mod;
  *((uint8_t*)&mod) = m;

#ifdef usb_debug
  Serial.print((mod.bmLeftCtrl   == 1) ? "C" : " ");
  Serial.print((mod.bmLeftShift  == 1) ? "S" : " ");
  Serial.print((mod.bmLeftAlt    == 1) ? "A" : " ");
  Serial.print((mod.bmLeftGUI    == 1) ? "G" : " ");

  Serial.print(" > ");
  PrintHex<uint8_t>(key, 0x80);
  Serial.print(" < ");
#endif

#ifdef usb_debug
```

```
  Serial.print("Key = ");
  Serial.println(key);
#endif

#ifdef usb_debug
  Serial.print((mod.bmRightCtrl   == 1) ? "C" : " ");
  Serial.print((mod.bmRightShift  == 1) ? "S" : " ");
  Serial.print((mod.bmRightAlt    == 1) ? "A" : " ");
  Serial.println((mod.bmRightGUI    == 1) ? "G" : " ");
#endif
}
```

The `KbdRptParser::OnKeyDown()` function indicates to the USB host library that a key has been pressed down. Due to how the USB keyboard's function, they will send information to the USB host that a key has been pressed, and again when the key is released.

```
void KbdRptParser::OnKeyDown(uint8_t mod, uint8_t key)
{
#ifdef usb_debug
  Serial.print("DN ");
#endif

  repeat_enabled = true;
  PrintKey(mod, key);

  uint8_t c = OemToAscii(mod, key);

  if (c)
    OnKeyPressed(c);

  key_code = key;
  ascii_key_code = c;

#ifdef usb_debug
  Serial.print("Before OEM to ASCII: ");
  Serial.print(key, HEX);
  Serial.print("   After OEM to ASCII: ");
  Serial.println(c, HEX);
#endif
}
```

The `KbdRptParser::OnControlKeysChanged()` function is used by the USB host library to get the key code information when a control key on the keyboard has been pressed or released.

```
void KbdRptParser::OnControlKeysChanged(uint8_t before, uint8_t after)
{

  MODIFIERKEYS beforeMod;
  *((uint8_t*)&beforeMod) = before;

  MODIFIERKEYS afterMod;
  *((uint8_t*)&afterMod) = after;

#ifdef usb_debug

  if (beforeMod.bmLeftCtrl != afterMod.bmLeftCtrl)
  {
    Serial.println("LeftCtrl changed");
  }
  if (beforeMod.bmLeftShift != afterMod.bmLeftShift)
  {
    Serial.println("LeftShift changed");
  }
  if (beforeMod.bmLeftAlt != afterMod.bmLeftAlt)
  {
    Serial.println("LeftAlt changed");
  }
  if (beforeMod.bmLeftGUI != afterMod.bmLeftGUI)
  {
    Serial.println("LeftGUI changed");
  }

  if (beforeMod.bmRightCtrl != afterMod.bmRightCtrl)
  {
    Serial.println("RightCtrl changed");
  }
  if (beforeMod.bmRightShift != afterMod.bmRightShift)
  {
    Serial.println("RightShift changed");
  }
  if (beforeMod.bmRightAlt != afterMod.bmRightAlt)
  {
    Serial.println("RightAlt changed");
```

```
  }
  if (beforeMod.bmRightGUI != afterMod.bmRightGUI)
  {
    Serial.println("RightGUI changed");
  }

#endif

}
```

The `KbtRptParser::OnKeyUp()` function is used to indicate to the USB host library that a key has been released.

```
void KbdRptParser::OnKeyUp(uint8_t mod, uint8_t key)
{

#ifdef usb_debug
  Serial.print("UP ");
  PrintKey(mod, key);
#endif

  prev_key_code = 0;
  prev_ascii_key_code = 0;
  repeat_enabled = false;
}
```

And finally, the `KbtRptParser::OnKeyPressed()` function will return the ASCII value of the key that is pressed.

```
void KbdRptParser::OnKeyPressed(uint8_t key)
{
#ifdef usb_debug
  Serial.print("ASCII value: ");
  Serial.print(key);
  Serial.print(" Character: ");
  Serial.println((char)key);
#endif
  // ascii_key_code = key;
}
```

Enhancement Ideas

As I mentioned earlier, the USB keyboard library only has a ten-character typeahead buffer. At the slower CW speeds, it would be easy for a fast typist to overrun the buffer. Expanding the buffer by as few as ten characters could add a great deal of flexibility to this project. If you remember, in the original PS/2 CW keyboard, there were five user-programmable memories. It should not be difficult to add programmable memory capability to this project, and even save the memory information to the Nano's EEPROM. And while we're saving things to the Nano's EEPROM, why not save the speed and CW mode settings as well? Another possible enhancement would be to use a rotary encoder with a pushbutton switch in the place of the keyboard arrow keys to set the sending speed, mode, and select/program memories. And finally, you could replace the Arduino Nano with a faster processor board, such as the STM32, ESP32, or the RP2040 Raspberry Pi Pico to enhance the speed and response of the project.

The sketch and libraries for this project are available at **kw5gp.com**.

20 Lightning Detector

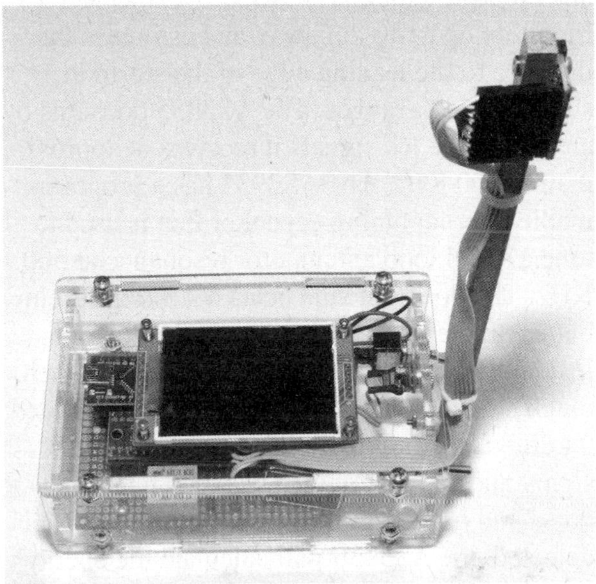

Figure 20.1 — The lightning detector.

Figure 20.2 — Sparkfun lightning detector.

The lightning detector (see **Figure 20.1**) has always been one of my personal favorite projects. It's fun to be able to revisit this project and update it with an Arduino Nano, a color TFT display, and an antenna disconnect relay. We'll also add a bypass switch, which will prevent the antenna disconnect relay from activating when lightning is detected in case you like to live dangerously. Like so many of my Arduino projects, the lightning detector project concept started out with my usual thought of, "wouldn't it be cool if…?" I think that's one of the things I like most about the Arduino. If you can think it up, there's a good chance you can find a sensor, module, or a circuit design that will do what you want it to do. That's how it was with the lightning detector. Since I'm on the road a lot, I either have to remember to disconnect my antennas before I leave or roll the dice and hope that lightning has no interest in feeding on my rigs. At one of the local ham breakfasts, I was brainstorming with Tim Billingsley, KD5CKP (SK), and I had the thought, "Wouldn't it be cool if you could make a lightning detector that would automatically disconnect your antennas when it detected lightning, and reconnect them when the storm was past?" The discussion ended with the consensus that a lightning detector that did all of that would be cool, but existing lightning detectors were too expensive.

That led to some research about lightning detectors, and eventually I stumbled across the Austria Microsystems AS3935 Franklin Lightning Sensor chip. I also found several fully assembled modules that will connect directly to the Arduino (see **Figure 20.2**). You can now get an AS3935 module from eBay, **Sparkfun.com**, **Embeddedadventures.com**, and others. The modules incorporate the AS3935 chip and all the support components that connect to the Arduino via either the SPI or I^2C bus, selected by two solder jumpers on the module. The AS3935 lightning sensor can detect lightning up to 40 km away and provide the distance and strength of the lightning strike. And the best part is that the module only costs about $25. Now that I had cleared the "too expensive" hurdle, it was time to get my hands on one of these modules and start playing.

Around this same time, I was invited by Craig Behrens, NM4T (SK), to present at a forum on the Arduino at the Huntsville, AL Hamfest. Craig asked me to come up with a couple of new and unique ham-related projects for the presentation. Since I had never seen a lightning detector anywhere, much less an affordable one, I figured this would be the perfect project to present at the forum.

Several years later I was at an ARRL Field Day weekend where nothing but constant thunderstorms were rolling through, one after the other. But our trusty lightning detector kept us safe throughout the event. We were able to safely disconnect and shut everything down as soon as lightning was detected in the area. As wave after wave rolled through, we were able to stay safe, protect the equipment, and still have some fun on Field Day.

The Austria Microsystems AS3935 Franklin Lightning Detector

The Austria Microsystems AS3935 Franklin Lightning Detector is a programmable lightning sensor chip that can detect the presence and approach of lightning, both cloud-to-ground and cloud-to-cloud. The AS3935 can detect lightning up to 40 km away and estimates the distance to the leading edge of the storm in steps of 16 km ranges. The AS3935 does this by analyzing the RF signals it receives at approximately 500 KHz. The AS3935 has a programmable internal tuning capacitor that is used to tune a small loop antenna for resonance at 500 KHz. This loop antenna bears a striking resemblance to the wire-wound ferrite core antennas inside the old AM transistor radios, only much, much smaller. The AS3935 uses a sophisticated on-chip algorithm to validate the incoming signal and can distinguish between man-made noise and lightning. If a valid lightning strike is detected, the AS3935 will calculate the energy of the strike, and then perform a statistical estimate of the distance of the strike. The AS3935

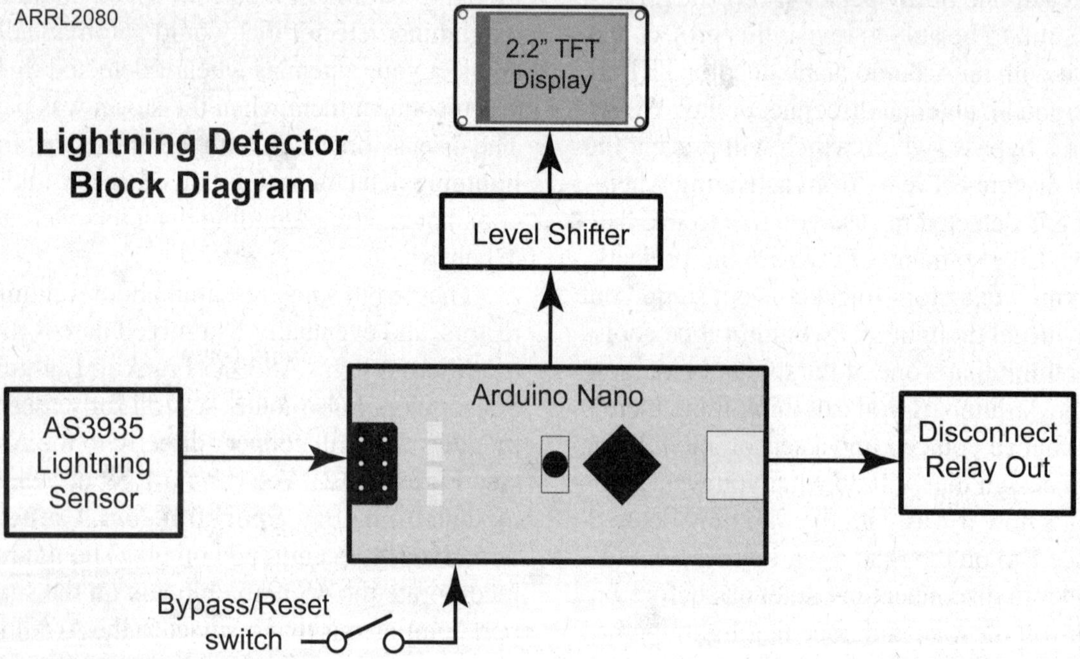

Figure 20.3 — The lightning detector block diagram.

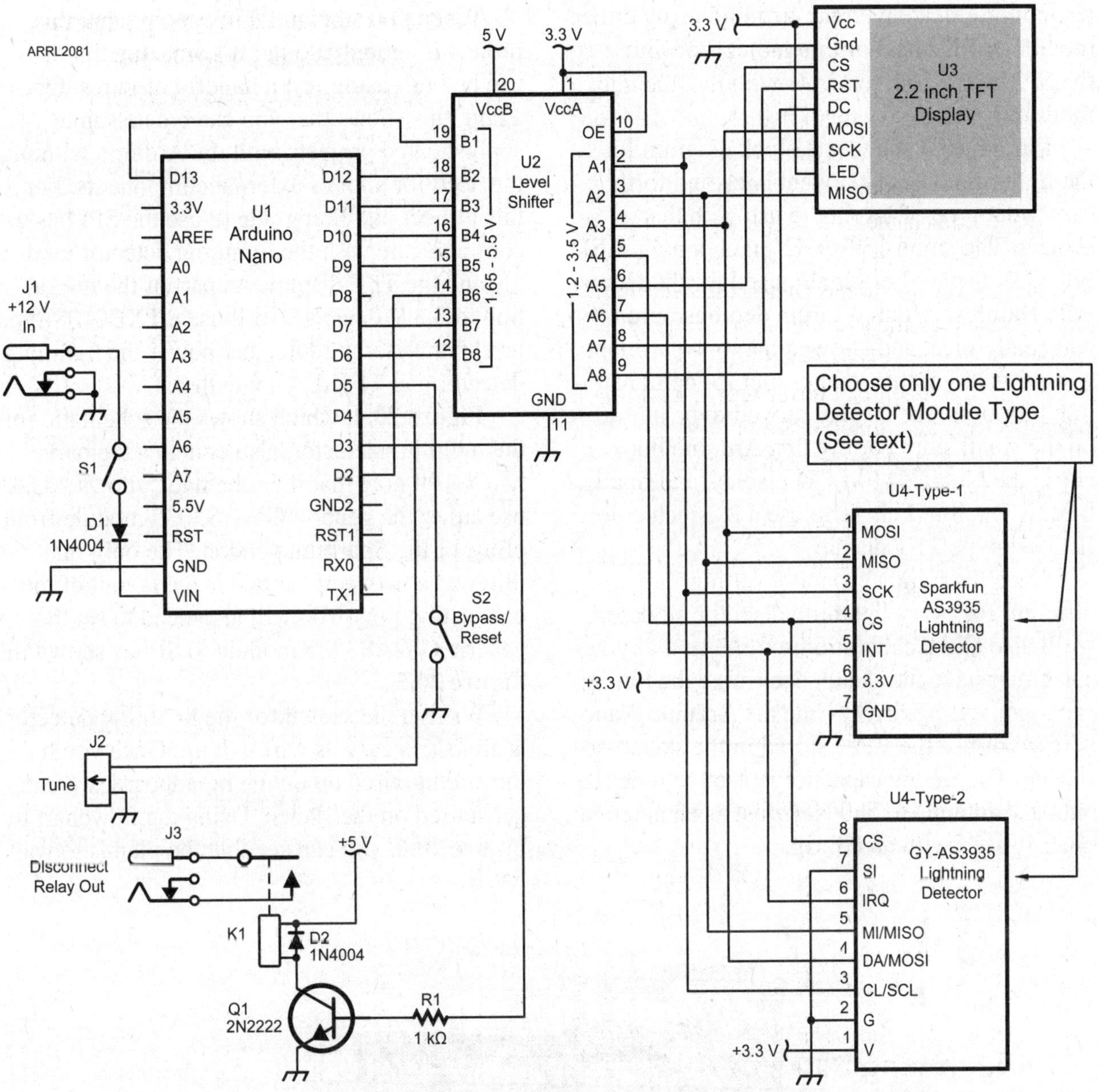

Figure 20.4 — The lightning detector schematic.

Part	*Value*
D1, D2	1N4004 diode
J1	DC power connector
J2	2-pin Dupont-style 2.54 mm header
J3	Mini-mono phone jack
K1	5 V dc relay
Q1	2N2222A NPN transistor
R1	1 KΩ resistor, 1/8 W
S1, S2	SPST switch

Part	*Value*
U1	Arduino Nano
U2	8-channel level shifter module (TXB0108 type)
U3	2.2-inch ILI9341-type TFT display
U4	AS3935 module
	Type 1 = Sparkfun AS3935 module
	Type 2 = GY-AS3935 module
Enclosure	Solarbotics Mega S.A.F.E.

can communicate with the Arduino using either the SPI or I^2C bus. For this project, we'll use the SPI bus to communicate with the AS3935 module.

Figure 20.3 shows the block diagram for the lightning detector. I went back and forth over what type of display to use with this project. The small 128- × 32-pixel organic LED (OLED) display I originally used for the Huntsville Hamfest Arduino forum demonstration was really nice, and allowed me to experiment with bitmap graphic images, but for demonstration purposes, the display proved to be a little on the small side. For my first Arduino book, I chose the Nokia 5110 LCD display, and finally, for this version, I chose to use a 2.2-inch color ILI9341-type TFT display.

Since the tuning capacitor settings are different for every lightning detector module, we'll also add a calibration routine so that you can either have the sketch determine the tuning capacitor settings every time the Arduino Nano is reset, or use the value stored in the sketch at startup. The tuning capacitor is used to tune the onboard antenna to 500 KHz just as an antenna tuner is used with an HF rig.

When you stop and think about what this project is actually doing, it's amazing that it can be built using just a handful of wires. Once again, this shows that you can create some sophisticated projects with the Arduino without the need for a lot of external components. For this project, we'll upgrade to use the SPI bus to communicate with the lightning detector module and the TFT display. As part of the migration to 3.3 V devices, we'll use a TXB0108-type level converter module, and power the lightning detector and TFT display with 3.3 V.

Figure 20.4, which shows the schematic for the lightning detector, also contains the parts list. You'll note that the schematic allows you to use either the generic GY-AS3935 module from eBay, or the Sparkfun version. The only major difference between the two is the layout of the connecting pins. You will also need to set the generic GY-AS3935 module to SPI as shown in **Figure 20.5**.

Writing the sketch for the lightning detector is almost as easy as wiring it up. Once you have the circuit wired up on the breadboard, we can get started on the sketch. Using the flowchart in **Figure 20.6**, you can see that the plan is to wait

Figure 20.5 — Soldering the GY-AS3935 solder jumper settings for SPI communication.

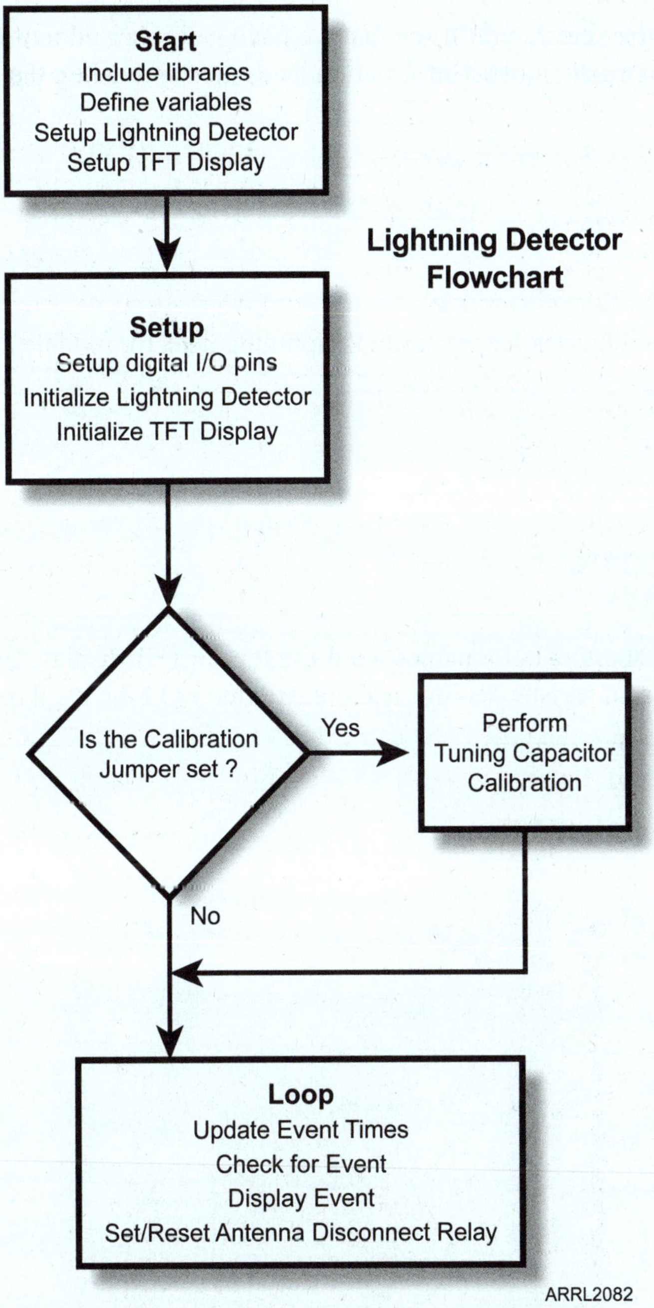

Figure 20.6 — The lightning detector flowchart.

for a lightning event and display the information about the lightning strike on the TFT display. We'll keep track of the time since the last lightning strike and display that also. We'll be using libraries for the AS3935 and the ILI9341-type TFT display, which will help to simplify the work we have to do to create this sketch. The sketch and libraries for the lightning detector project can be found at **kw5gp.com**.

As we start out with the sketch, you'll see that we have incorporated multiple levels of debugging to provide us with extra diagnostic information as we test and debug the sketch.

```
//#define debug // Startup, calibration and basic debug
//#define debug1  // Primarily timekeeping and display debug
//#define debug2  // Lightning detection and interrupt processing debug
```

Next, we'll include the libraries for the AS3935 lightning detector module and the TFT display.

```
#include <SPI.h>
#include <SparkFun_AS3935.h>
#include <Adafruit_GFX.h> // The Adafruit Graphics Library
#include <Adafruit_ILI9341.h> // The ILI9341 TFT Display Library
```

Next, we'll define the shortcut color names we'll use for the TFT display, along with various lightning detector module internal register settings and the Arduino I/O Pins we'll be using:

```
#define tft_cyan ILI9341_CYAN
#define tft_black ILI9341_BLACK
#define tft_red ILI9341_RED
#define tft_green ILI9341_GREEN
#define tft_yellow ILI9341_YELLOW

#define INDOOR 0x12
#define OUTDOOR 0xE
#define LIGHTNING_INT 0x08
#define DISTURBER_INT 0x04
#define NOISE_INT 0x01
#define ANTFREQ 3
#define LD_Tune_jumper 4
#define LD_relay 3
#define IRQ_pin 2
#define TFT_CS 9
#define TFT_DC 6
#define LD_bypass 5
#define tft_delay 100
```

```
#define settle 100  // Time to allow the AS3935 IRQ line to settle after a read
#define clear_alert_time 3600 // # of ticks before clearing an alert
#define start_delay 10000 // milliseconds to delay start to avoid false trigger-
ing while initial statistics are gathered
```

In the last part of the initialization portion of our sketch, we'll define all the variables we'll need and instantiate the AS3935 and TFT objects.

```
long hold_down; // Timer used to prevent false triggers after startup and bypass
switch changes
int hold_down_time = 1000; // Timer to ignore interrupts in milliseconds

long last_event = 0;  // How long ago the last event occurred
long last_scan = 0;
bool active_alert = false;
bool new_event = false;
bool disconnected = false;
bool LD_bypassed = false; // The status of the bypass/reset switch
bool bypass_active = false;
String event_time = "";
int tft_event_background = tft_black;
int text_spacing; // Use to space the timestamp line
byte distance;  // Lightning Distance
long lightEnergy; // Lightning Energy
bool auto_clear = true; // Automatically clear alert after clear_alert_time

// The lightning detector defaults to an indoor setting (higher
// gain/sensitivity)

bool LD_mode = INDOOR;

byte Interrupt_type = 99; // Since 0 is a valid interrupt type, we use 99 to
tell us there are no interrupts

byte intVal = 0; // Value of the Interrupt Register
String int_text = "";

// "Disturbers" are events that are false lightning events. If you find
// yourself seeing a lot of disturbers you can have the chip not report those
// events on the interrupt lines.
```

```
bool LD_Disturber = true; // False = Enabled, True = Disabled (Masked)

// Values for modifying the Lightning Detector's settings.
// All of these values are set to default values.
// Noise floor setting from 1-7, one being the lowest.
// Default setting is two.

byte noiseFloor = 3;

// Watchdog threshold setting can be from 1-10, one being the lowest.
// Default setting is two.

byte watchDogVal = 2;

// Spike Rejection setting from 1-11, one being the lowest. Default setting is
// two. The shape of the spike is analyzed during the chip's
// validation routine. You can round this spike at the cost of sensitivity to
// distant events.

byte spike = 2;

// This setting will change when the lightning detector issues an interrupt.
// For example you will only get an interrupt after five lightning strikes
// instead of one. Default is one, and it takes settings of 1, 5, 9 and 16.

byte lightningThresh = 1;

#define font_size 4  // The CW text display font size
#define status_font_size 2  // The font size for the status line

// Here you give the value of the capacitor you want turned on. It accepts up
// to 120pF in steps of 8pF: 8, 16, 24, 32 etc.The change in frequency is
// somewhat modest. At the maximum value you can lower the frequency up to 22kHz.

// As a starting point, the Sparkfun products designed in house ship
around 496kHz
// (though of course every board is different) putting you within one percent
// of a perfect resonance; the datasheet specifies being within 3.5 percent as
// optimal. You can leave it at zero if using auto-tune jumper and let the sketch
```

```
determine the
// proper capacitor value.

int tuning_cap = 16; // Set the value to the correct capacitance in pF if not us-
ing auto-tune jumper - must be a valid capacitance

long freq = 0;
long end_time = 0;
long tick = 0;

bool osc_fail;

// Chip select for Lightning Detector SPI on pin ten.
int spiCS = 10;
int tuneVal;
int best_cap = 16;
float best_diff = 0;
float diff;
long last_sim = 0; // timer for simulated interrupts
long test_tick = 0; // 10 second tick for simulated interrupts
int trigger_time = 60; // Set the sim trigger time to 60 seconds
bool first_pass = true;

// SPI
SparkFun_AS3935 lightning;

// Use hardware SPI (on Uno/Nano, #13, #12, #11) and the above for CS/DC
Adafruit_ILI9341 tft = Adafruit_ILI9341(TFT_CS, TFT_DC);  // Create the TFT
instance
```

In the `setup()` loop, we'll start the TFT display and show a brief startup message. We'll also setup the I/O pin modes and then we'll run a check on the lightning detector module. We'll set the lightning detector module settings, and then check the calibration of the tuning capacitor if the tuning calibration jumper is in place, otherwise we'll use the tuning capacitor value defined in the initialization area of the sketch. Next, we'll run a basic communications check with the lightning detector module to ensure that it is operating properly.

```
void setup()
{

#if defined (debug) || defined(debug1) || defined (debug2)  // Only start the Se-
rial port if we're debugging
  Serial.begin(baud_rate);
  Serial.println("Starting Serial Port for debugging");
#endif

  start_tft();

  pinMode(IRQ_pin, INPUT); // The Detector IRQ input pin
  pinMode(LD_bypass, INPUT_PULLUP); // The disconnect relay bypass/alert reset
switch
  pinMode(LD_relay, OUTPUT); // The Relay drive output pin
  pinMode(LD_Tune_jumper, INPUT_PULLUP);  // The Tuning_jumper input pin

  SPI.begin(); // Start the SPI bus

  check_LD_module();  // Check to see if the lightning detector module is basi-
cally functional

  check_calibration();  // Run/Set the LD module calibration

  clear_display();

  update_status();

  // Read the Lightning detector to clear any interrupts before starting
  intVal = lightning.readInterruptReg();
  intVal = 99;

#ifdef debug2
```

```
  Serial.println("Reading Detector IRQ to reset");
#endif

#if defined(debug) || defined(debug2)
  Serial.println("**** Detector Ready ****");
#endif

  tft.setTextSize(2);

  bypass_active = true; // Used to force bypass check and LD bypass reset pro-
cess on first pass
  active_alert = false;
  hold_down = millis() + hold_down_time + start_delay;
}
```

In the `main loop()`, we wait for the interrupt pin on the lightning detector module to go active and update the TFT display every minute. The background of the TFT display will indicate the current lightning and antenna bypass status. If it's green, there are no lightning events detected within the 3,600 seconds (60 minutes) range. If the bypass/reset switch is enabled, the TFT background will be yellow, and the antenna switch will not trigger the antenna disconnect relay.

```
poll_interrupt(); // Check the interrupt pin
  // read the bypass switch
  LD_bypassed = !(digitalRead(LD_bypass));  // Read the bypass/reset switch

  if (LD_bypassed)
  {
    // Bypass switch is enabled
    tft_event_background = tft_yellow;
    if (!bypass_active || first_pass)
    {
      hold_down = millis() + hold_down_time;
      if (disconnected)
      {
        // Reconnect the Antenna
        digitalWrite(LD_relay, LOW);
        disconnected = false;
      }
      // Update the screen
      tft_event_background = tft_yellow;
      clear_text_area(tft_event_background);
```

```
      tft.setTextColor(tft_black, tft_event_background);
      tft.setCursor(35, 70);

      tft.print("No Lightning Detected");
      tft.setCursor(80, 120);

      tft.print("Bypass Enabled");
      bypass_active = true;
      active_alert = false;
      first_pass = false;

#ifdef debug2
      Serial.println("first pass off");
#endif
    }
  } else
  {
    if (!bypass_active)
    {

    } else
    {
      // The bypass switch has been turned off, reset alerts, etc
#ifdef debug2
      Serial.println("Bypass turned off");
#endif
      hold_down = millis() + hold_down_time;
      bypass_active = false;

      tft_event_background = tft_green;
      clear_text_area(tft_event_background);
      tft.setTextColor(tft_black, tft_event_background);
      tft.setCursor(35, 70);

      tft.print("No Lightning Detected");
      tft.setCursor(60, 140);

      tft.print("Antenna Connected");
      active_alert = false;
```

```
#ifdef debug2
     Serial.println("first pass off");
#endif
    }
    first_pass = false;
  }

  if (!disconnected)
  {
    tft.setCursor(60, 140);

    tft.print("Antenna Connected");
  }

  if (!active_alert)  // If we don't have an active alert and nothing new we do
nothing
  {
```

If the TFT display background is red, it indicates that there has been a lightning event detected and the antenna disconnect relay is activated. After no lightning events have been detected for 60 minutes, TFT background will turn green, and the antennas will automatically be reconnected.

```
       // Now we wait for an interrupt and process it

    if (Interrupt_type != 99) // Note: 99 is an impossible type. Since 0 is
valid, we use this to test for a real one
    {
      switch (Interrupt_type)
      {

        case 1:
          // It's noise
          int_text = "High Noise";
          tft_event_background = tft_yellow;
          tft.setTextColor(tft_black, tft_yellow);
          break;

        case 2:
          // It's a disturber
```

```
        int_text = "Disturber";
        tft_event_background = tft_yellow;
        tft.setTextColor(tft_black, tft_yellow);
        break;

      case 3:
        // It's lightning
        // Activate the Antenna Disconnect relay asap
        int_text = "Lightning";
        tft.setTextColor(tft_yellow, tft_red);
        tft_event_background = tft_red;
        if (!LD_bypassed)
        {
          digitalWrite(LD_relay, HIGH);
          disconnected = true;
        }
        break;
    }
    if (LD_bypassed)
    {
      tft.setTextColor(tft_red, tft_yellow);
      tft_event_background = tft_yellow;
    }

#ifdef debug2
    Serial.println(int_text);
#endif

    clear_text_area(tft_event_background);

    if (Interrupt_type == 3)
    {
      // Add the distance and energy if it's lightning
      tft.setCursor(110, 65);

      if (distance == 1)
      {
```

```
          tft.print("Overhead");
        }
        else
        {
          if (distance != 63)
          {
            tft.print(distance); tft.print(" km away");
          } else
          {
            tft.setCursor(80, 65);
            tft.print("Out of Range");
          }
        }
        tft.setCursor(100, 90);

        tft.print("Energy: "); tft.print(map(lightEnergy, 1, 16777216, 1, 100));
tft.print("%");

        if (LD_bypassed)
        {
          tft.setCursor (80, 115);

          tft.print("Bypass Enabled");
        }
        if (disconnected)
        {
          tft.setCursor(40, 140);

          tft.print("Antenna Disconnected");
        }
        tft.setCursor(50, 40);
      }
      else
      {
        tft.setCursor(50, 90);
      }
```

```
      tft.print(int_text);
      tft.print(" Detected");
      if (LD_bypassed)
      {
        tft_event_background = tft_yellow;
      }
      else
      {
        if (Interrupt_type == 3)
        {
          tft_event_background = tft_red;
        } else
        {
          tft_event_background = tft_yellow;
        }
      }
      event_time = "Now";
      tick = 0;
      print_timestamp();
      active_alert = true;  // Set the active alert flag
      Interrupt_type = 99; // Reset the interrupt type
      new_event = false;
      last_event = millis();
    }
  }
  else
  {
    // Calculate how long ago and update the event time line
    if (millis() - last_event >= 10000) // Timestamp if not a new event
    {
      timestamp();
    }
  }
}
```

A brief on and off flip of the bypass/reset switch can be used to clear a lightning event and re-connect the antennas.

If there is an active lightning event, the most recent detection event will be displayed on the TFT display, with distance and strength information, and how long ago the event occurred, as shown in **Figure 20.7**.

Now, here's the hardest part of the whole project, finding lightning to detect. When I finished building my lightning detector prototype on the breadboard, I spent weeks waiting for a lightning storm to pass through. I used the debug simulation function to test things, but it's just not the same without real lightning. The day finally came when I woke up to lightning, and up popped all the lightning events. Once I had proof that the thing actually worked, it was time to move everything into a nice enclosure to show off.

The finished project was built on a piece of 6 × 8-cm copper-clad perfboard (see **Figure 20.8**), and then mounted in a Solarbotics Arduino Mega S.A.F.E. enclosure. Then it was back to waiting for the next thunderstorm to test it. Unfortunately, it was weeks before the next storm, and therefore the discovery of the last little glitch in the project.

I had originally placed everything inside the enclosure, mainly to protect the TFT display and the fragile-looking antenna on the AS3935 module. Let's back up and review how the AS3935 actually works. For all intents, it is a narrowband receiver tuned for 500 KHz that is looking for a signal with a signature that is indicative of lightning. In my initial design, I had mounted the AS3935 module inside the enclosure near the Arduino Nano. As it turns out, the Nano generates your typical low-grade RF noise from its 16 MHz oscillator and various clock signals, including the SPI bus. The color TFT display also generates some RF noise as it scans and displays the pixels. These factors combined to interfere with the reception on the AS3935 module, and it would not detect lightning very well at all. The performance was nowhere near as good as it had been when it was on the breadboard. I tried various types of displays and configurations, having to wait for thunderstorms after each new design was mounted in the enclosure.

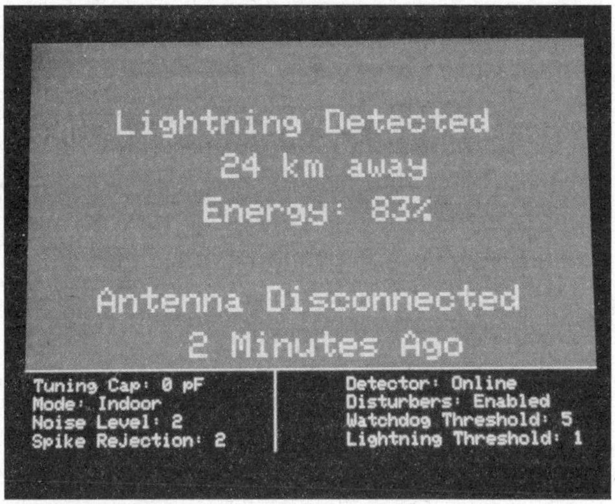

Figure 20.7 — The lightning detector TFT display.

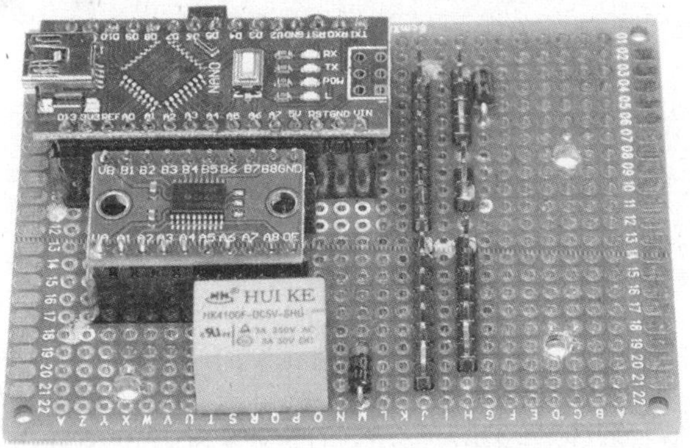

Figure 20.8 — The assembled lightning detector board.

As a last resort, I moved the AS3935 outside the Arduino enclosure onto a small wooden mast to provide some added distance from the noisemakers inside the enclosure. At one point, I had three different versions of the lightning detector sitting around the house waiting for a storm. As luck would have it, a rare winter thunderstorm came rolling through one morning, and I was finally able to verify my suspicions about the RF noise and determine which design worked best. The winning design can be seen in Figure 20.1 at the beginning of this chapter. As a final touch, you could use a ping-pong or plastic golf ball to provide some measure of protection and make the project look like it has a little radome to add cuteness.

Enhancement Ideas

Now that we have a working lightning detector, there are all sorts of interesting features we can add. After building this project, I discovered the various text-to-speech modules, including the DFRobot gravity speech synthesizer module. Adding voice output to the lightning detector project opens a world of possibilities. You can drive a PA system at ARRL Field Day to announce lightning in the area, or you can just have it on your desk to tell you when lightning is nearby. If you didn't want to use a voice module, you could have the lightning detector sound an alert tone when it detects lightning.

The antenna disconnect relay is a standard relay that is only energized when the lightning detector is powered on and a lightning even has occurred. This relay only provides a relay contact output that you could use to drive an actual antenna disconnect relay. Another possible solution would be to use this project to drive a latching relay in order to reduce power consumption.

You can find the sketch and libraries for the project at **kw5gp.com**.

21 RTTY Reader

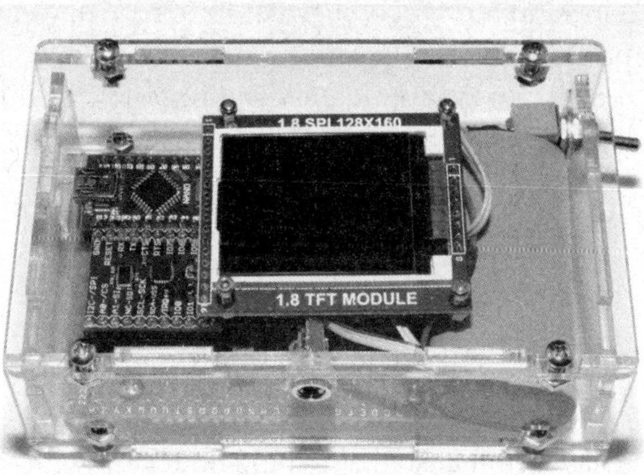

Figure 21.1 — The RTTY reader.

Radioteletype, more commonly known as RTTY (see **Figure 21.1**), was one of the main incentives for upgrading my Novice license to General back when I first got into ham radio. My local ham buddies had a 2-meter RTTY net, and when I first saw that noisy, clanking, model 15 teletype, and smelled that machine oil, I was hooked. I'm absolutely certain that my parents wished I had chosen a different hobby, especially one that did not include having that loud beast going off in my bedroom at all hours. Yes, I was enjoying the heck out of operating CW as a novice, but there was something about the RTTY that was fascinating to me, and I just had to upgrade it as soon as possible.

If you don't count CW, RTTY was one of the first digital modes used by hams. Instead of the standard 8-bit ASCII character set, amateur radio RTTY generally uses a 5-bit data word using the Baudot character set, transmitted at 45.45 baud. That may seem like an unusual speed at first, but this actually works out to 60 words per minute. RTTY is considered to be a keyboard-to-keyboard mode, meaning that operators essentially type to each other over the air, and is similar in operation to the more modern PSK31 digital mode. RTTY has none of the error correction, handshaking, or time synchronization that is needed with modes such as JT65, FT8, and FT4. For "canned" messages, some RTTY machines had paper tape readers and punches instead of the computer software and memories used today. In the old days, hams enjoyed collecting and sharing typewritten "pictures" via RTTY that were stored on paper tape. (An online search for "RTTY art" will turn up numerous examples. Some are quite intricate.)

RTTY consists of a frequency shift keyed (FSK) signal that is modulated between two carrier frequencies, called the mark frequency and the space frequency. The difference between the mark and space frequencies is called the FSK shift, usually 170 Hz for an amateur RTTY signal. Typically, these frequencies are 2125 Hz for mark, and 2295 Hz for space, a frequency shift of 170 Hz. These are called high tones, you may also see references to low tones, which are typically 1275 Hz for mark, and 1445 Hz for space with a 170 Hz shift. If your transceiver does not have a dedicated RTTY (FSK) mode, it will use audio frequency shift keying (AFSK) with the transceiver set to LSB (lower sideband). On the air, FSK and AFSK signals are indistinguishable. AFSK RTTY can also be sent on VHF/UHF using an FM transceiver.

Back in the day, a piece of equipment known as a terminal unit was used to convert the mark and space tones into the current drive for the "loop" circuit that powered the signal relay inside the teletype machine. This signal relay was used to control all the gears and synchronous motor timing to select and type the individual characters. In today's RTTY environment, the tone decoding is done via software, and the terminal unit and current loop are long gone — unless you are still using a mechanical teletype machine.

Because the Baudot RTTY data word is only 5-bits, you can only have 32 different characters. To get around this limitation, Baudot incorporates the letters (LTRS) and figures (FIGS) shift functions, allowing access to a second set of 32 characters. The LTRS shift consists mainly of alphabetic characters, and the FIGS shift consists mainly of decimal numerals and punctuation marks. These two unique Baudot characters are used by the sender to command the decoder to switch between these two character sets.

The old mechanical teletypewriter keyboards have two keys to send the LTRS and FIGS characters. If you're going to have a computer translate between 5-bit Baudot and 8-bit ASCII, you'll need to take the FIGS and LTRS shift operation into account and handle it within your software. If you miss a FIGS or LTRS shift due to noise or fading while receiving, the output will appear to be garbled since you're receiving on the opposite character set from what you should be. For example, "599" will be received as "TOO." Most RTTY programs perform what is known as unshift on space (USOS) to restore the LTRS shift every time a space character is received.

Back in my early ham years, I was also tinkering around with the RCA 1802 COSMAC Elf CPU, the world's first CMOS microprocessor. While all my friends and co-workers were playing with the new Apple, Radio Shack, and 8080/Z-80-based systems, I was playing with the RCA 1802. I liked the 1802 because it was the closest thing you could get to a microcontroller back then, and I was doing a lot of microcontroller-style projects at that time. The 1802 found a niche in the early embedded computer environment and was widely used in the industrial and aerospace worlds. The 1802 was also used in NASA's Galileo space probe to Jupiter in 1989, and other space projects, including the Hubble Space Telescope. The 1802 also found a home in ham radio satellites, as the onboard internal housekeeping unit (IHU) for AMSAT's Phase 3-D, as well as OSCAR 10 and 13. On a more personal note, the very first magazine article I ever had published was about building an ASCII to Baudot RTTY converter for an RS-232 video display terminal so that I could do RTTY without all the noise and motor vibration of the big clunky machines.

Figure 21.2 — The SC16IS750 UHART module.

Since those early days, RTTY has always been one of my favorite operating modes, and today, virtually all RTTY operation on the ham bands uses PC-based RTTY programs such as the *Ham Radio Deluxe (HRD)* DM780, the open-source *MMTTY*, and *FLDIGI* programs.

While thinking up projects for this book, I came across a new Arduino module that really caught my eye, and my first thought was to use an Arduino to create a handheld RTTY decoder and display as shown in Figure 21.1. While I've never been a big fan of using the Arduino's software serial library to handle all the serial bit timing, this new module, the NXP SC16IS750 universal asynchronous receiver-transmitter (UART) module (see **Figure 21.2**), would be the perfect thing to receive the 5-bit RTTY serial data and convert it to the 8-bit ASCII used by the Arduino and most other modern computers.

What is a UART?

A UART is an IC that is used to send and receive serial data. A typical UART can handle serial communication in either half-duplex or full-duplex mode. In its original design, the UART would convert the parallel data from a computer into a serial RS-232-compatible data stream. The UART manages all the functions necessary to transmit and receive the serial data, including handshaking, bit timing, data buffering, and error detection. A UART can be configured to handle a data word from 5 – 8-bits long, at a programmable baud rate based on the input clock frequency. Typically, the input clock would be 16 × the baud rate, and either a crystal or NE555 oscillator circuit would provide this clock input.

The UART also supports even, odd, or no parity, and 1, 1-1/2, or 2 stop bits. Handshaking and data flow control can be done using the RTS (request to send), CTS (clear to send), DTR (data terminal ready), and other standard serial hardware flow control functions. Many UARTs also incorporated software flow control using the Xon/Xoff characters in the data stream.

Because the UART is designed to handle virtually all the serial communication functions, very little extra is often needed to add hardware serial communication functions to your program. Simply put, the UART is the one-stop-shop for all things serial communications-wise.

The NXP SC16IS750 UART

The NXP SC16IS750 UART is an update to the original bi-directional parallel to serial data UART design. Instead of a discrete 8-bit digital parallel interface, the SC16IS750 can be configured to use either the I^{2}C or SPI bus to transfer the UART data to the Arduino. This frees up several I/O pins for the UART, and in fact, the UART interface to this project only requires eight wires, including the power and ground connections.

Apart from the I^{2}C or SPI bus selection, and the I^{2}C device address, all the UART configuration functions are fully programmable and managed via software. Additionally, the SC16IS750 incorporates a 64-bit transmit and receive FIFO (first in first out) buffer. For baud rate selection, the UART module uses an external 14.7456 MHz crystal on the board, in conjunction with a programmable divider inside the UART, to create the internal clock rate of 16 × the desired baud rate. The SC16IS750 has several internal configuration registers, however, the main one we're interested in is the line control register (LCR), which controls the data word length, parity, and the number of stop bits.

The SC16IS750 UART chip itself is designed for 2.5 or 3.3 V operation, however all the I/O pins are 5 V tolerant, and the module itself is powered by 5 V, with a 3.3 V power regulator onboard. I^{2}C or SPI operation is selected by an input pin on the module, and one of 16 different I^{2}C bus addresses can be selected using the A0 and A1 pins. Additionally, the SC16IS750 also has 8 programmable I/O pins available for your use.

While this feature is not used in this project, the SC16IS750 can be configured to use interrupts to provide status information. For example, it could signal that a character has been received, the transmit buffer is empty, a communication error was detected, or the state of one of the onboard I/O pins has changed.

The RTTY Reader Project

Figure 21.3 shows the block diagram for the RTTY reader project. For this project, we'll use an Arduino Nano, the SC16IS750 UART, an ST7735-type 1.8-inch color graphic TFT display, and a tone decoder circuit to convert the RTTY audio tones from the receiver to logic-level signals for the UART receive input. The finished project will be mounted in a Solarbotics Mega S.A.F.E. enclosure, along with a 9 V battery for portable operation. We'll be using the SPI bus to communicate with the TFT display, and the I^{2}C bus for the UART. The I^{2}C bus configuration for the UART was chosen because while the SC16IS750 library supports the SPI configuration, the I^{2}C configuration within the sketch is easier to understand and use.

For the tone decoder portion of the circuit, we'll create two different decoder circuits to highlight the differences between two of the most popular tone decoder chips, the Exar XR-2211 FSK demodulator/tone decoder, and the ever-popular LM567 tone decoder. Back in the 1970's and 80's, the XR-2211, along with its counterpart, the XR-2206 tone generator, were my go-to chips for any RTTY or FSK audio decoding projects. The XR-2211 is an inexpen-

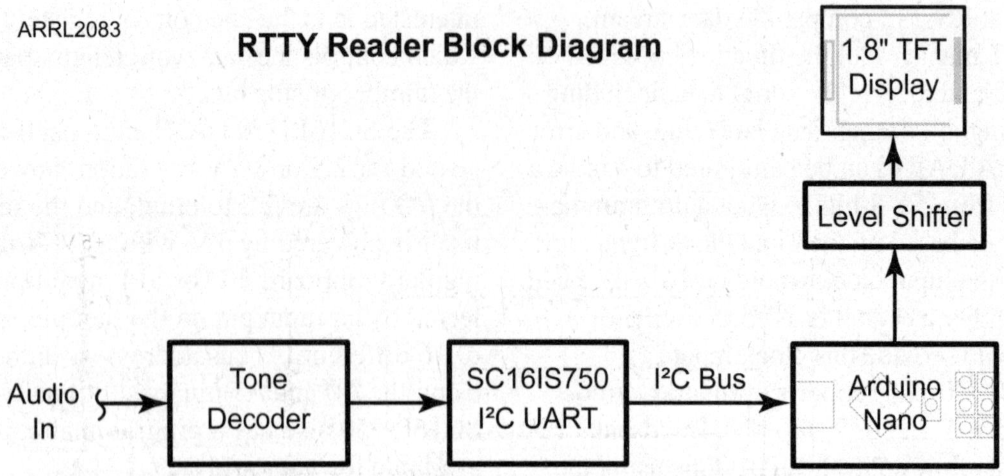

Figure 21.3 — The RTTY reader block diagram.

sive, and very good, decoder. The hardest part about the XR-2211 is figuring out the values of the various tuning and filtering components, but once you get those values worked out, the actual construction is very simple.

The LM567 is a single tone decoder, but is also very easy to work with, and at 100 mA, has a much higher open-collector output sinking current that the XR-2211's 5 mA limit. Both decoder designs were kept simple and basic, although there is a lot of extra filtering and tuning functionality left in both chips for you to explore in your own designs. Because both decoders have their own specific set of advantages and disadvantages, I decided to design both decoder options and let you choose which one you prefer to use in your project.

For this project, the XR-2211 version is designed for 1000 Hz mark and 1170 space frequencies. This was chosen as these frequencies are in the middle of the normal HF sideband audio range, without having to use the RTTY receive setting on your transceiver (if it has one). This makes the RTTY audio much easier to tune and easier on the ears. By changing just three components, you can change the XR-2211 version to operate at the standard 2125/2295Hz mark and space tone frequencies if you prefer to do it that way. The LM567 version was designed to only use the 1000/1170 Hz frequencies, although it too could be switched to the 2125/2295 frequencies by changing just a component or two.

The XR-2211 FSK Decoder Circuit

Figure 21.4, which shows the schematic for the XR-2211 FSK decoder unit, also includes the parts list. While the XR-2211 can also be used as a single frequency tone decoder, for this circuit, we'll be using it as an FSK decoder. Potentiometer VR1 is used to adjust the input level. The XR-2211 input signal level can be anything between 10 mV and 3 V RMS. The reason for this wide range is that the XR-2211 has an input preamplifier that also functions as a voltage limiter, providing a constant signal level at the output. While this project only uses the FSK decoder to provide a logic-level low signal on the output pin, the XR-2211 is actually tuned to a frequency between the two tones and can output a signal indicating whether the received frequency is higher or lower than the center frequency of the circuit. Since we're just using the XR-2211 as a basic FSK decoder, we're only interested in identifying the 1000 Hz mark frequency.

There are several additional components you can add to this circuit to improve the signal filtering and decoding response, but for this project, we'll just be using a basic FSK decoder without any additional filtering components.

The basic calculations for determining the components used to determine the XR-2211's center frequency and bandwidth are relatively straightforward. For 1000 Hz, the component values work out to 18 kΩ for resistor R1, 100 kΩ for resistor R4, and .047 μF for capacitor C4. You

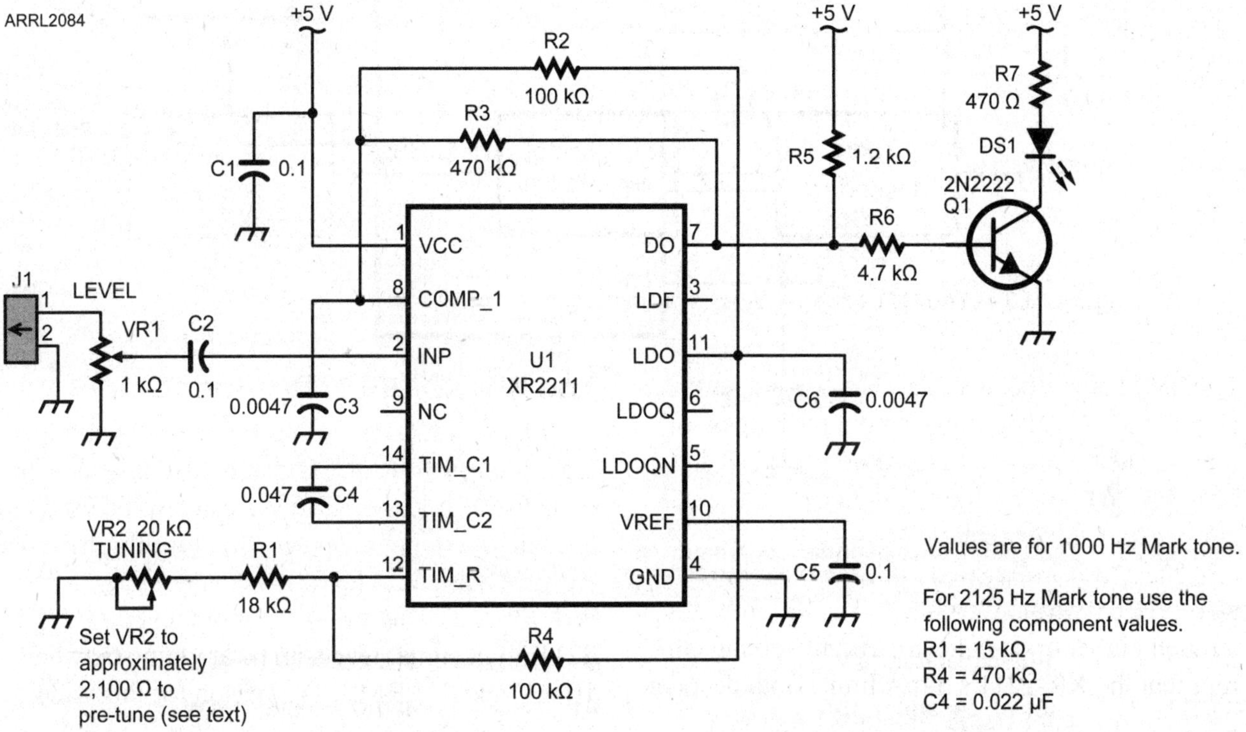

Figure 21.4 — The XR-2211 FSK decoder schematic.

Part	*Value*	*Part*	*Value*
C1, C2, C5	**0.1 μF capacitor, 16 V**	**R3**	**470 kΩ resistor, 1/8 W**
C3, C6	**.0047 μF capacitor, 16 V**	**R5**	**1.2 Ω resistor, 1/8 W**
C4	**.047 μF capacitor, 16 V**	**R6**	**4.7 kΩ resistor, 1/8 W**
DS1	**5 mm red LED**	**R7**	**470 Ω resistor, 1/8 W**
J1	**2-pin Dupont male header**	**U1**	**XR2211 FSK demodulator/ tone decoder**
Q1	**2N2222 NPN transistor**	**VR1**	**1 kΩ 10-turn potentiometer**
R1	**18 kΩ resistor, 1/8 W**	**VR2**	**20 kΩ 10-turn potentiometer**
R2, R4	**100 kΩ resistor, 1/8 W**		

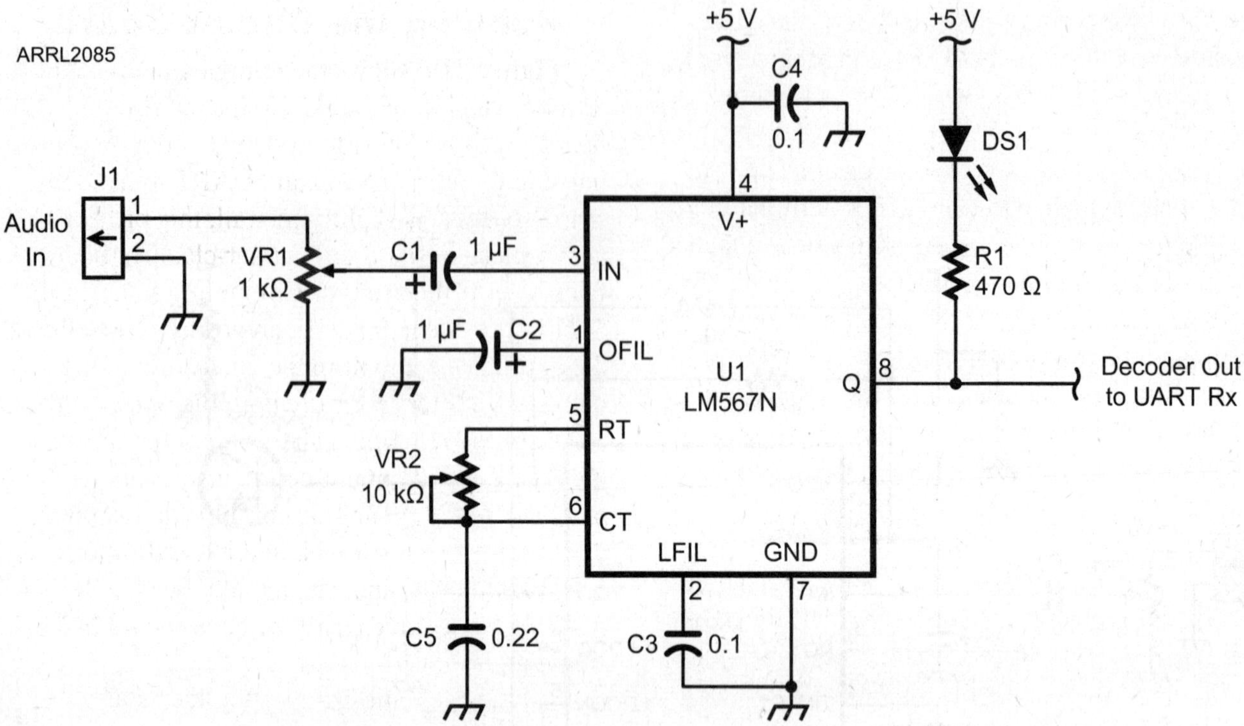

Figure 21.5 — The LM567 tone decoder schematic.

Part	*Value*	*Part*	*Value*
C1, C2	1 µF capacitor, 16 V	Q1	2N2222 NPN transistor
C3, C4	0.1 µF capacitor, 16 V	R1	470 Ω resistor, 1/8 W
C5	0.22 µF capacitor, 16 V	U1	LM567 tone decoder
DS1	5 mm red LED	VR1	1 kΩ 10-turn potentiometer
J1	2-pin Dupont male header	VR2	10 kΩ 10-turn potentiometer

can easily modify the XR-2211 decoder circuit for the standard mark and space tones of 2125 and 2295 Hz by changing the values to 15 kΩ for resistor R1, 470 kΩ for resistor R4, and .022 µF for capacitor C4. You'll notice that the pullup resistor on the output is a much lower value than is typically used for a pullup resistor to drive a 2N2222 transistor. This is because the XR-2211 can only sink 5 mA of current, and without a lower value of pullup resistor, the output can't drive the UART receive input and the transistor used to light the LED on the board. Because the transistor stage inverts the output signal, you'd need a second 2N2222 transistor to invert the output signal again for the UART if you wanted to provide more drive to the UART receive input. When idle, the output of the XR-2211 is at logic-level high, which is also what the UART is expecting to see when no characters are being received.

The LM567 Decoder Circuit

Figure 21.5, which shows the schematic for the LM 567 tone decoder circuit, also contains the parts list. The LM567 is a simple, easy-to-use general purpose single tone decoder chip. With modern RTTY-type devices, we no longer need to decode both the mark and space frequencies unless we want to implement a higher level of eliminating false decoder outputs. A simple way to do this would be to implement two 567-tone decoders and not allow the receive input to the UART to change unless one or the other decoder was active. This is a bit of overkill, since fading or noise that would cause the loss of the space tone would most likely affect more than a single character anyway. So, in this case, a minimal RTTY tone decoder is adequate for the job.

The LM567 is very easy to design and use. The tone frequency to decode is determined by a single resistor (VR2) and capacitor (C5) attached to pins 5 and 6 of the decoder chip. As with the XR-2211, you can add levels of phase-locked loop and output filtering as desired, simply by the addition of a component or two. Since we're using a minimal tone decoder for this project, we don't have any filtering added to the design, but an online search will yield some ideas.

In addition to the tone frequency, the bandwidth is also selectable using a single capacitor (C3) and/or changing the input signal level. Unlike the XR-2211 decoder, the LM567 is much more sensitive to different signal levels, with a maximum design specification of 200 mV RMS on the audio input pin. For this project, I chose to use the 1000 Hz mark frequency for the LM567 decoder to make it easier to tune on the radio's receiver. Information on determining the component values for tone frequency and bandwidth may be found in the LM567 datasheet.

Figure 21.6 — The RTTY reader circuit board.

Building the Circuit Board

Figure 21.6 shows the completed XR-2211 decoder version of the RTTY reader circuit board. Because an Arduino UNO and a prototyping shield with the Nano and UART in header strip "sockets" is slightly too tall, this project uses a piece of 6 × 8-cm copper-clad perfboard with holes drilled to fit the mounting holes in the Solarbotics Mega S.A.F.E. enclosure. This allows the entire project to fit in the enclosure, along with a 9 V battery for truly portable operation.

Figure 21.7, which shows the schematic for the RTTY reader circuit board, also contains the parts list. Once again, this circuit is an example of how simple, functional, and interesting projects with the Arduino can actually be. The RTTY reader project could easily be constructed in a single afternoon.

In the schematic, you will notice that the UART's I^{2}C/SPI pin (Pin 1) is connected to +5 V, causing the UART to be in the I^{2}C operating mode. UART addressing pins A0 and A1 (Pins 2 and 3) are connected to ground, selecting I^{2}C bus address BB. All the remaining UART configuration and operation is performed using the I^{2}C bus.

You will also see J2 in the schematic, this is a three-pin header jumper that allows you to either select normal operation or loopback, connecting the UART TX output pin to the UART RX input pin. This was added as part of the design and initial testing process and left in the design as a troubleshooting aid. This allows you to perform complete end-to-end testing of the project using the serial monitor to verify that your project is functioning correctly.

Next, we'll discuss the sketch used for the RTTY reader. We'll cover testing and using the completed RTTY project once the sketch has been uploaded to the finished project.

The RTTY Reader Flowchart

Figure 21.8 shows the RTTY reader flowchart. As usual, we'll start out by including the libraries needed for the ST7735 TFT display and the SC16IS750 UART. Then, we'll define the necessary constants and I/O pins needed. Next, we'll define the Baudot-to-ASCII translation

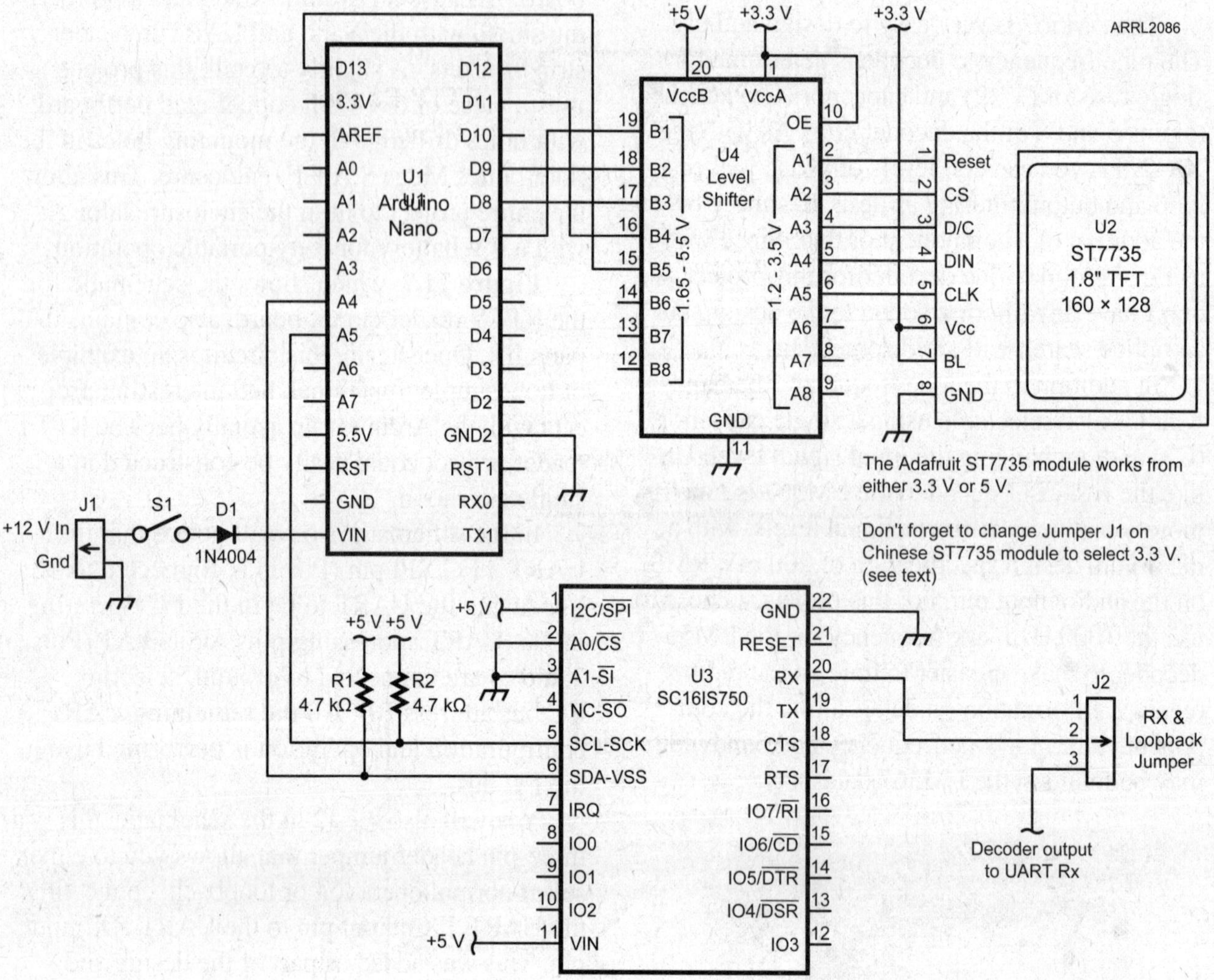

Figure 21.7 — The RTTY reader schematic.

Part	*Value*	*Part*	*Value*
D1	**1N4004 diode**	**U2**	**ST7735-type 1.8-inch color TFT display**
J1	**2-pin Dupont male header**	**U3**	**SC16IS750 I2C UART module**
J2	**3-pin Dupont male header**	**U4**	**8-channel level shifter module (TXB0108 type)**
R1, R2	**4.7 kΩ resistor, 1/8 W**	**Enclosure**	**Solarbotics Mega S.A.F.E.**
S1	**SPST toggle switch**		
U1	**Arduino Nano**		

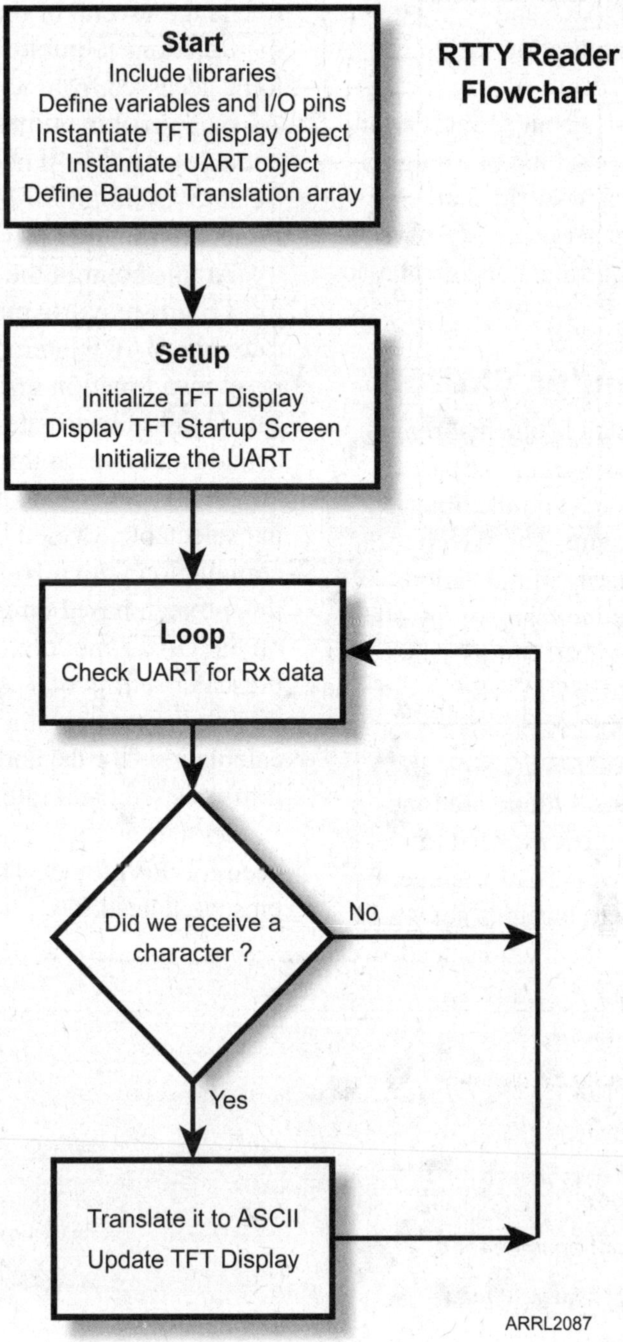

Figure 21.8 — The RTTY reader flowchart.

array, and then declare the remaining variables needed by the sketch. Finally, we'll instantiate the TFT display and UART objects.

In the `setup()` section, we'll initialize the TFT display and post a brief startup message. Next, we'll initialize the UART and verify that it is online, then set the configuration we'll need to receive the Baudot RTTY data.

The `main loop()` is about as short as an Arduino sketch can get, consisting of a single function call to check for a received character from the UART. If a character is received, we'll translate it to its ASCII equivalent and display it on the TFT.

The RTTY Reader Sketch

The RTTY reader sketch is one that really demonstrates the power of Arduino libraries. You can download the sketch and the libraries for this project at **kw5gp.com**. The sketch itself is short and consists primarily of the various building blocks we've used in many of the other sketches in this book. The heart of this sketch is the library for the SC16IS750 UART. As of writing this chapter, the libraries available for the SC16IS7xx UART modules are still a bit on the scarce side. The ones I found are functional and have all the features we need to make the RTTY reader project work. Unfortunately, the best library I was able to locate is not well documented, and many of the features needed to fully customize the UART were set up as private functions inside the library, meaning these library functions cannot be called from an external sketch. I had to modify the function declarations within the SC16IS750.h library file to change several of the library's private internal functions to public functions, allowing them to be accessed from within the Arduino sketch. As such, until the time that a better library comes along, you'll need to install the custom SC16IS750 library from **kw5gp.com** for this project to function properly.

At this point in the sketch, we'll also define the TFT display I/O pins and the RTTY baud rate. The UART internal baud rate clock setting is set by a function inside the UART library. The UART's baud rate clock is set to 16 × the actual baud rate, in this case 45 × 16 = 720 Hz. To get this clock rate, the UART uses an internal selectable divided by 1 or 4 prescaler, in conjunction with a 16-bit divisor register. For this project, based on the 14.7456 MHz crystal on the UART module, the library calculates a prescaler setting of 1, and a divisor setting of 20480. Fortunately, the library handles all these calculations for us, and all we have to do is give it the desired baud rate.

To start out, we'll include the libraries we'll need for this project. Then we'll define the I/O pins we'll need.

```
#include <Adafruit_GFX.h>    // Core graphics library
#include <Adafruit_ST7735.h> // Hardware-specific library
#include <SPI.h>  // SPI library required for SPI devices
#include <Wire.h> // Wire library required for I2C devices
#include <SC16IS750.h>  // SC16IS750 UART library

#define TFT_CS     9  // Assign the TFT CS to pin 10
#define TFT_RST    7  // Assign the TFT RST to pin 7
#define TFT_DC     8  // Assign the TFT DC to pin 8
```

Since the RTTY Baudot character set is comprised of 5-bit data words, and the Arduino uses the 8-bit ASCII character set, we'll need to translate between the two. Since Baudot uses two sets of 32 different characters, based on whether we're in LTRS or FIGS shift, we can create two 32-character arrays, and by keeping track of the LTRS/FIGS shift, select which of the two arrays we'll use to translate the Baudot character to its ASCII equivalent. By using an array to translate between the two character sets, we can also substitute ASCII or null values for the Baudot characters we don't want to print, such as the bell, LTRS, and FIGS shift characters.

```
// Assign the translation values for the Baudot Letters character set
char baudot_ltrs[] = {'\0', 'E', ' ', 'A', ' ', 'S', 'I', 'U', ' ', 'D', 'R',
'J', 'N', 'F',
                    'C', 'K', 'T', 'Z', 'L', 'W', 'H', 'Y', 'P', 'Q', 'O',
'B', 'G', '\0',
                    'M', 'X', 'V', '\0'
                   };

// Assign the translation values for the Baudot Figures character set
char baudot_figs[] = {'\0', '3', '\0', '-', ' ', '\0', '8', '7', '\0', '\0', '4',
'\0', ',', '!',
                    ':', '(', '5', '+', ')', '2', '$', '6', '0', '1', '9',
'?', '&', '\0',
                    '.', '/', ';', '\0'
                   };
```

Next, we'll declare the variables we'll need in the sketch itself.

```
String status_msg;  // Variable for the current USB status message to display
String rtty_data = "";  // String variable to hold rtty text data
String rtty_char = "";  // Single character string variable to hold converted
rtty char
bool system_status ;  // Flag indicating current system status - true = ok,
false = bad

bool  figs = false; // The Figures/Letters toggle flag
bool unshift_on_space = true; // Enable shifting back to LTRS when a space char-
acter is received
int rx; // The received character
byte reg_data; // Variable for UART LCR Register data
And finally, we'll instantiate (create) the UART and TFT display objects.
SC16IS750 uart = SC16IS750(SC16IS750_PROTOCOL_I2C, SC16IS750_ADDRESS_BB); //
Create the UART instance
Adafruit_ST7735 tft = Adafruit_ST7735(TFT_CS,  TFT_DC, TFT_RST);  // Create the
TFT display instance
```

In the `setup()` section, we'll initialize the TFT display and display a brief start-up message.

```
  tft.initR(INITR_18BLACKTAB);   // initialize a 1.8" TFT with ST7735S chip,
black tab
  delay(tft_delay);
  tft.fillScreen(ST7735_BLACK); // Clear the display - fill with BLACK background
  delay(tft_delay);
  tft.setRotation(1); // Set the screen rotation
  delay(tft_delay);
  tft.setTextWrap(true); // Turn off Text Wrap
  delay(tft_delay);
  tft.setTextSize(3); // Set the Font Size
  delay(tft_delay);
  tft.setTextColor(ST7735_GREEN); //Set the Text Color
  delay(tft_delay);
  tft.setCursor(40, 10);  //Set the Cursor and display the startup screen
  delay(tft_delay);
  tft.print("KW5GP");
  delay(tft_delay);
  tft.setTextSize(2);
  delay(tft_delay);
  tft.setCursor(15, 65);
  delay(tft_delay);
  tft.print("RTTY Reader");
```

Next, we'll start the UART at 45-baud, then verify that the UART is operational. Then, we'll update the TFT display with the UART status.

```
  uart.begin(baud_rate);  //Initialize SC16IS750 UART with defined baudrate set-
ting
  if (uart.ping() != 1) // Check for a response form the UART
  {

#ifdef debug
    Serial.println("SC16IS750 UART not found");
#endif

    status_msg = no_uart; // set the status message to no UART found
    system_status = false;  // set the system status ok flag to false
    update_status();  // update the status line on the TFT display
    while (1);
```

```
  } else
  {
#ifdef debug
    Serial.println("SC16IS750 UART found");
#endif

    status_msg = uart_rdy; // set the status message to UART ok
    system_status = true;  // set the system status ok flag
    update_status();  // update the status line on the TFT display
  }
```

Finally, we'll set the UART's LCR for 5-bit data words with no parity and one stop bit. The UART module is now properly configured to receive 45-baud RTTY data.

```
#ifdef debug
  Serial.println("Start Serial Communication");
  reg_data = uart.ReadRegister(SC16IS750_REG_LCR);  // Read the UART Line Con-
trol Register (LCR) register
  Serial.print("LCR Register = ");
  Serial.println(reg_data, BIN);
#endif

  uart.SetLine(5, 0, 0); // Set the UART Line Control Register (LCR) register
fot 5 bit word, no parity, 1 stop bit

#ifdef debug
  reg_data = uart.ReadRegister(SC16IS750_REG_LCR);  // Read the UART Line Con-
trol Register (LCR) register to be sure it's set
  Serial.print("Modified LCR Register = ");
  Serial.println(reg_data, BIN);
#endif
}
```

The `main loop()` for the RTTY reader sketch is about as simple as you can get. It's comprised of a single function, the `get_rx_char()` function. This function will handle the remainder of the operations to display the received RTTY data on the TFT display.

```
void loop()
{
  get_rx_char();  // Get the received character and convert to ASCII
}
```

The clear_display() function is used to clear the TFT display screen. There is no real clear display command for the TFT displays used with the Arduino. Instead, the display is filled with the background color, thereby erasing the display data.

```
void clear_display()  // Function to clear the TFT display
{
  tft.fillScreen(ST7735_BLACK); // Clear the display - set to use a black back-
ground
  delay(tft_delay);
}
```

The update_status() function is used to update the status information displayed on the bottom line of the TFT display. The TFT display for this project is set to display 6 lines of 13 characters each, with the bottom line of the TFT display reserved for status information such as the UART status and the current baud rate.

```
void update_status() // Prints status messages on bottom line of TFT display
{
  if (system_status)  // True if good, false if bad
  {
    //We're all good - update status in green
    tft.setTextColor(ST7735_GREEN);  // Green on black background color required
to "clear" previous text
  } else
  {
    // There is a system error
    tft.setTextColor(ST7735_RED);  // Red on black background color required to
"clear" previous text
  }
  tft.setTextSize(status_font_size);  // Set the font size to the status line
font size
  tft.setCursor(5, 118); // Move to the bottom line of the display
  tft.print(status_msg);  // Display the status

  tft.setCursor(115, 118);  // Move to the end of the status line
  tft.print("Baud:");
  tft.print(baud_rate);  // Display the baud rate

  tft.setTextSize(rtty_font_size);  // Reset the text size for the rtty text
area
  tft.setCursor(0, 5);  // Move to the top line of the display
}
```

The `update_text()` function will update the received RTTY text on the TFT display. The RTTY text area is 6 lines of 13 characters each, with the received RTTY characters scrolled within this 6-line window.

```
void update_text() // Updates the RTTY text on the display
{
  tft.setTextSize(rtty_font_size);  // Set the rtty text font size
  delay(tft_delay);
  if (rtty_data.length() >= 78)  // Scroll bottom line up a line if at end of
line 6
  {
    rtty_data = rtty_data.substring(13);  // Remove the first 13 characters (the
top line)
    tft.setCursor(0, 5); // Move the cursor to the top line
    delay(tft_delay);
    tft.fillRect(0, 0, 159, 105, ST7735_BLACK); // Clear the rtty data area
    delay(tft_delay);
    tft.print(rtty_data); // Print the rtty data string variable
    tft.setCursor(0, 85);  // Set the cursor to the start of line 6
  }

  rtty_data = rtty_data + rtty_char;  // Add the character to the rtty data
string
  tft.print(rtty_char); // display the received character

#ifdef debug
  if (rtty_data.length() != 0)
  {
    Serial.print("rtty_data string length: ");
    Serial.print(rtty_data.length());
  }
#endif
}
```

The `get_rx_character()` is called from the `main loop()` and is the only statement in the `loop()`. This function will check the UART to see if there is a received character, and if there is one, it will process and display the received character. Otherwise, the sketch will continue to loop until a character is received.

If a character is received, we'll check to see if it's a space, FIGS, or LTRS character. This sketch is designed to implement the unshift on space (USOS) RTTY feature, which will return the character set to the LTRS shift whenever a space character is received.

```
void get_rx_char()  // Get a received character from the UART
{
  if (uart.available() != 0)
  {
    // Read the received character
    rx = uart.read();
    if (rx != 0)
    {
      switch (rx)
      {
        case 4: // It's a space character. Unshift if flag is enabled
          if (unshift_on_space)
          {
            figs = false; // Set FIGS/LTRS flag to LTRS

#ifdef debug
            Serial.print("   Space - unshift");
#endif

          }
          break;

        case 27: // It's a FIGS key
          figs = true;  // Set the FIGS/LTRS flag to FIGS

#ifdef debug
          Serial.print("  FIGS");
#endif

          break;

        case 31:  // It's a LTRS key
          figs = false; // Set FIGS/LTRS flag to LTRS

#ifdef debug
          Serial.print("  LTRS");
#endif

          break;
      }
```

Next, we'll check for the FIGS or LTRS character and set the FIGS flag accordingly. Finally, we'll translate the received RTTY character to its matching ASCII counterpart using one of the two Baudot translation arrays selected by the FIGS flag, and then call the `update_text()` function to add the received character to the TFT display output.

```
if (figs) // Choose which character decode array to use
      {
        rtty_char = baudot_figs[rx];  // Use the Figures array
      } else
      {
        rtty_char = baudot_ltrs[rx];  // Use the LTRS arry
      }

      // Decode it
#ifdef debug
      Serial.print("   Decode: ");
      Serial.println(rtty_char);
      //    delay(300);
#endif
      update_text();  // update the TFT display
    }
```

Testing the RTTY Reader

The nice thing about this project is that there are several ways to tune and test the finished project. For initial testing of the decoder circuit, I used a USB sound dongle plugged into a workstation, with the headphone output of the USB dongle plugged into the audio input on the RTTY reader. To do basic tone decoder testing, I used the online tone generator at **szynalski.com/tone-generator** to generate an audio signal so that I could adjust the audio input level and tune the decoder. You can measure the audio level going to the decoder circuit at the wiper of the input audio level potentiometer using a voltmeter on the AC setting. Remember, for the LM567 tone decoder, you'll want a maximum of 200 mV RMS of audio. For the XR-2211 decoder, the audio input level can be anywhere from 10 mV – 3 V. Once you have the audio level set, you can adjust the decoder tuning potentiometer.

For the LM567 decoder circuit, since we're only decoding the mark tone frequency, you can set the online tone generator to 1000 Hz and adjust the tuning potentiometer until the LED on the board goes out. Remember that this LED is powered by a transistor on the output of the decoder and is the inverse of the actual decoded signal. Turn the potentiometer until you have the tuning centered, and you should be good to go. For the XR-2211 version of the decoder, the tuning is slightly different. You'll want to tune for the center frequency, which would roughly be 1085 Hz. In either case, the tuning range is quite broad, and you have a tuning range of about 2 – 3 turns on the tuning potentiometer where the tone will be properly decoded.

Once you have the tone decoder set, you can now try decoding an actual RTTY signal. I used the free MMTTY program available for download from **hamsoft.ca/pages/mmtty.php** for testing. After installing MMTTY in your PC, set the mark frequency to 1000 Hz with the NET button selected in the demodulator settings. With these settings, MMTTY will transmit the 1000 Hz mark and 1170 Hz space tones used by the RTTY reader. Be sure to set the MMTTY soundcard option to transmit to the sound card audio output you have connected to the RTTY reader, and you should be able to select the RY macro button to continuously send a series of RY's to the RTTY reader (The RY character sequence was originally used to stress test mechanical RTTY machines, and it carried over to more modern RTTY decoder testing. Occasionally you still may see RYs transmitted on the air.) You can do any final tuning of the decoder while sending

RY's and stop the test by selecting the "TXOFF" control button. You can also press the TX button and type directly to the RTTY reader. Be sure that you have the correct sound card setting for these tests, otherwise if you have your workstation connected to a transceiver, you could accidentally transmit out over the air instead of to the RTTY reader. **Figure 21.9** shows the output of the RTTY reader tested using MMTTY.

Once you have the RTTY reader tuned, you're ready to connect it to the audio output of your receiver and tune in a RTTY signal. Remember that on the HF bands, RTTY is generally tuned using the LSB (lower sideband) setting on your receiver. You can use the LED on the circuit board as a tuning aid.

A reliable source of on-air RTTY signals that you can use for testing the decoder and tuning practice is the W1AW RTTY bulletins that are transmitted daily during the week. See **arrl.org/w1aw-operating-schedule** for details.

Enhancement Ideas

The RTTY reader is one of those fun projects that I could keep tinkering and refining indefinitely. There are so many tweaks that I came up with that it would take me forever to implement them all. But I had to stop somewhere or else the project would never get finished, and there wouldn't be anything left for you to add yourself.

One possible enhancement would be to add in the USB keyboard project and modify the RTTY reader to send as well as receive. A chip such as the XR-2206 monolithic function generator could be used to generate the mark and space audio tones to feed to your transmitter. This would give you a portable PC-less RTTY station that you could take anywhere.

Another enhancement would be to display more information on the status line, such as whether you're in LTRS or FIGS shift. You could also change the font size and put more RTTY text on the screen and use more lines of the TFT display. You could even jazz up the display by using different colors and adding lines between the text and status area. You could also increase the size of the TFT to a 2.2-inch TFT, or larger. There's really no limit to the size of the TFT display you could use with this project.

On the XR-2211 version of the decoder, you could add a switch and an option to select either the 1000 Hz or 2125 Hz mark tone that would use small reed relays to switch the resistors and caps needed to make the change. You could also use this same method to change the frequency shift from 170 Hz to any other frequency shift you may choose. You could even go so far as to use the digital potentiometer chips we used in the Yaesu FH-2 keypad project to change the resistor values via software.

You can add tighter filtering on the decoder circuits. With the addition of a few components, you can tighten up the decoder filter so that it is less affected by adjacent signals. The XR-2211 and LM567 datasheets are good resources for design ideas.

I also think that adding a pushbutton switch to manually perform the LTRS shift, and another to clear the display, could be nice additions this project.

Finally, you could add a piezo buzzer to one of the Arduino's I/O pins and play a tone whenever a BELL character is received (The old mechanical teleprinters had bells that rang to alert the operator of an important incoming message).

As with all the projects in this book, the sketch and libraries for this project can be found at **kw5gp.com**.

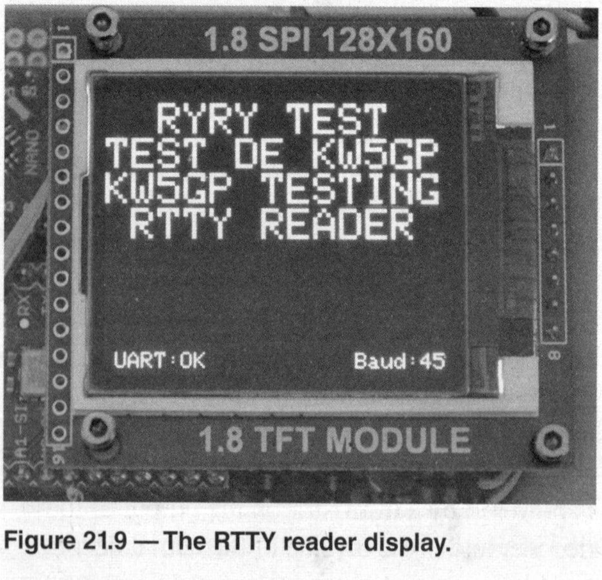

Figure 21.9 — The RTTY reader display.

22 40-Meter QRP CW Transceiver

Figure 22.1 — Mini-Cube 40-meter CW transceiver.

Figure 22.2 — Frogs Calling QRP CW transceiver kit.

I've had a lot of fun building the Arduino-powered 40-meter QRP CW and JT65 transceiver projects. The original CW transceiver version (see **Figure 22.1**), that I called the "40-meter cube," used an Arduino UNO along with a Nokia 5110 LCD display and fit inside an acrylic softball trophy case. When needing a new project for the QRP-ARCI's Homebrew contest at their Four Days in May (FDIM) event a few years back, I upgraded the Arduino Uno to a Nano and replaced the Nokia 5110 LCD display with a 1.8-inch color TFT display. The project was renamed the "mini-cube" and was sized down to fit inside an acrylic golf ball display case.

While the Arduino isn't fast or powerful enough to be a full-fledged Software Defined Radio (SDR), there are quite a few radio functions that it does do well. A few years ago, Tom Medlin, W5KUB, introduced me to the Frog Sounds and Frogs Calling 40-meter QRP CW transceiver boards. These boards are very inexpensive 40-meter CW transceiver kits that can be built in an afternoon. I got mine from eBay for about $11 each. For this project, we'll use the Frogs Calling kit shown in **Figure 22.2**, but you could just as easily use the Frog Sounds board by using the modifications shown in the next chapter.

These boards are based on the NoCal Forty-9er by Wayne Burdick, N6KR, co-founder of Elecraft. They are crystal-controlled CW transceivers that put out about 1.8 W and use an NE-602-based direct conversion receiver. The current kits that are available on eBay, and other online retailers such as **Banggood.com** and AliExpress, come with a 7.023 MHz crystal. Tom showed me his Frogs Calling board and asked

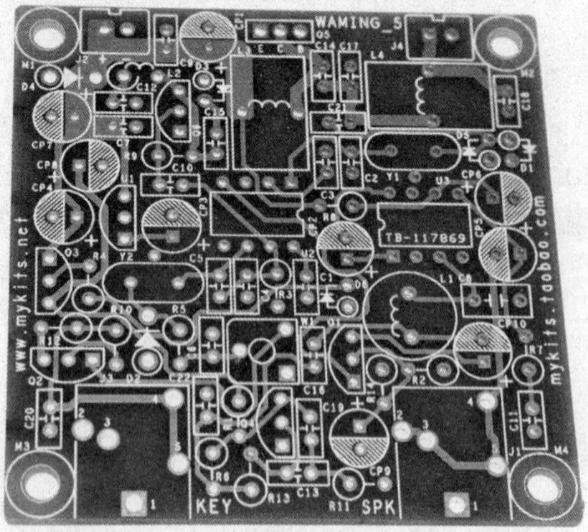

Figure 22.3 — The 40-meter QRP transceiver circuit board.

if we could add an Arduino-based VFO to have it tune the entire 40-meter band. That triggered the "wouldn't it be cool if…" part of me, so I quickly ordered my kit and started playing with it on the lab bench.

I have no idea where the Frog Sounds and Frogs Calling names came from, but while easily built without instructions, for a chuckle, you simply must take the time to read the instructions.

The translation from Chinese to English ranges from strange to downright hilarious, with examples like, "…the headset with the beat Hair drops pattered generated audio," and, "avoid prolonged shorting key launch, or D882 and dummy load fever large, carefully damaged." I'll come back to the "carefully damaged" part in a bit.

But don't let the instructions scare you away. The boards are well made, and very clearly silk-screened, with the component numbers and their proper placement easily read, as shown in **Figure 22.3**.

Using the 1 – 30 MHz DDS VFO project to serve as a signal generator, I found that you could easily convert the Frogs Calling board to use a DDS instead of a crystal simply by removing the receive filter crystal on the transceiver board and adding in a couple of components with a few circuit board trace cuts, and a couple of wires. We'll cover these modifications in detail in the construction portion of this chapter.

Now that we have our proof of concept, we can design a fully featured Arduino-powered 40-meter QRP CW transceiver.

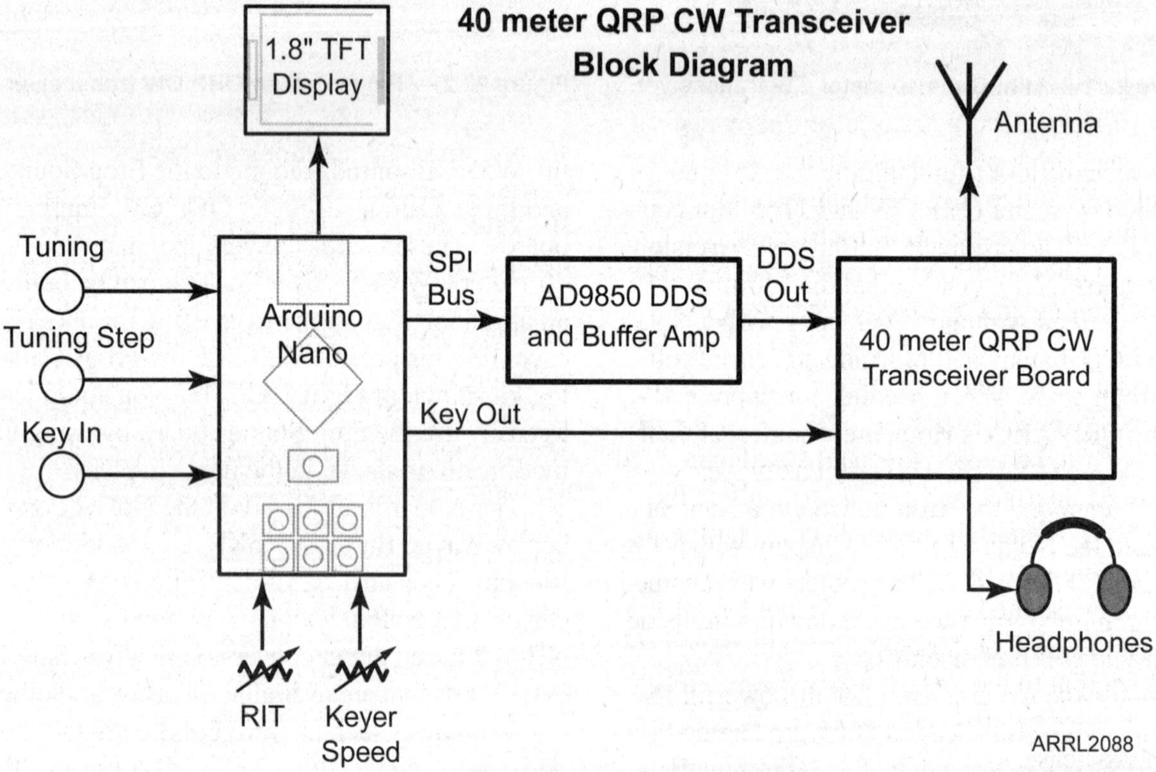

Figure 22.4 — The 40-meter CW transceiver block diagram.

Design Goals

Using the Frogs Calling board as our transceiver platform, we'll upgrade the original Arduino UNO to a Nano and use an AD9850 DDS module to act as a VFO, allowing the transceiver to cover the entire 40-meter band. Since we have the power of the Arduino available, we're not going to stop with just adding a VFO. We'll turn this project into a full-featured QRP 40-meter CW transceiver, with a built-in CW keyer with adjustable speed, dual-range receive incremental tuning (RIT), and we'll add dual speed tuning using a rotary encoder with a built-in pushbutton switch. We'll upgrade the Nokia 5110 LCD display module to a ST7735-type color TFT display that we'll use to show the transmit and receive frequencies, along with the RIT setting, the tuning rate, and the CW keyer speed. We'll connect a voltage divider to the incoming 12 V power and display that on the Nokia display as well. To jazz things up, we'll mount everything inside an acrylic golf ball trophy case. At the end of the day, we'll have a fun little 40-meter QRP CW transceiver, and it will cost us about $40.

The 40-Meter QRP CW Transceiver Block Diagram

Figure 22.4 shows the block diagram for the 40-meter QRP CW transceiver project. We'll use a small piece of copper-clad perfboard to mount the Arduino Nano and other components to the inside of the golf ball trophy case. A rotary encoder with a built-in pushbutton switch will be used for tuning and selecting the tuning frequency step rate. Potentiometers will be used as analog inputs to the Arduino to control the RIT and the CW Keyer speed.

The Arduino prototyping shield will contain all the interface components, including the AD9850 DDS module and the AD826 buffer amplifier, along with the transmitter keying circuit that will be isolated using an optocoupler. We'll connect the DDS buffer amplifier output and CW keying output to the Frogs Calling 40-meter CW transceiver board that is mounted to the base of the display case.

The 40-Meter QRP CW Transceiver Flowchart

Figure 22.5 shows the flowchart for the 40-meter QRP CW transceiver project. Surprisingly enough, we don't need many libraries for this project. All we'll need to include are the M0XPD DDS library, the Adafruit TFT display libraries, and the rotary encoder library.

Next, we'll define the I/O pins used in the project, then declare and initialize the variables we'll need in the sketch. Then we'll create (instantiate) the DDS, TFT display, and rotary encoder objects.

In the last part of the initialization area, we'll define all of the variables needed by the keyer software written by Steven Elliott, K1EL. You may remember that this is the same keyer code we used in the iambic keyer project earlier in this book. Then we'll create an enumerated data type that defines all of the necessary keyer state machine elements we'll be using in the keyer portion of the sketch.

In the `setup()` loop, we'll set the pin modes for the Arduino's I/O pins, and initialize the TFT display. After a brief startup message, we'll start the DDS, setup the K1EL keyer, and display the operating information on the TFT display.

Because many of the operations in the sketch are performed by functions, the `main loop()` is actually quite small. If we are not processing any keyer activity, we'll read and set the RIT frequency and keyer speed. Next, we'll read the rotary encoder, and if the position has changed, we'll update the transmit and receive frequencies accordingly. We'll check the status of the rotary encoder's pushbutton switch to select either fast or slow tuning modes, and also read and display the battery voltage. If there is any keyer activity, we'll check the keyer input, and key the Frogs Calling transmitter accordingly.

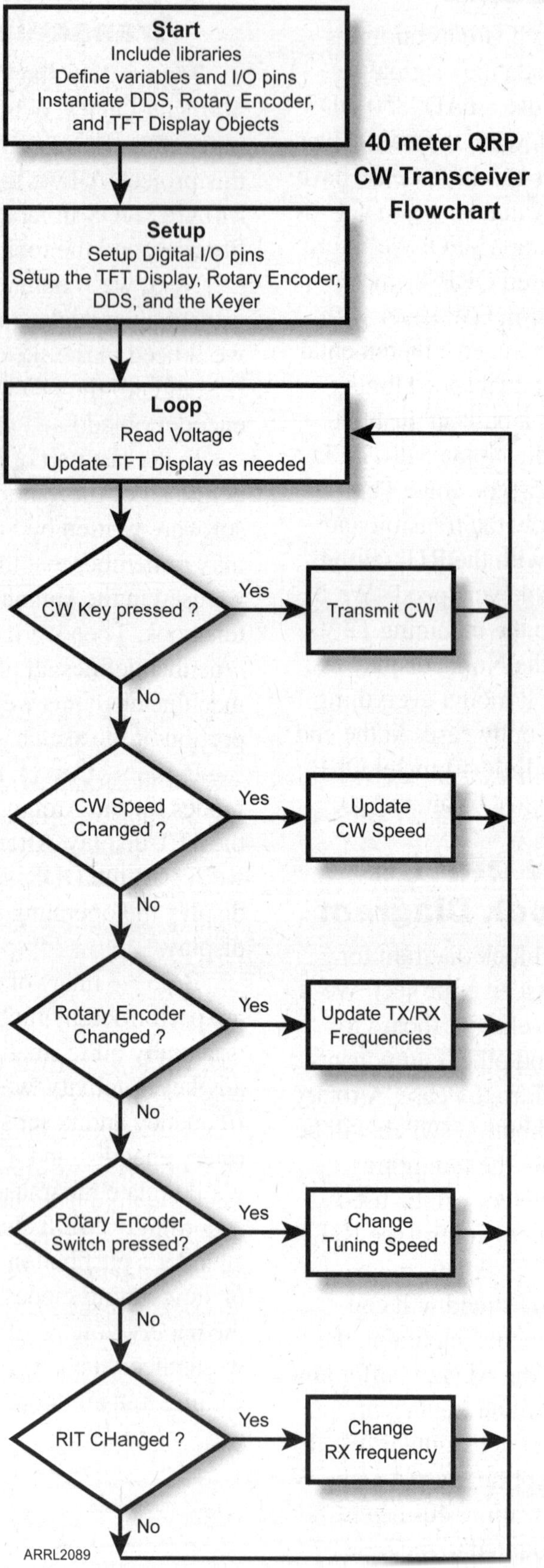

Figure 22.5 — 40-meter QRP CW transceiver flowchart.

Constructing the 40-Meter QRP CW Transceiver

The first thing you'll want to do is build your Frogs Calling (or Frog Sounds) 40-meter QRP CW transceiver kit. You'll want to build and test your Frogs Calling kit to make sure it works before we modify it for the DDS VFO. Don't be concerned if your kit comes with an NE612, SA602, or SA612 IC for the receive mixer/oscillator IC. They're all essentially the same chip, other than the SA series which has a higher temperature specification. There's a long story about the NE602 versus the NE612, which involves an accidental fire destroying the original masters for the NE602, and today's NE602's are basically relabeled NE612's.

Figure 22.6 shows the completed Frogs Calling transceiver board. One thing that is not clearly discussed in the instructions is the construction of the two ferrite core inductors for the transceiver board. L4 is the black ferrite core used to make the 22 μH coil, and you'll need to wind 11 turns of the solid enameled wire around this core. L5 uses the red ferrite core to make the 1 μH coil, and you'll need to wind 15 turns of the enameled wire around this core. Be sure to get the ends cleaned of all the enamel insulation before soldering to the board. I used a match to burn off the enamel insulation, then 400-grit sandpaper to be sure I had removed it.

Figure 22.6 — The completed Frog QRP transceiver.

Once you have a working and tested transceiver board, the next thing we'll do is modify it for use with the Arduino and DDS VFO. Since the Frogs Calling board uses a 7.023 MHz crystal (Y1) for the receive filter, we'll first need to remove that crystal, and add a broader LC-based receive filter. Rather than reinvent the wheel, I chose to adapt the 40-meter receive input filter used in "The MRX-40 Mini Receiver," by Steve Bornstein, K8IDN, that was published in the September 1997 issue of *QST*. **Figure 22.7**, which shows the schematic for the receiver modification of the Frogs Calling board, also contains the parts list.

Figure 22.8 shows the receive filter mod on the Frogs Calling board. Using the "dead bug" style of construction, most of the receive filter modification can be built above the board and connected between the solder pads of the removed Y1 crystal. I placed L1, the 3.3 μH inductor, C2, and the 470-pF capacitor on the bottom side of the board as shown in **Figure 22.9**.

Because we replaced the very narrow bandwidth crystal receive filter with a wider LC-based receive filter, you may encounter broadcast interference (BCI) on your receiver. While my board did not need it, if you do encounter BCI, you can add a BCI filter ahead of the LC-based receive filter modification. **Figure 22.10**, which shows the schematic for the BCI filter, also contains the parts list.

The next modification we need to make to the Frogs Calling board is to add in the connections for the DDS output. First, we'll need to cut two circuit board traces on the Frogs Calling board as shown in **Figure 22.11**. The first trace to cut is between C5 and pin 6 on the NE602, as shown in the bottom arrow in Figure 22.11. Next, we'll need to cut the circuit board trace between pin 7 and C10, as shown in the top arrow of the figure. Referring to Figure 22.9, we attach the output from the DDS buffer amplifier to pin 6 of the NE602 and add a 680 Ω resistor between pin 6 and the pin 7 side of C10. This completes the modification of the Frogs Calling board, and we're now ready to build the rest of the proj-

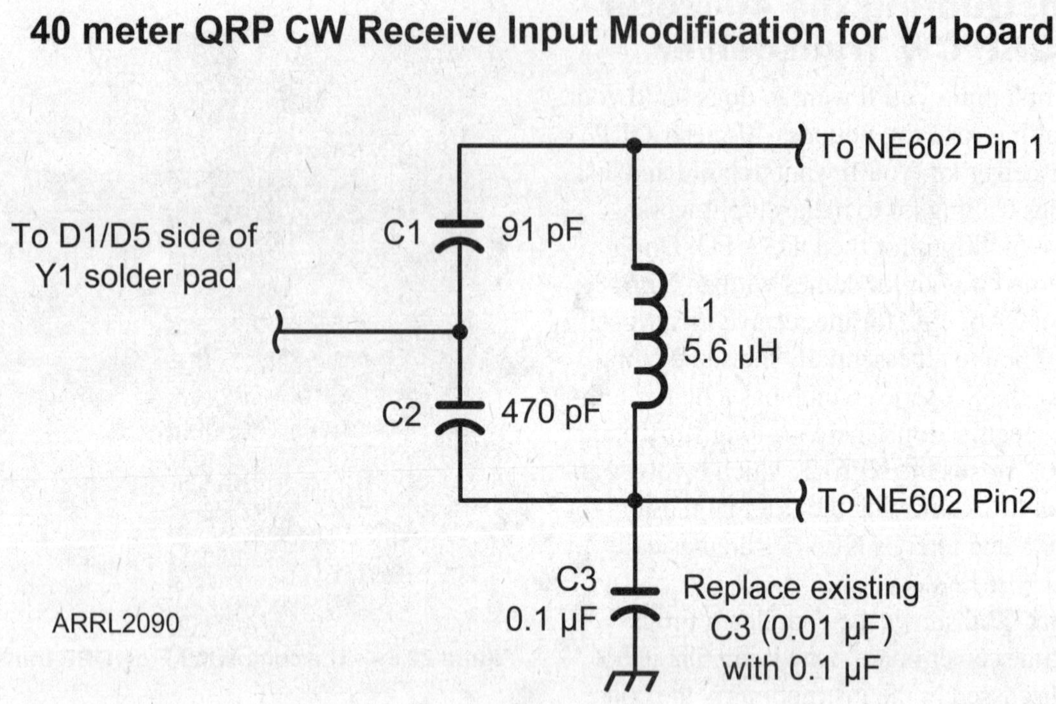

Figure 22.7 — 40-meter QRP CW receiver modification schematic.

Part	*Value*	*Part*	*Value*
C1	91 pF ceramic capacitor, 50 V	C3	0.1 µF ceramic capacitor, 50 V
C2	470 pF ceramic capacitor, 50 V	L1	5.6 µH inductor

Figure 22.8 — The completed 40-meter QRP CW transceiver receive filter modification.

Figure 22.9 — The bottom view of the receive filter modification.

Optional Broadcast Interference Filter for V1 Board

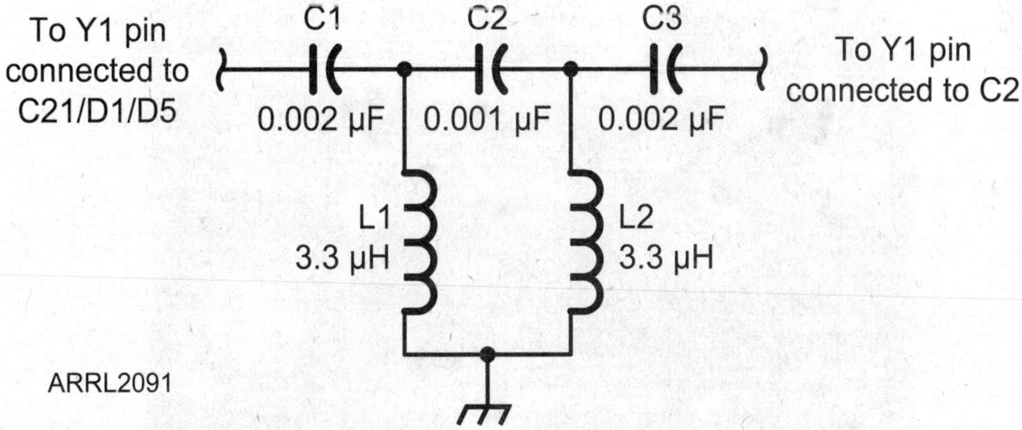

Figure 22.10 — The 40-meter CW transceiver BCI modification.

Part	***Value***
C1, C3	**.002 µF ceramic capacitor, 50 V**
C2	**.001 µF ceramic capacitor, 50 V**
L1, L2	**3.3 µH inductor**

ect on the Arduino prototyping shield that will mount on top of the Arduino Uno.

The final modification to the Frogs Calling board is to add the volume control. To do this we need to cut the LM386 amplifier output on pin 3 as shown in **Figure 22.12** and add in the connections to the volume control potentiometer on the bottom of the board as shown in **Figure 22.13**. One side of the volume control potentiometer connects to the LM386 Pin 3, and the other side goes to any ground connection. The wiper of the potentiometer goes to the other side of the trace we cut on LM386 pin 3 to complete the audio circuit path that we also cut.

Now that you have a modified Frogs Calling board, it's time to build the rest of the circuit for the 40-meter QRP CW transceiver project. **Figure 22.14**, which shows the schematic for the 40-meter QRP CW transceiver project, also contains the parts list. This design uses resistors in the place of a level converter module. If you would prefer to use a TXB0108-type level converter module, please refer to the parts list included with **Figure 22.15** of the schematic for the transceiver using level converters.

We'll mount the Arduino perfboard assembly to the side of the display case, and the Frogs Calling board in the base. We'll use Dupont-style 2.54 mm headers to connect from the perfboard to the transceiver board, TFT display, and the other transceiver controls. We'll connect a voltage divider to the incoming 12 V power and connect that to an analog I/O pin on the Arduino Nano. This will allow us to read and display the battery voltage on the TFT display.

We'll mount the potentiometer for the keyer speed, along with the stereo mini phone jacks for the headphones and keyer paddles, and a BNC connector for the antenna on the rear and side of the display case. The Powerpole connectors used for the 12 V power input and the power switch are mounted to the rear side of the display case. The rotary encoder, RIT potentiometer, and audio level potentiometer are mounted on the front of the display case. The TFT display is mounted in the top of the display case using 2 mm hardware.

Figure 22.11 — The circuit board trace cuts needed to connect the DDS module.

Figure 22.12 — The circuit board trace cuts to add the volume control potentiometer.

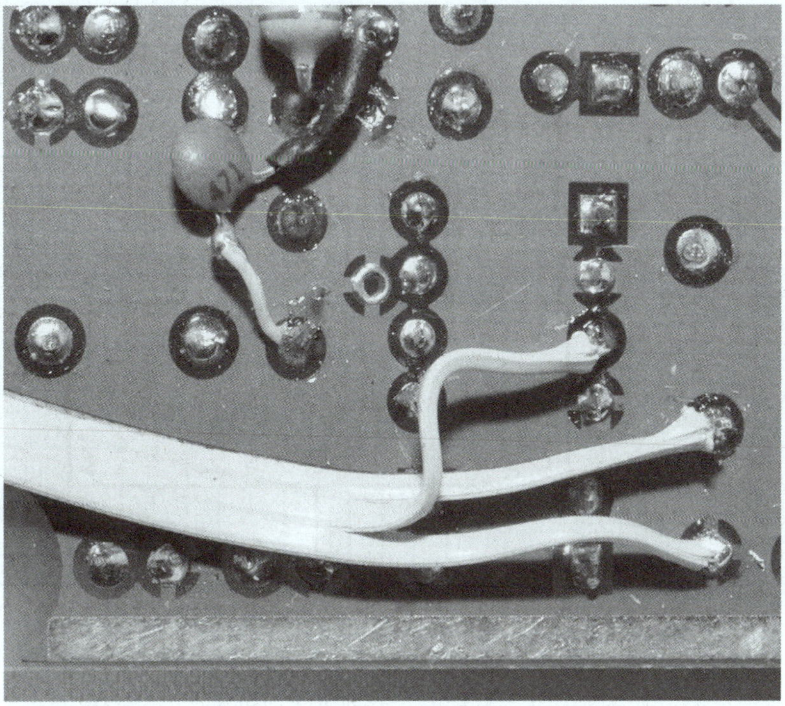

Figure 22.13 — Adding the volume control potentiometer connections.

40 meter QRP CW Transceiver

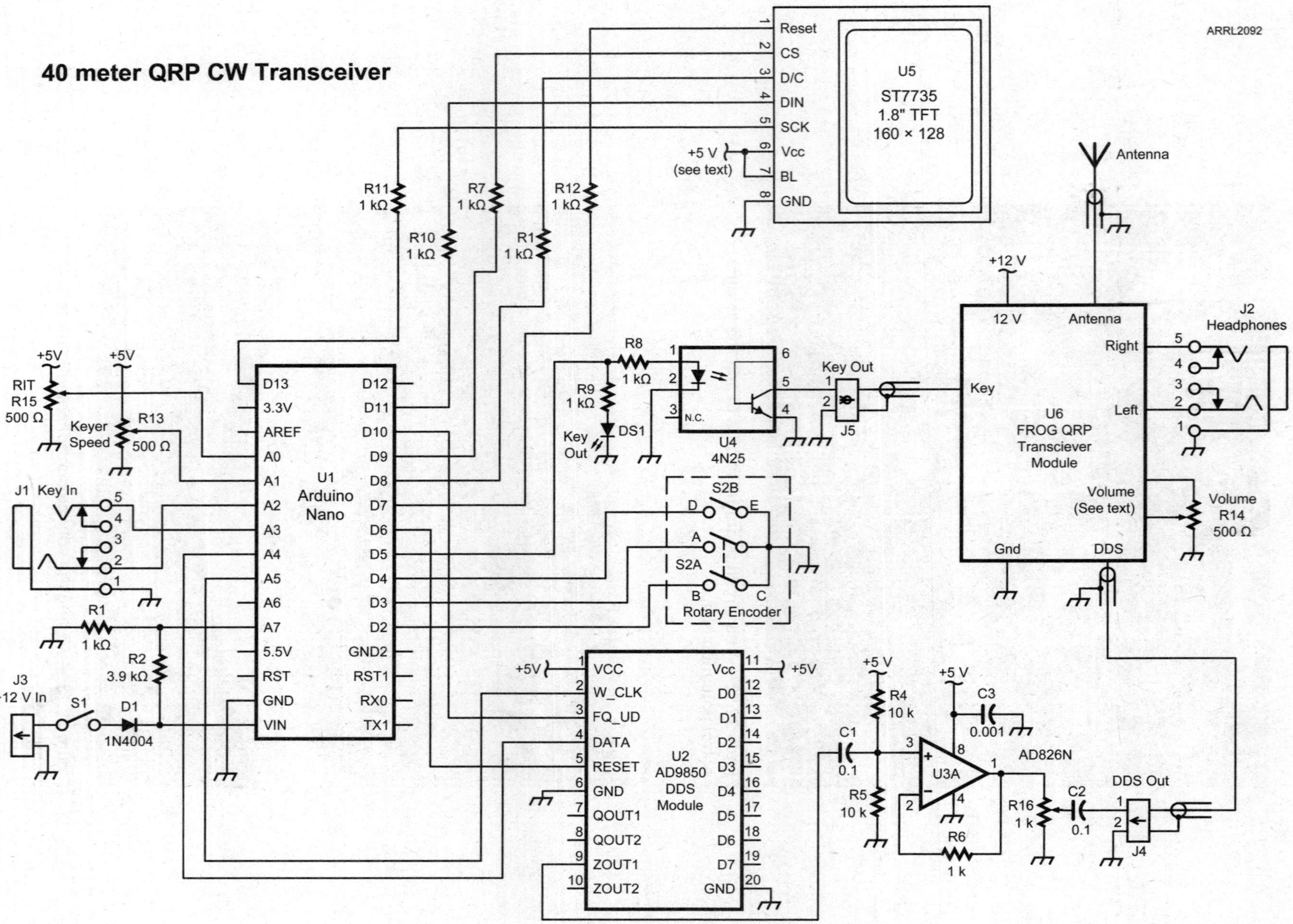

Figure 22.14 — 40-meter QRP CW transceiver schematic.

Part	*Value*
C1, C2	.1 μF ceramic capacitor, 50 V
C3	.001 μF ceramic capacitor, 50 V
D1	1N4004 diode
DS1	5 mm LED
J1, J2	stereo mini phone jack
J3	Powerpole connector
J4, J5	2-pin Dupont 2.54 mm header and connector
R1, R3, R6–R12	1 KΩ resistor 1/8 W
R2	3.9 KΩ resistor, 1/8 W
R4, R5	10 KΩ resistor, 1/8 W
R13, R14, R15	500 Ω potentiometer
R16	1 KΩ potentiometer
S1	SPST switch
S2	Rotary encoder with pushbutton switch
U1	Arduino Nano
U2	AD9850 DDS module
U3	AD826N high speed dual operational amplifier IC
U4	4N25 optocoupler IC
U5	ST7735-type 1.8-inch TFT display (see text)
U6	Frogs Calling QRP CW transceiver board (eBay)
Enclosure	Softball display case

The AD 9850 DDS is mounted to the copper-clad perfboard using two rows of 2.54 mm female header strips, and the AD826 instrumentation amplifier and the 4N25 optocoupler are mounted using standard 8-pin DIP sockets. While the 4N25 optocoupler is a 6-pin device, it fits just fine in an 8-pin DIP socket. You could also get a 6-pin DIP socket if you prefer. I only used 8-pin sockets because I had some laying around. The keying output of the 4N25 is connected in parallel with the key jack on the Frogs Calling board. The height of the AD9850 socket allows us to place most of the DDS buffer amplifier circuit underneath the AD9850 DDS module. The DDS gain potentiometer is mounted on the edge of the perfboard so that we can drill a small hole in the display case and adjust the DDS output level as necessary. To avoid confusion, the potentiometer on the DDS module itself is only used to adjust the duty cycle of the square wave output and has no effect on the sine wave output of the DDS module.

You may notice that we don't use the transistor final output section of the DDS buffer amplifier for this project like we have in previous projects. The AD826 instrumentation amplifier has enough output in the 40-meter band to provide what we need for the VFO input to the Frogs Calling board.

One final note on constructing the 40-meter QRP CW transceiver project. You may want to add a heat sink to the D882 final output transistor. Under normal operation at full 1.8 W output, the D882 final output transistor tends to run warm.

40 meter QRP CW Transceiver Using Level Shifter on TFT Display

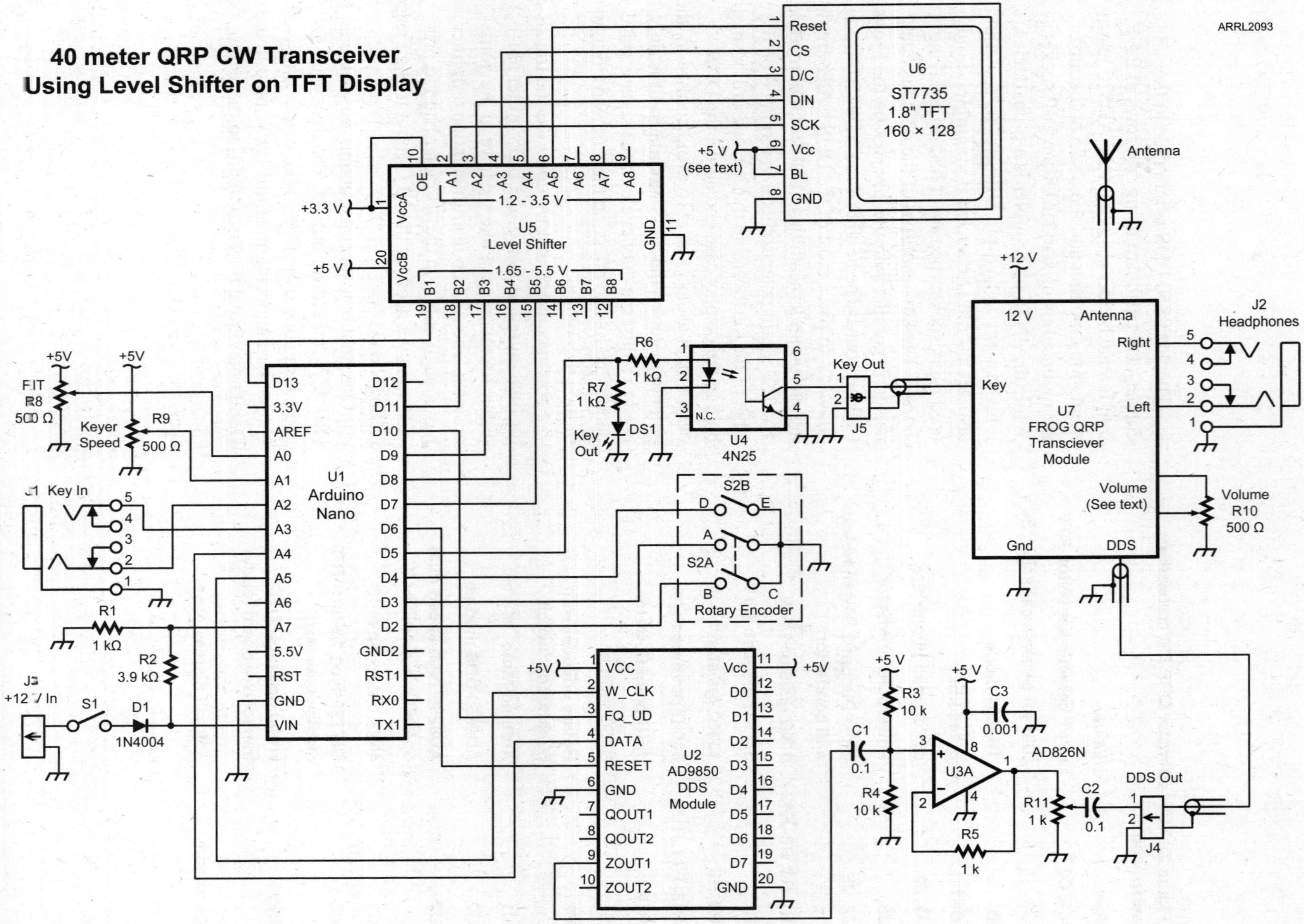

Figure 22.15 — 40-meter QRP CW transceiver using level converter schematic.

Part	*Value*
C1, C2	.1 µF ceramic capacitor, 50 V
C3	.001 µF ceramic capacitor, 50 V
D1	1N4004 diode
DS1	5 mm LED
J1, J2	Stereo mini phone jack
J3	Powerpole connector
J4, J5	2-pin Dupont 2.54 mm header and connector
R1, R3, R5–R7	1 KΩ resistor 1/8 W
R2	3.9 KΩ resistor, 1/8 W
R3, R4	10 KΩ resistor, 1/8 W
R8-R10	500 Ω potentiometer
R11	1 KΩ potentiometer
S1	SPST switch
S2	Rotary encoder with pushbutton switch
U1	Arduino Nano
U2	AD9850 DDS module
U3	AD826N high speed dual operational amplifier IC
U4	4N25 optocoupler IC
U5	8-channel level shifter module (TXB0108 type)
U6	ST7735-type 1.8-inch TFT display (see text)
U7	Frogs Calling QRP CW transceiver board (eBay)
Enclosure	Softball display case

Figure 22.16 — What "dummy load fever large, carefully damaged" means.

Under the "prolonged shorting key launch" (CW key down) conditions for the same testing with the JT65 transceiver project, I managed to discover the meaning of "D882 and dummy load fever large, carefully damaged,' as shown in **Figure 22.16**. The moral of this story, the Frogs Calling transceiver board is designed for CW, and does not like to be keyed down for an extended period of time at full power output. Be careful in your testing to avoid extended key-down periods and always have an antenna or dummy load connected when transmitting to keep your board from catching "dummy load fever."

The 40-Meter QRP CW Transceiver Sketch

It took a while, but we're now ready to work on the sketch for the 40-meter QRP CW transceiver project. We'll start by including the M0XPD DDS, ST7735-type color TFT display, and rotary encoder libraries.

```
#include <Encoder.h>
#include <Adafruit_GFX.h>    // Core graphics library
#include <Adafruit_ST7735.h> // Hardware-specific library
#include <SPI.h>
```

Next, we'll define the I/O pins needed for the AD9850 DDS, and the ST7735-type TFT display module. You may notice that the W_CLK pin for the DDS module is connected to digital I/O pin 19, and the voltage divider for the battery voltage input is attached to analog input pin 7. This is the one major difference between the Arduino Nano and the UNO. The Nano has 8 analog input pins, A0-A7, also labeled as I/O pins 14 through 21, which we can use. In this case, digital I/O pin 19 is the A5 analog pin on the Nano that we use as a digital output pin for the DDS W_CLK signal.

```
const int W_CLK = 19;
const int FQ_UD = 10;
const int DATA = 18;
const int RESET = 6;

// For the breakout, you can use any 2 or 3 pins
// These pins will also work for the 1.8" TFT shield
#define TFT_CS     9
#define TFT_RST    7  // you can also connect this to the Arduino reset
// in which case, set this #define pin to 0!
#define TFT_DC     8

const int rit_pot = A0;  // Set the RIT Pot pin to A0
const int encoder_PinA = 2;  // Pin A of the Rotary Encoder
const int encoder_PinB = 3;  // Pin B of the Rotary Encoder
const int encoder_Switch_Pin = 4;  // Encoder Pushbutton Switch input pin
const int speed_Pot = A1;  // CW Speed Pot Pin
const int key_dit = A3;  // Keyer Dit input pin
const int key_dah = A2;  // Keyer Dah input pin
const int key_pin = 5;  // Key Out pin
const int battery_pin = A7; // Battery Voltage Input pin
```

Now we'll declare and initialize the variables we'll use in the sketch. The dual range RIT tuning range is also controlled by the pushbutton switch on the rotary encoder, and we'll define its tuning range at +/-500 Hz and +/-5000 Hz respectively. We'll define the I/O pins used by the rotary encoder, the keyer speed potentiometer, the keyer inputs, and the battery voltage input. We'll also set the dual-range tuning steps to 1Hz and 100Hz.

```
const int rit_fast_step = 5000;  // The RIT fast step rate
const int rit_slow_step = 500;  // The RIT slow step rate
const int key_mode = 1; // Set to 0 for Iambic A mode. Set to 1 for Iambic B,
Set to 2 for Straight Key
const int debounceInterval = 500;  //disable step change for 500 ms between en-
coder pushbutton press
const int fast_step = 100;  // Rotary Encoder Fast Tuning Step
const int slow_step = 1;  // Rotary Encoder Slow Tuning Step
```

Next, we'll define the high and low end of the 40-meter band and set the transmit frequency offset to 600 Hz. This offset will allow us to hear a sending station on our frequency with a CW tone of 600 Hz. You can adjust this to whatever CW tone frequency you are most comfortable listening to. We'll also preset the RIT frequency, rotary encoder position, and keyer speed with default values, but these values will be set to their operating values once the `main loop()` starts. We'll set the starting transmit and receive frequencies to 7.030 MHz with the fast 100 Hz tuning rate. Then we'll set the calibration values we'll need to read the battery voltage.

```
const double band_low = 7000000; // Low end of Band
const double band_high = 7125000; // High end of CW portion of band

int tx_offset = 600;
int rit;
int old_rit = 0;
int rit_freq;
int rit_range;
int encoder_Pos = 0;
int old_encoder_Pos = 0;
int cw_speed = 5;
int old_speed = 0;

int tick = 0;
int last_voltage = -1;
int old_rx_freq = 0;
```

```
int step_size = fast_step;

double freq = 7030000;   // Starting freq
double tx_freq = freq;   // set the TX freq to the starting freq
double rx_freq = freq;   // set the RX freq to the starting freq
double dds_freq = rx_freq;

boolean encoder_Switch;
boolean current_step = HIGH;   // Default Tuning Step is Fast

unsigned long int delay_time = 0;

float battery_volts;
float volt_drop = 0.89;
int volt_cal = 2490;
```

Next, we'll create (instantiate) the DDS, TFT display, and rotary encoder objects.

```
// Instantiate the DDS
DDS dds(W_CLK, FQ_UD, DATA, RESET);

// Instantiate the Rotary Encoder
Encoder Enc(encoder_PinA, encoder_PinB);

//Instantiate the TFT
Adafruit_ST7735 tft = Adafruit_ST7735(TFT_CS,  TFT_DC, TFT_RST);
// Option 2: use any pins but a little slower!
#define TFT_SCLK 13   // set these to be whatever pins you like!
#define TFT_MOSI 11   // set these to be whatever pins you like!
//Adafruit_ST7735 tft = Adafruit_ST7735(TFT_CS, TFT_DC, TFT_MOSI, TFT_SCLK, TFT_
RST);
```

In the last section of the initialization area, we'll define and setup the K1EL keyer we'll be using in this sketch.

```
// Keyer Variables
unsigned long ditTime;  // Number of milliseconds per dit
char keyerControl;
char keyerState;
```

```
static long ktimer;
////////////////////////////////////////////////////////////////////////////////
/////////
//  keyerControl bit definitions
//
#define     DIT_L      0x01     // Dit latch
#define     DAH_L      0x02     // Dah latch
#define     DIT_PROC   0x04     // Dit is being processed
#define     PDLSWAP    0x08     // 0 for normal, 1 for swap
#define     IAMBICB    0x10     // 0 for Iambic A, 1 for Iambic B
#define     ULTIMATIC  0x20     // 1 for ultimatic
#define     STRAIGHT   0x80     // 1 for straight key mode
////////////////////////////////////////////////////////////////////////////////
////////
//
//  State Machine Defines

enum KSTYPE {IDLE, CHK_DIT, CHK_DAH, KEYED_PREP, KEYED, INTER_ELEMENT };
```

In the first part of the `setup()` loop, we'll start by initializing the TFT display and displaying a brief startup screen before continuing.

```
void setup()
{

  // Use this initializer if you're using a 1.8" TFT
  tft.initR(INITR_BLACKTAB);   // initialize a ST7735S chip, black tab

  tft.fillScreen(ST7735_BLUE); // Clear the display
  tft.setRotation(1); // Set the screen rotation
  tft.setTextWrap(false);
  tft.setTextSize(3);
  tft.setTextColor(ST7735_GREEN);
  tft.setCursor(40, 10);
  tft.print("KW5GP");
  tft.setTextSize(1);
  tft.setCursor(55, 40);
  tft.print("Mini-Cube");
  tft.setCursor(40, 55);
```

```
  tft.print("Arduino Powered" );
  tft.setCursor(20, 70);
  tft.print("40 meter QRP CW XCVR");

  tft.setCursor(40, 90);
  tft.print("Dual Range RIT");

  tft.setCursor(30, 100);
  tft.print("Dual Speed Tuning");

  tft.setCursor(7, 110);
  tft.print("5-35 WPM Iambic A/B Keyer");
```

Next, we'll initialize the AD9850 DDS, set the pin modes for the rotary encoder and keyer inputs, and enable the internal pullup resistors.

```
// start up the DDS...
  dds.init();
  // (Optional) trim if your xtal is not at 125MHz...
  dds.trim(125000322); // enter actual osc freq


  // Set up the Rotary Encoder
  pinMode (encoder_PinA, INPUT);
  pinMode (encoder_PinB, INPUT);
  pinMode (encoder_Switch_Pin, INPUT);
  // Enable the Internal Pull-up resistor on the Encoder Inputs
  digitalWrite(encoder_PinA, HIGH);
  digitalWrite(encoder_PinB, HIGH);
  digitalWrite(encoder_Switch_Pin, HIGH);

  // Set up the Keyer
  pinMode (key_dit, INPUT);
  pinMode (key_dah, INPUT);
  pinMode (key_pin, OUTPUT);
  // Enable the Internal Pull-up resistor on the Keyer Inputs
  digitalWrite(key_dit, HIGH);
  digitalWrite(key_dah, HIGH);
```

Then, we'll set the K1EL keyer state to IDLE, and set the pin mode for the analog battery input pin. We'll delay five seconds, and then clear the TFT display screen.

```
// Setup the Keyer
 keyerState = IDLE;
 keyerControl = 0;


 // Setup the Battery Voltage input
 pinMode (battery_pin, INPUT);

 tft.fillScreen(ST7735_BLUE);
```

In the next part of the `setup()` loop, we'll set up the TFT display with a template to show the transmit and receive frequencies, along with the tuning step rate, and the battery voltage.

```
 tft.setCursor(0, 35);
 //  tft.setTextSize(2);
 //  tft.print("RX:");
 tft.setTextSize(1);
 tft.setCursor(0, 70);
 tft.print("Step:");
 tft.setCursor(0, 80);
 tft.print("RIT:");
 tft.setCursor(0, 90);
 tft.print("Batt:");
 tft.setCursor(0, 120);
 tft.print("KW5GP 40m Mini-Cube CW Xcvr");

 read_volts();
```

Now, we'll set the DDS to the default receive frequency of 7.030 MHz, and display the CW keyer speed. Note that you can set the `key_mode` variable to 2 and use the keyer inputs with a straight key if you'd prefer.

```
  // start the oscillator...
  set_DDS(rx_freq);

  tft.setCursor(0, 100);

  if (key_mode != 2)
  {
    tft.print("Speed: ");
    tft.print(cw_speed);
    tft.print(" wpm");
  } else {
    tft.print("Straight Key");
  }
```

In the final part of the `setup()` loop we'll display the current tuning speed, along with the default starting transmit and receive frequencies.

```
  // print the current RIT and encoder position
  tft.setCursor(35, 70);
  tft.print(fast_step);
  tft.print(" Hz");
  tft.setCursor(35, 80);
  tft.print( (char) 240);
  tft.print(" ");
  tft.print(rit_fast_step / 1000);
  tft.print(" KHz");
  tft.setCursor(10, 0);
  tft.setTextSize(3);
  tft.print(tx_freq / 1000, 3);
  tft.setTextSize(2);
  tft.setCursor(0, 40);
  tft.print("RIT: ");
  if (rit_freq > 0)
  {
    tft.print("+");
  }
  tft.print(rit_freq / 1000, 3);
  tft.print(" Hz");

} // End Setup Loop
```

Because we perform so many of the transceiver operations by using functions, the `main loop()` is quite short. This helps the radio controls respond quickly and smoothly.

Starting out in the `main loop()`, we first check to see that the CW keyer is not active. If the keyer is active, we process any keyer functions before dealing with any radio control functions. Assuming that the keyer is not active, we start with reading the RIT setting by calling the `read_rit() function`, this will read the current position of the RIT potentiometer. If the RIT potentiomenter position has been changed, we'll change the RIT frequency setting based on the RIT tuning range set by the rotary encoder's pushbutton switch. We'll then update the DDS receive frequency and the TFT display with the new receive frequency.

```
void loop()
{
  if (keyerState != KEYED)
  {
    read_rit();  // Read and set the RIT

The read_rit() function

void read_rit()
{
  // Read and set the RIT frequency
  rit = analogRead(rit_pot);  // Read the RIT pot
  if (rit != old_rit)  // Set the RIT if it has changed
  {

    if (current_step)
    {
      rit_range = rit_fast_step;  // Set the RIT Range for Fast Step
    } else {
      rit_range = rit_slow_step;  // Set the RIT Range for Slow Step
    }
    rit_freq = map(rit, 10, 1020, rit_range * -1, rit_range); // Map the RIT
frequency to the appropriate range
    if (abs(rit_freq) > rit_range)  // Constrain the RIT to the RIT range if
necessary
    {
      if (rit_freq > 0)
      {
        rit_freq = rit_range;
      } else {
```

```
      rit_freq = rit_range * -1;
    }
  }
  if (abs(rit_freq) < rit_range / 50)
  {
    rit_freq = 0;
  }
  rx_freq = tx_freq + rit_freq;
  set_DDS(rx_freq);

  if ( (int) rit / 15 != (int) old_rit / 15)
  {
    tft.fillRect(0, 40, 155, 25, ST7735_BLUE); // Erase the lower portion of
the TFT display
    tft.setTextSize(2);
    tft.setCursor(0, 40);
    tft.print("RIT: ");
    if (rit_freq > 0)
    {
      tft.print("+");
    }
    tft.print(rit_freq);
    tft.print(" Hz");
    old_rx_freq = rx_freq;
  }
  old_rit = rit;
 }
}
```

The next thing we do in the `main loop()` is read and set the CW keyer speed by calling the `set_keyer_speed()` function, which will read the current position of the CW speed potentiometer.

```
  set_keyer_speed(); // Read and set the Keyer Speed
```

```
The set_keyer_speed() function


void set_keyer_speed()
{
  if (key_mode != 2)
  {
```

```
    // Read and set the CW Keyer Speed
    cw_speed = map(analogRead(speed_Pot), 10, 1020, 5, 35); // Read the CW Speed
Pot
    if (cw_speed != old_speed)  // Update if the speed has changed
    {
      old_speed = cw_speed;
      //        cw_wpm = map(cw_speed,10,1020,5,35);  // Map the CW Speed pot
range to 5-35 wpm
      if (cw_speed > 35)
      {
        cw_speed = 35;
      }

      if (cw_speed < 5)
      {
        cw_speed = 5;
      }

      tft.fillRect(40, 100, 90, 10, ST7735_BLUE); // Erase the lower portion of
the TFT display
      tft.setTextSize(1);
      tft.setCursor(40, 100);
      tft.print(cw_speed);
      tft.print(" wpm");

      loadWPM(cw_speed);   // Set the keying speed
    }
  }
}
```

If the speed potentiometer position has been changed, we'll update the CW keyer speed to the new setting by calling the `loadWPM()` function in the K1EL keyer code, which will calculate the CW element (dit or dah) timing used in the K1EL keyer code and update the CW keyer speed based on the words-per-minute (wpm) parameter that we pass to the function. We'll then update the TFT display to reflect the new CW keyer speed. For this sketch, the CW keyer speed can be set from 5 – 35 words per minute (wpm). You can change these settings as desired.

```
void loadWPM (int wpm)
{
  ditTime = 1200 / wpm;
}
```

Next, we'll check to see if the rotary encoder position has changed by calling the `read_encoder()` function.

```
read_encoder(); // Read the encoder and set the Tx/Rx frequencies

The read_encoder() function
void read_encoder()
{
  // Read the Encoder
  encoder_Pos = Enc.read() / 4; // divide by 4 to match encoder detent
  if (encoder_Pos != old_encoder_Pos) // If the Encoder has changed update freq
  {
    if (encoder_Pos < old_encoder_Pos) // If we're increasing frequency
    {
      if (tx_freq >= band_high) // Limit to top end of band
      {
        tx_freq = band_high;
      } else {
        tx_freq = tx_freq + step_size;
      }

    } else {
      if (tx_freq <= band_low)// We're decreasing frequency, limit to low end of band
      {
        tx_freq = band_low;
      } else {
        tx_freq = tx_freq - step_size;
      }
    }

    //    dds.setFrequency(tx_freq); // Set the TX frequency
    old_encoder_Pos = encoder_Pos;
    rx_freq = tx_freq + rit_freq;
    set_DDS(rx_freq);

    tft.fillRect(10, 0, 150, 25, ST7735_BLUE); // Erase the lower portion of the
TFT display
    tft.setTextSize(3);
    tft.setCursor(10, 0);
    tft.print(tx_freq / 1000, 3);
  }
}
```

If the rotary encoder has been turned, the transmit and receive frequencies will be changed based on the current tuning speed setting by calling the `set_DDS()` function.

```
void set_DDS(double set_freq)
{
  dds.setFrequency(set_freq);  // Set the DDS to the starting freq
  if (set_freq >= band_high)
  {
    set_freq = band_high;
  }
  if (set_freq <= band_low)
  {
    set_freq = band_low;
  }
  //  // glcd.printNumI(set_freq,30,16,7);  // Display the DDS Frequency
}
```

Any RIT settings will be retained as the tuning frequency changes. The new transmit and receive frequencies will then be displayed on the TFT display. The `set_DDS()` function will set the DDS frequency to the frequency specified in the parameter passed to the function when it is called by using the M0XPD library's `dds.setFrequency()` function call. Next, we'll check to see if the pushbutton switch on the rotary encoder has been pressed by calling the `read_pushbutton()` function.

```
    read_pushbutton();   // Read the Encoder Pushbutton switch and set Fast/Slow
tuning mode

The read_pushbutton() function
void read_pushbutton()
{
  // Read the Encoder Pushbutton Switch
  encoder_Switch = digitalRead(encoder_Switch_Pin);
  if (encoder_Switch == LOW && millis() > delay_time) // Check to see if pressed
  {
    // if it's changed, toggle the step size but don't allow again for debounce
time
    delay_time = millis() + debounceInterval; // if it's changed, toggle the
step size but don't allow again for debounce time
    current_step = !current_step; // Change the Step rate
```

```
    if (current_step)
    {
      rit_range = rit_fast_step;  // Set the RIT Range for Fast Step
    } else {
      rit_range = rit_slow_step;  // Set the RIT Range for Slow Step
    }

    tft.fillRect(35, 70, 90, 20, ST7735_BLUE); // Erase the lower portion of the
TFT display
    tft.setTextSize(1);
    tft.setCursor(35, 80);
    tft.print((char) 240);
    tft.print(" ");
    if (current_step)
    {
      step_size = fast_step;
      tft.print(rit_range / 1000);
      tft.print(" KHz");
    } else {
      step_size = slow_step;
      tft.print(rit_range);
      tft.print(" Hz");
    }
    tft.setCursor(35, 70);
    tft.print(step_size);
    tft.print(" Hz");
  }
}
```

If it has, we'll debounce the switch in software, and then toggle the fast and slow tuning step rate between 1 Hz and 100 Hz. This also toggles the RIT tuning range at the same time, switching between the 0 to 500 Hz range and the 0 to 5000 Hz range.

As the last step in the `main loop()` when the keyer is not active, we'll read and update the battery voltage display by calling the `read_volts()` function, which will read the power supply voltage divider attached to the Arduino's analog input pin and convert the A/D reading into volts. This voltage reading is then adjusted to account for the voltage drop across the polarity protection diode, and then the battery voltage is displayed on the TFT display.

```
read_volts();  // Read and display the battery voltage
```

```
The read_volts() function
void read_volts()
{
  int battery = analogRead(battery_pin);
  float battery_voltage = map(battery, 0, 1023, 0, volt_cal); // Map the Battery
A/D value to voltage

  if (battery_voltage != last_voltage)
  {
    tft.fillRect(35, 90, 30, 10, ST7735_BLUE); // Erase the lower portion of the
TFT display
    tft.setTextSize(1);
    tft.setCursor(35, 90);
    tft.print((battery_voltage / 100) + volt_drop, 1);
    tft.print(" V");
    last_voltage = battery_voltage;
  }
}
```

The last thing we do in the `main loop()` is check if either of the CW keyer paddles are pressed, or if the straight key is pressed if we're in straight key mode, by calling the `check_key()` function.

```
  check_key();   // Check the Key and Transmit CW
```

This function will first check to see if we're in straight key or keyer mode. If the `key_mode` variable is set to 2, we're using a straight key, if it's set to 0, we'll use the keyer in iambic mode A, and if it's set to 1, we'll set the keyer to use iambic mode B. For this sketch, the default `key_mode` is set to iambic B mode. You can change this to any of the three available modes by changing the `key_mode` variable in the initialization area of the sketch. If we're in straight key mode, we'll key the transmitter based on the key down time. When the CW key is pressed, we'll set the DDS frequency to the transmit frequency before transmitting. When the key is released, the DDS is set back to the receive frequency until the CW key is pressed again.

If we're in keyer mode, we'll execute the K1EL keyer code. The various keyer states are represented by the status of the `keyerState` variable. This variable is used to determine what action the keyer code is performing. The possible keyer states are: IDLE, CHK_DIT, CHK_DAH, KEYED_PREP, KEYED, and INTER_ELEMENT. We'll examine how these states are used in the Switch…Case code block that follows. If we're in the IDLE state, the keyer code checks to see of either of the keyer paddles has been pressed, and if they have, we call the `update_PaddleLatch()` function before setting the `keyerState` to CHK_DIT. The `update_PaddleLatch()` function will determine whether the Dit or Dah paddle was pressed, and set the `keyerControl` variable to DIT_L or DAH_L depending on which paddle was pressed.

The next step in the keyer process is to process the CHK_DIT state. If the `updatePaddleLatch()` function has set the DIT_L status, we'll set the `ktimer` to the `ditTime` setting, the `keyerState` variable to KEYED_PREP, and the `keyer_control` variable to DIT_PROC. If the `updatePaddleLatch()` function has set the DAH_L status, we'll set the `keyerState` to CHK_DAH, and continue processing. Then we'll process the CHK_DAH state. If the `updatePaddleLatch()` function has set the DAH_L status, we'll set the `ktimer` to the ditTime multiplied by three to equal the dah timing, and the `keyerState` variable to KEYED_PREP. If neither the Dit or Dah paddle was pressed but we still somehow managed to get to this keyer state, we'll return the keyer to the IDLE state.

Next, we'll process the KEYED_PREP state. This state will set the transmit frequency and key the transmitter for the duration specified by the `ktimer` variable. Once the CW element (dit or dah) has been sent, it will reset the DIT_L and DAH_L paddle latches and set the KEYED state. Then we'll process the KEYED state. At this point the transmitter has been keyed and we're waiting for the `ktimer` to expire, indicating that we have sent the desired CW element (dit or dah). When the timer expires, we'll set the keyer state to INTER_ELEMENT, unkey the transmitter, and set the DDS back to the receive frequency. Depending on the iambic mode, we'll either clear the paddle latch to indicate the last CW element has been sent if we're using iambic mode A, or, if we're in iambic mode B, leave the latch set to send the same element we just sent.

If we're processing the INTER_ELEMENT state, we'll set a delay between CW elements (dit or dah), setting the keyer state to CHK_DAH and clearing the DIT_L and DIT_PROC flags if we just sent a dit, or clearing the DAH_L latch and setting the keyer state to IDLE if we just sent a dah.

The good news about using the K1EL keyer code inside our sketch is that we really don't need to know how it works. It's just another one of those "black box" type functions that you can insert into your code block and have it just work. That's why we didn't dig deep into the K1EL keyer functions. Since this is essentially a black box block of code, we just let it do its thing. That's part of the power and simplicity of having the resources of the open-source community available to you as you create your own sketches and projects.

Figure 22.17 shows the output of the TFT display on the QRP CW transceiver project in operation.

The sketch and libraries for this project can be downloaded from **kw5gp.com**.

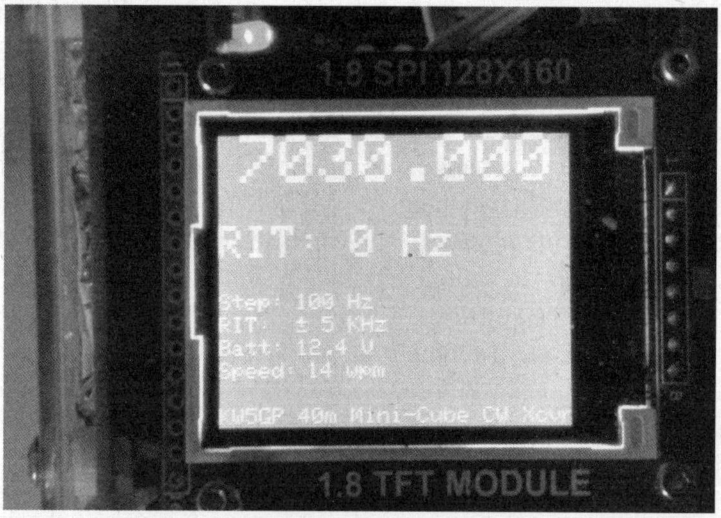

Figure 22.17 — The QRP CW transceiver display.

23 Conclusion

It was so fun being able to revisit and upgrade some of my earlier Arduino projects. This journey has shown me how far I've personally grown with the Arduino. It's amazing to think how that first Arduino book has turned into a whole series, and just keeps going. I hope you have as much fun building, using, and modifying these projects as I've had with creating them. There could have easily been a dozen more projects included in this book, but I had to stop somewhere. Don't let this book be the end of your own Arduino adventure, though. Think up your own Arduino projects and make them happen, that's where the fun really begins.

If you like this book and would like to see more Arduino books in the future, be sure to write to ARRL at pubsfdbk@arrl.org and let them know. And of course, always feel free to email me at kw5gp@arrl.net with your comments or suggestions.

Hopefully, this book has provided you with the knowledge, confidence, and skills that you need to create and modify your own Arduino projects. You now have all the pieces to create and modify your antenna rotator controller and give it all kinds of new features, regardless of its age or manufacturer. I have several very old rotator controllers that don't even have position sensors. I'm sure you remember those old rotator controllers with the dial that you turned, and the controller that had a solenoid that clicked while it rotated to move the position indicator on the box. A couple of those are up next on my lab workbench, just for fun.

I think the biggest question is, where do I go next? I have a whole file folder full of ideas for new Arduino projects that I've been wanting to tackle. Let's take a peek at some of those scribblings on napkins and sticky notes:

Speech recognition projects with the Arduino have always been at the top of my list for project ideas. I really like the DFRobot gravity speech synthesis V2.0 module for projects that use text-to-speech.

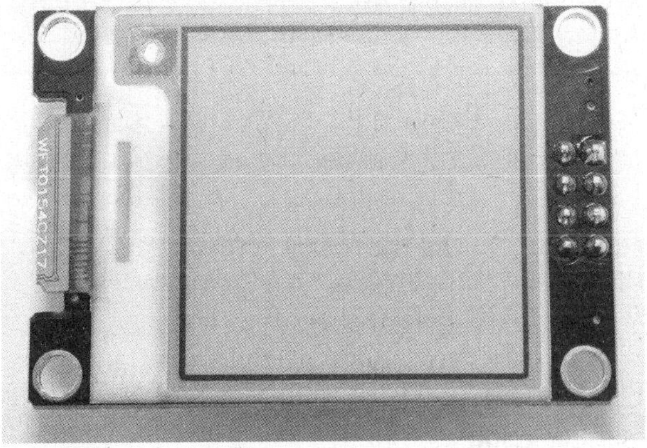

Figure 23.1 — A 152 × 152-pixel E-Ink display.

E-Ink displays, such as the one shown in **Figure 23.1**, are rapidly gaining in popularity, with a number of them being combined with processors such as the STM32, and the Raspberry Pi Pico RP2040. I think that E-Ink displays will prove quite valuable to use in low power projects, since the display only draws power when changing the information.

The Arduino IDE just got a major upgrade with the release of the Arduino IDE 2.0. In addition to having an overall upgrade to the library manager, board manager, and many other standard IDE features, it also includes a built-in debugger which will be a big help in troubleshooting your Arduino sketches. IDE 2.0 also incorporates synchronization with Arduino Cloud, the new Arduino Cloud-based IDE. You can learn more about IDE 2.0 at **docs.arduino.cc/software/ide-v2**.

There are also moves to use the Python programming language with the Arduino. Adafruit, Sparkfun, Raspberry Pi, and others are develop-

Figure 23.2 — The Raspberry Pi Pico.

ing a Python-based line of microcontrollers using CircuitPython and MicroPython. This is aimed at the Internet of Things (IoT) application side of microcontrollers. The Raspberry Pi Pico gives you the best of both the Arduino and Python worlds. With a simple press of the bootsel button, you can have the Pico support either the Arduino ecosystem or the MicroPython and CircuitPython ecosystem. All of this is available in just a $6 board (see **Figure 23.2**).

Of course, the Arduino processor boards just keep getting better and better. The PJRC Teensy 4.1 (see **Figure 23.3**) is a 32-bit ARM Cortex M7 CPU with 8 M of flash, 1 M of RAM, and 4 k of emulated EEPROM, running at 600 MHz with 55 digital I/O pins, 18 analog input pins, 8 serial ports, 3 SPI ports, 2 I²C ports, a USB host port, a 10/100 ethernet port, and much more.

The STMicroelectronics STM32 (see **Figure 23.4**) series of processors is becoming very popular as well. In addition to being inexpensive, the STM32 "Blue Pill" version boasts a 72 MHz ARM Cortex-M3 processor with 64 K of flash and 20 K of SRAM. It has 32 digital I/O pins, 14 analog input pins, 3 UART ports, 2 SPI ports, and 2 I2C ports, all for around $3.

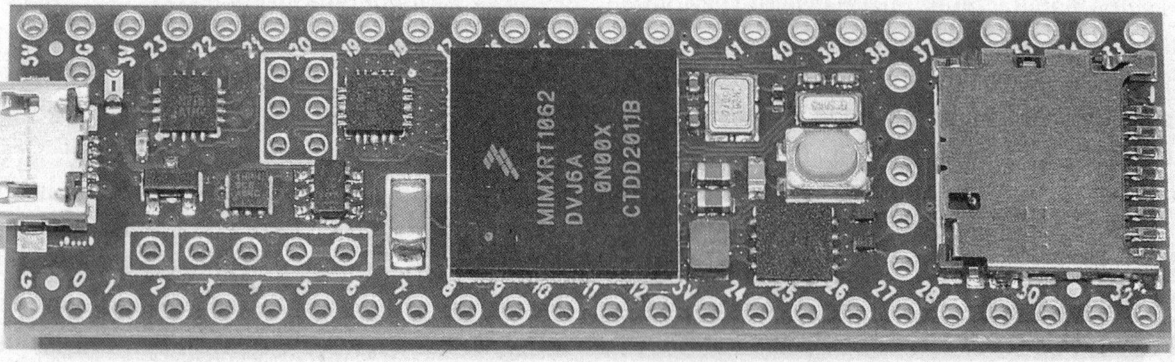

Figure 23.3 — The PJRC Teensy 4.1.

Figure 23.4 — The STM32.

The Espressif ESP32 family (see **Figure 23.5**) is the upgrade for the ESP8266. Also costing around $3, it uses a Tensilica 32-bit CPU running up to 240 MHz with up to 16 M of flash, 77 digital I/O pins, 18 12-bit D/A pin 28-bit D/A pins, integrated Bluetooth, Bluetooth BLE, and much more.

Next is the Raspberry Pi Pico and RP2040-compatible boards. The Raspberry Pi Pico uses a dual-core, 133 MHz ARM Cortex M0+ CPU with 2 M of flash, 256 K of RAM, 26 digital I/O pins, 3 12-bit A/D input pins, 2 SPI ports, 2 I2C ports, USB host capability, and more. The Pi Pico-W adds 2.4 GHz 802.11 b/g/n Wi-Fi capability, all for $6. Several RP2040 CPU-based boards are becoming available, including integrated E-Ink and OLED displays such as the LILYGO I-Display RP2040.

There's even a nifty CPU board that has both an RP2040 CPU and an ESP32 CPU on the same board, along with an integrated 125 × 240-pixel E-Ink display. You select which CPU to use by flipping over the USB connector to the board. The LILYGO T-PicoC3 (see **Figure 23.6**) is available from suppliers such as **Banggood.com** for just $13.

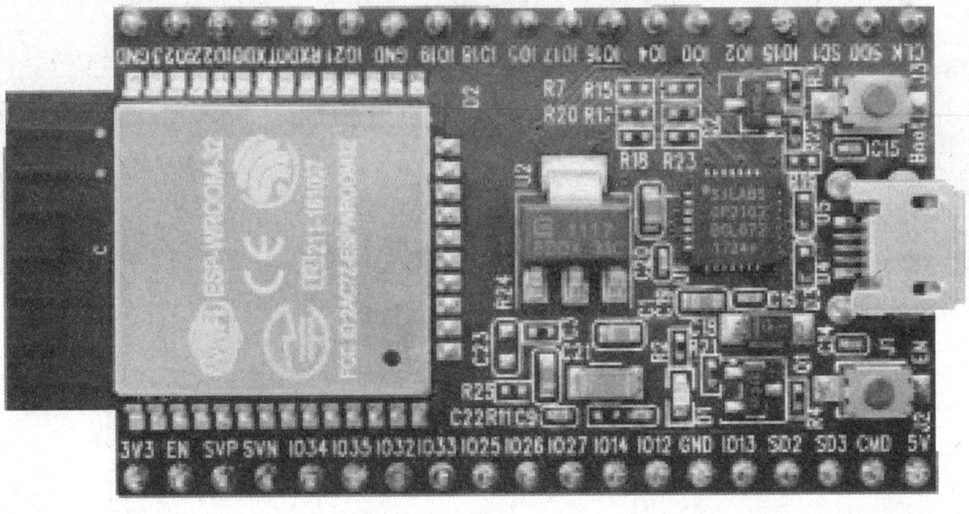

Figure 23.5 — The Espressif ESP32.

Figure 23.6 — The LILYGO T-PicoC3 board.

The Arduino is at the heart of one of the newer ham radio-based hobbies, picoballooning. Using something as simple as a Mylar party balloon, or a larger ultra-lightweight high-altitude plastic balloon, hams are sending Arduino-powered experimental payloads (see **Figure 23.7**) high into the sky, sometimes as high as 45,000 feet, and tracking them using WSPR and APRS as they circumnavigate the world.

So, the future is looking wonderfully bright for the entire Arduino family. I'm already planning out the next Arduino book, focusing on the STM32, ESP32, and the Raspberry Pi Pico projects, and yes, I'm still having tons of fun.

73,

Glen Popiel, KW5GP

Spring 2024

Figure 23.7 — The solar-powered Skytracker Arduino-powered balloon payload by Bill Brown, WB8ELK. (Photo courtesy of Bill Brown, WB8ELK)

Appendix A

Table A

Azimuth/Elevation Rotor Controller Commands Chart
(Subset of Yaesu GS-232A Commands)

Command	*Action*
A	Stop azimuth rotation
B	Return current elevation value in degrees (format +0eee)
C	Return current azimuth value in degrees (format +0aaa)
C2	Return current azimuth and elevation (format +0aaa+0eee)
D	Rotate elevation down
E	Stop elevation rotation
F	Set azimuth full scale calibration
F2	Set elevation full scale calibration
L	Rotate azimuth counter-clockwise
Maaa	Rotate azimuth to aaa degrees
O	Set azimuth aero calibration
O2	Set elevation zero calibration
R	Rotate azimuth clockwise
S	Stop all rotation
U	Rotate elevation up
Waaa eee	Rotate azimuth to aaa degrees and elevation to eee degrees

Notes

Notes

Notes

Notes

Notes